S0-CAH-424

Isotopes in the Atomic Age

HARI JEEVAN ARNIKAR

Professor Emeritus in Chemistry
University of Poona
Pune, India

JOHN WILEY & SONS
New York Chichester Brisbane Toronto Singapore

First Published in 1989 by
WILEY EASTERN LIMITED
4835/24 Ansari Raod, Daryaganj
New Delhi-110 002, India

Distributors:

Australia and New Zealand:
JACARANDA-WILEY LTD., JACARANDA PRESS
JOHN WILEY & SONS, INC.
GPO Box 859, Brisbane, Queensland 4001, Australia

Canada:
JOHN WILEY & SONS CANADA LIMITED
22 Worcester Road, Rexdale, Ontario, Canada

Europe and Africa:
JOHN WILEY & SONS LIMITED
Baffins Lane, Chichester, West Sussex, England

South East Asia
JOHN WILEY & SONS, INC.
05-05 Block B, Union Industrial Building
37 Jalan Pemimpin, Singapore 2057

Africa and South Asia:
WILEY EASTERN LIMITED
4835/24 Ansari Road, Daryaganj
New Delhi 110 002, India

North and South America and rest of the world:
JOHN WILEY & SONS, INC.
605 Third Avenue, New York, N.Y. 10158, USA

Library of Congress Cataloging in Publication Data

ISBN 0-470-21046-X John Wiley & Sons, Inc.
ISBN 81-224-0058-2 Wiley Eastern Limited

Printed in India at Siba Exim Pvt. Ltd., Delhi-110051

To Professors

Marius Chemla
Toshijo Nakaya

FOREWORD

Till about fifty years back, there was little need for separating isotopes, except in minute amounts for purposes of research. All this changed with the harnessing of atomic energy for daily use. Modern (even) peacetime life requires the isolation of certain isotopes in bulk amounts, some in a high degree of *isotopic* purity. All isotopic separations and even significant enrichments, are tedious, difficult and extremely expensive. The present book describes in an easily readable manner, the classical important, as well as the more recently developed techniques of high sophistication for the separation of isotopes. Conveniently enough, the stable, radioactive and nuclear fuel isotopes are treated separately, succinctly and lucidly. Theoretical principles involved are discussed briefly and the limitations of each method listed. Major applications of radioisotopes in the service of mankind are highlighted.

Presently there are two excellent works on the subject but dealing mainly with classical methods of separation. The scope of the ISOTOPES IN THE ATOMIC AGE is much wider, covering strategically important stable, as well as radioisotopes within its covers. It is a valuable, welcome contribution.

Today terms like nuclear energy, radiation hazards and isotope enrichment, are very much in common parlance. However, much avoidable confusion, often bordering on panic, is caused by glib misuse of these expressions. India is one of the most advanced countries in the field of atomic energy. We can ill-afford to let this confusion to continue. In this context, Dr. Arnikar's book is timely and instructive.

Bombay
2nd September, 1988

M.V. Ramaniah
formerly Director, Radiological Group
Bhabha Atomic Research Centre

PREFACE

A story is told of a maid serving in a German family who was warned, soon after the first atom bomb, that every piece of china in the house was made up of "atoms", since then not a piece of crockery ever broke in the house! No doubt the common man today knows a lot more of the atom; all the same perhaps not enough to appreciate the role of diverse isotopes in the present age. Some fifty years back, everything was happily well-regulated in an isotopically homogeneous cosmos. Isotopes have always existed, no doubt, in nearly invariant proportions but it was never necessary to effect an isotopic separation, nor even a partial enrichment over the natural abundance, for human well-being, or in the pursuit of his intellectual activities. This was due to the comfortable assurance that the properties of all elements depended only on the number of protons their atoms contain and this is same for all the isotopes of the same element.

All this changed rapidly with the discovery of isotope shifts in the atomic spectra of elements which led to the setting up of more precise standards of length, mass and time. The discovery that it is the rarer isotope ^{235}U, 0.7202% in natural uranium, which is susceptible to fission by slow neutrons, ushered in, for good and for bad, the atomic age with its bounty of nuclear energy as well as the curse of the bomb. Increasing demand for radioisotopes for use in research, medicine and industry has made it necessary to bring about enrichments of numerous isotopes, some in a high state of purity and some in massive amounts, despite the technical difficulties and high prices involved.

The object in writing this book was to make the reader familiar with the subject of isotope separations and their applications in the present atomic age. Books available on the subject being limited to (1) *Separation of isotopes,* Ed. H. London (George Newnes Ltd. London 1961) (2) *Isotope Separation*, by Stelio Villani (American Nuclear Society, 1976), and (3) *La Separation des Isotopes* par M. Chemla et J. Perie (Presses Universitaires de France, 1974) (in French), it was natural for the author to have derived much inspiration and material from these publications:

Besides classical electromagnetic techniques, including the calutron, and statistical reversible and irreversible processes, the review covers recent developments on the use of lasers and biological and microbial agencies in effecting isotope fractionations. For a better understanding of two sets of particles (isotopes) when present in a large assemblage, some knowledge of statistical thermodynamics would be of value. With this object an attempt has been made to present the basic concepts of the subject in Chapter 2 in a non-mathematical way.

Some of the author's work on the separation of the isotopes of alkali metals, specially of lithium by "simple" electromigration in aqueous agar gel and in fused salt media are described and the results compared with those obtained by more elaborate countercurrent electromigration.

The author places on record his deep sense of gratitude to the late Dr. London for the inspiration provided by his book. To the different publishers who have accorded me permission to use some of their materials, I offer my grateful thanks. Due credit lines appear at the appropriate places in the text. I like to express in the warmest words my indebtedness to Professors Stelio Villani, Marius Chemla, and Jacques Périé, for letting me draw freely from their well-known books on isotope separation.

I take this opportunity to extend my special thanks to Professors L.K. Nash, H.J.M. Bowen, K.L. Kampa, F.S. Becker, A. Klemm, A. Lunden, G.W.A. Newton, V.J. Robinson, J.F. Ready and B.B. Snavely for some of the materials drawn from their publications which are acknowledged in the text. I am much beholden to Dr. M.V. Ramaniah for writing the *Foreword*.

The continued guidance and encouragement extended to me by Professor J.E. Willard and Drs. M. Tarmes. G.A. Brinkman and S.A. Chitambar have gone a long way in the preparation of the book. I offer my thanks to Shri K.S. Bapat, and B.B. Pawar for their help in the preparation of the manuscript. In the end I place on record my sincere thanks to Wiley Eastern Ltd. for their cooperation in bringing out the book so well.

University of Poona
September 1988

H.J. ARNIKAR

CONTENTS

Chapter 1

ISOTOPY

1.1 DISCOVERY OF ISOTOPES

Our knowledge of the phenomenon of isotopy grew through developments in two areas, more or less concurrently. These refer to studies of the decay products of radioactivity and the behaviour of positive ions of gases under electric discharge. The relevant developments are presented in what follows.

1.1.1 Positive Ion Analysis

In 1886 Eugen Goldstein observed that by using a perforated cathode in a discharge tube, a new kind of rays emerged which he named *kanalstrahlen* (canal rays). These were later shown by Wien to be simply positive ions of the gas under discharge. It was Sir J. J. Thomson[1] who in 1907 analysed these rays by applying external electric and magnetic fields.

To follow Thomson's experiment, let us consider a single positive ion of the beam coming out of the discharge tube from point A (Fig. 1.1). It strikes the screen or film placed normally in its path at O. If a uniform electric field of X volts/cm is applied across the plates p^+ / p^-, the ion would be deflected to the left and meet the screen at E, the deviation x, equalling OE, is given by

$$x = k_1 \frac{Xe}{mv^2} \tag{1.1}$$

where e, m and v are the charge, mass and velocity of the ion and k_1 a constant depending on the dimensions of the apparatus. If now, instead of the electric field, a uniform magnetic field of strength B is applied, the ion gets deflected perpendicularly, i.e. upwards, and meets the screen at M. The deviation y, equalling OM, is given by

$$y = k_2 \frac{Be}{mv} \tag{1.2}$$

with k_2 as another constant of the apparatus. Finally if, both electric and magnetic fields are applied simultaneously, the ion would be subjected to two forces mutually at right angles and it would strike the screen at a point P whose coordinates are x and y. By eliminating v between Eqs. 1.1 and 1.2, we arrive at

$$\frac{y^2}{x} = k\,\frac{B^2}{X}\cdot\frac{e}{m} \tag{1.3}$$

where k is a constant equal to k_2^2/k_1. It is thus easy to see how the different positive ions of the beam travelling with different initial velocities would be subject to varying x and y deviations, but always such that

$$\frac{y^2}{x} = \text{constant} \tag{1.4}$$

as required by Eq. 1.3, since e/m would be the same for all the ions of a given gas under low excitation and B and X are held constant. Since Eq. 1.4 stands for a parabola, the trace on the screen or film due to the positive ions would be a parabola SS.

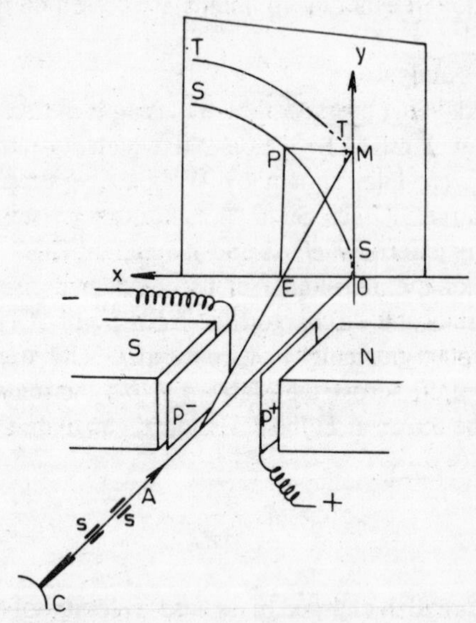

Fig. 1.1 **Principle of Thomson's positive ion analysis:** C—cathode with a hole for positive ions; ss — slits; NS— magnet poles; $\overset{+}{p}\ \overset{-}{p}$ —electric field plates; SS, TT—positive ion parabolas

Lastly, if the gas were a mixture of two components, there would result two parabolas SS and TT, corresponding to $(e/m)_1$ and $(e/m)_2$ of the two gases. If we now consider the ordinates y_1 and y_2 on the two curves SS and TT, corresponding to a given abscissa x, we have

$$\left(\frac{y_1}{y_2}\right)^2 = \frac{m_2}{m_1} \tag{1.5}$$

since the stage of ionization of the two gases under a constant discharge would usually be the same, i.e. e would be the same. Thus, by using crossed electric and magnetic fields, the beam of positive ions from a gas is resolved into a mass spectrum, i.e. particles of different masses arriving at different points on the screen. This is the principle of Thomson's parabolic mass spectrograph, which provided the first evidence that atoms of the same element may have different masses, contrary to Dalton's theory of the atom. Thus, in 1912-13 Thomson obtained clear evidence of atoms of masses 20 and 22 for neon, though he could not, at that time, explain the results correctly. Later, Francis W. Aston[2] constructed a greatly improved version of the mass spectroscope (§ 5.2.1). With the help of this, he showed that over 70 elements consist of atoms of more than one mass. Modern mass spectrographs can detect a mass difference of 1 part in 10^5. These are used for measuring atomic masses with great accuracy and in chemical analysis of different substances.

1.1.2 Radioactivity

With increasing understanding of radioactivity, it soon became clear that there can be more than one nuclide of the same atomic number, but of different masses. Thus, uranium I (^{238}U) and its third descendent following one α and two β^- decays, namely, uranium II (^{234}U) both have $Z = 92$ and hence must be chemically identical with uranium, though of different atomic masses of 238 and 234. Similarly, uranium—X_1 (^{234}Th) and ionium (^{230}Th), the parent of radium, as well as the naturally occurring thorium (^{232}Th), all have $Z = 90$ and hence must be chemically identical with thorium. We find numerous examples of this kind amongst the decay products of the radioactive series of elements.

1.2 SODDY'S CONCEPT OF ISOTOPES

Frederick Soddy[3] (1911) conceived of *isotopes* as nuclides of the same atomic number but of different atomic weights (rather mass numbers). Literally, isotope means (occupying) the *same place* in the Periodic Table, quite consistent with the by then well recognized idea that chemical characteristics of an element are governed by its atomic number and not atomic weight. Hence, a given element and its isotopes, have only one place in the Periodic Table determined by its atomic number. Thus, if isotopes of the same element are mixed up, their separation from one another would be extremely difficult. This was the experience of Boltwood, in 1906, on a mixture of thorium (^{232}Th) and ionium (^{230}Th), and later of others on other similar mixtures.

1.2.1 Symbolic Representation of Isotopes

In terms of the contemporary concept of atomic structure, nuclei of isotopes of an element have the same number (Z) of protons but a different number (N) of neutrons, hence, different mass numbers A ($= Z + N$), also referred to as the isotope numbers. The way of indicating these numbers is illustrated by the examples of isotopes of hydrogen, chlorine and uranium ${}^{1}_{1}H_{0}$, ${}^{2}_{1}H_{1}$, ${}^{3}_{1}H_{2}$; ${}^{35}_{17}Cl_{18}$, ${}^{37}_{17}Cl_{20}$ and ${}^{238}_{92}U_{146}$, ${}^{235}_{92}U_{143}$ and ${}^{234}_{92}U_{142}$. Usually the neutron number is not indicated as it is always apparent from A and Z. Also, since Z is a unique characteristic of the chemical element, it need not be shown along with the chemical symbol of the element. Thus, these isotopes are adequately represented by the symbols ${}^{1}H$, ${}^{2}H$, ${}^{3}H$; ${}^{35}Cl$, ${}^{37}Cl$ and ${}^{238}U$, ${}^{235}U$ and ${}^{234}U$.

1.2.2 Isotope Mass Number and Atomic Weight

The actual *atomic weight** of a naturally occurring element is the average weight of all the isotopic atoms which are present homogenously mixed up in it. Thus, the atomic weight of natural chlorine is 35.543 being made up of 75.77 per cent of ${}^{35}Cl$ and 24.23 per cent of ${}^{37}Cl$ atoms. Similarly, the atomic weight of natural lithium is 6.94, made up of 7.5% of ${}^{6}Li$ and of 92.5% of ${}^{7}Li$ atoms.

1.2.3 Mono- and Multiisotopic Elements

Of the 85 stable elements occurring in nature, only 20 are monoisotopic** (19 with odd-Z and one ${}^{9}_{4}Be$ with even-Z). The rest 65 have two or more stable isotopes each. Generally, elements of even atomic numbers, and especially if this number were to be a 'magic number'***, have a larger number of stable isotopes over their neighbours in the Periodic Table (*vide* Table 1.1).

Table 1.1 Number of Stable Isotopes of Elements of Z around Magic Numbers

Region around	Z = Magic No. −1	= Magic No.	= Magic No. +1
Calcium (Z = 20)	2 (${}^{39,41}K$)	6 (${}^{40,42-44,46,48}Ca$)	1 (${}^{45}Sc$)
Tin (Z = 50)	2 (${}^{113,115}In$)	10 (${}^{112,114-120,122,124}Sn$)	2 (${}^{121,123}Sb$)
Lead (Z = 82)	2 (${}^{203,205}Tl$)	4 (${}^{204,206-208}Pb$)	1 (${}^{209}Bi$)

Apart from these naturally occurring nuclides, a much larger number of isotopic nuclides have been prepared by adopting various nuclear reactions.

* The atomic weight is not the same as the mass number (A) though it is close to it. A is invariably an integer, being the sum of Z + N, while the atomic weight of an atom is the ratio of the mass of the atom to that of a ${}^{12}C$ atom taken equal to 12. Thus, the atomic weight of a ${}^{31}P$ atom is 30.973 763 u while its mass number is 31, (1 u = 1.660 53 × 10^{-27} kg).

** These are ${}^{9}Be$, ${}^{19}F$, ${}^{23}Na$, ${}^{27}Al$, ${}^{31}P$, ${}^{45}Sc$, ${}^{55}Mn$, ${}^{59}Co$, ${}^{75}As$, ${}^{89}Y$, ${}^{93}Nb$, ${}^{103}Rh$, ${}^{127}I$, ${}^{133}Cs$, ${}^{141}Pr$, ${}^{159}Tb$, ${}^{165}Ho$, ${}^{169}Tm$, ${}^{197}Au$, and ${}^{209}Bi$.

*** 2, 8, 20, 50, 82, 126, are the magic numbers (see also foot note on p. 8).

All these are invariably radioactive. For example, of the 12 isotopes of Cl from ^{32}Cl to ^{43}Cl only ^{35}Cl and ^{37}Cl are naturally occurring and are stable; the rest 10 are man-made and are radioactive.

1.3 CHEMICAL IDENTITY OF ISOTOPES

It is well established that over 99.95% of the total mass of an atom is concentrated in the tiny core or the *nucleus* of the atom, occupying but 10^{-39} cm^3 at the centre of the atom of volume of the order of 10^{-24} cm^3. The vast space surrounding the nucleus is occupied by only electrons which are equal in number to the number of protons (Z) in the nucleus. It is easy to picture, on such a model for the atom, that all reactions, except nuclear reactions, involve only electrons, with the nucleus playing little or no part. In other words, in all atomic and molecular chemical reactions, isotopes of a given element react in an almost identical manner, in respect of the rate, the products and their yields, etc. All the isotopes of an element have the same number of electrons distributed in the same way in the outer atomic space. Thus, isotopic molecules as $H^{35}Cl$ and $H^{37}Cl$, or $C_2H_5{}^{127}I$ and $C_2H_5{}^{131}I$ are expected, and actually observed, to react in a nearly identical manner with a given reactant, indicating the absence of an isotope or mass effect. This situation is also reflected by identical isotopic constitution of all terrestrial substances, except in very light elements, notably hydrogen, and rarely in elements heavier than chlorine. We may be said to be living in an *isotopically uniform cosmos.*

1.3.1 Isotope Effects in Light Elements

The limits of the assumption that in chemical reactions isotopes behave in an *identical* manner cannot be overlooked, specially in the case of very light elements. For instance, the mole fraction of deuterium or heavy hydrogen (D or ^2H) in the total hydrogen content of waters of different sources, as rain, rivers, lakes, of the deep seas of equitorial and polar regions, etc., differs significantly from 0.011 9 to 0.015 6. These variations in the ^2H content from the commonly observed value of 0.014 8, unmistakably point to an isotopic fractionation, to varying extents under different conditions, resulting in the preferential loss of the lighter or the heavier isotope. This is evidence of a difference in the behaviour of the two isotopes of hydrogen, arising out of some property(ies) dependent on the mass of the particle, as diffusivity, mobility, reactivity, or other thermodynamic characteristics, or a nuclear property like spin. Such effects are termed *isotope effects*, which are considered in Sec. 1.4.

These effects are expected to be the largest in case of the isotopes of hydrogen (^1H, ^2H and ^3H), as illustrated by some of their properties listed in Table 1.2.

Table 1.2 Some Properties of Isotopic Molecules of Hydrogen[4,5]

Property	H_2	HD	D_2	T_2
Boiling point/K	20.39	22.13	23.67	25.04
Freezing point/K	13.95	16.60	18.65	—
Critical temperature/K	32.99	35.41	38.96	—
Triple point/K	13.96	16.60	18.73	20.62
Liquid density	0.08	—	0.17	0.18
Enthalpy H^0/kJ mol^{-1}	0	+0.63	0	0
Entropy S^0/JK^{-1} mol^{-1}	131.17	144.44	145.40	165.44
Free energy G^0/kJ mol^{-1}	0	−1.51	0	0

1.3.2 Isotope Effect in Atomic Hydrogen Spectrum : Discovery of Heavy Hydrogen

An isotope effect of a large magnitude between the hydrogen isotopes, H and D, is in the frequency shift of their line spectra. On Bohr's theory, the frequency ν, or the wave number $\bar{\nu}$ (= ν/c), of the hydrogen atomic spectrum, is given by

$$\bar{\nu} = R\left(\frac{1}{n_1^2} - \frac{1}{n_2^2}\right), \qquad (1.6)$$

where n_1 and n_2 are simple integers characteristic of the electron transition, with $n_2 \geq (n_1 + 1)$, and R the Rydberg constant given by

$$R = \frac{2\pi^2 e^4}{ch^3}\mu, \qquad (1.7)$$

where c is the velocity of light, h the Planck's constant and μ the *reduced mass** of the system (of electron of mass m and the nucleus of mass M in this case).

Expressing all masses in terms of the electron mass (*i.e.* $m_e = 1$, $m_H = 1836$ and $M_D = 2M_H = 3672$), we have for the reduced masses of H and D atoms

$$\mu_H = \frac{m.M_H}{M_H + m} = 0.999\ 456$$

and

$$\mu_D = \frac{m.M_D}{M_D + m} = 0.999\ 728$$

Hence for a given transition $n_2 \rightarrow n_1$

$$\frac{\nu_H}{\nu_D} = \frac{R_H}{R_D} = \frac{\mu_H}{\mu_D} = \frac{M_H}{M_D} \cdot \frac{M_D + m}{M_H + m} = 0.999\ 728 \qquad (1.8)$$

The relative shift in the given spectral line is, thus:

$$\frac{\bar{\nu}_D - \bar{\nu}_H}{\bar{\nu}_D} = \frac{\Delta\bar{\nu}}{\bar{\nu}} = 1-0.999\ 728 = 0.000\ 272$$

This can also be expressed as a shift in the wavelength $\frac{\Delta\lambda}{\lambda}$. Thus, the red H$\alpha$ line of 656.28 nm of hydrogen in the Balmer series ($n_1 = 2$ and $n_2 = 3$)

* The reduced mass μ of a system of two particles of masses m_1 and m_2 is $\mu = \frac{m_1 m_2}{m_1 + m_2}$

shifts by 0.178 5 nm towards the shorter wavelength side for deuterium. It was Urey, Brickwedde and Murphy[6] who found in 1932 a similar weak line on the ultraviolet side of each of the Balmer lines in the hyperfine structure of the hydrogen spectrum of a sample of liquid hydrogen reduced by evaporation from 4 litres to 1 ml, which they identified as belonging to the, till then unknown, isotope of hydrogen of mass 2. Later, a fine structure due to isotope effect had also been observed in the elements, Li, Ne, Zn, Hg, Tl and Pb.

1.3.3 **Isotopic Molecules**

Differences in the properties of the isotopes of an element are carried over into their *isotopic molecules* as well. This refers to molecules of similar chemical constitution and structure but having one or more isotopically different atoms. Here are examples of isotopic molecules :

Water : H_2O, HDO, D_2O, HTO, T_2O, DTO, ...*
Carbon dioxide : $^{12}CO_2$, $^{13}CO_2$, $^{14}CO_2$, etc...
Acetic acid : $^{12}CH_3\,^{12}COOH$, $^{12}CH_3\,^{14}COOH$, $^{12}CH_3COOD$, etc...
Hydrochloric acid : $H^{35}Cl$, $D^{35}Cl$, $H^{37}Cl$, $D^{37}Cl$, etc.
Ammonia : NH_3, NH_2D, NHD_2, ND_3, etc...

The properties of water isotopic molecules H_2O and D_2O, are listed in Table 1.3.

Table 1.3 Some Properties of Water and Heavy Water [5,7–9]

Property	H_2O	D_2O
Molecular weight	18	20
Density at 20 °C/g cm^{-3}	0.998 23	1.105 30
Maximum density/at temp.	1.000 at 3.96 °C	1.106 at 11.6 °C
Boiling point /°C	100	101.42
Freezing point /°C	0	3.8
Latent heat of vaporization /kJ mol^{-1}	44.163	45.545
Latent heat of fusion /kJmol^{-1}	6.031	6.363
Dielectric constant	80.1	79.8
Viscosity at 25 °C/ m poise	8.91	10.99
Electrode potential H_2/H^+ at 25 °C/V	0	-0.003 5 (in D_2O)
H^+ ion conductance /Ω^{-1} cm^{2}mol^{-1}	350	213.7
OH ion conductance /Ω^{-1}cm^2 mol^{-1}	197	119
Surface tension/dynes cm^{-1}	72.75	67.8
Refractive index (for 589 nm)	1.332	1.328
Specific heat/J g^{-1}K^{-1}	4.17	4.10
Δ H° / kJ mol^{-1} (gas)	–242.76	-248.85
Δ S° / JK^{-1}	189.55	199.00
ΔG° / kJ mol^{-1}	–229.45	-234.11

* There can be 18 isotopic molecules of water if the oxygen isotopes ^{16}O, ^{17}O and ^{18}O are also incorporated in the above six.

1.3.4 Isotope Effect in Molecular Spectra : Discovery of Oxygen and Carbon Isotopes

Giauque and Johnston[10] observed faint new lines in the molecular spectra of atmospheric air close to those of oxygen, that led to the discovery of ^{17}O and ^{18}O as additional stable isotopes of oxygen, their abundances being 0.04 and 0.2 atom per cent, (^{16}O = 99.76). The existence of the isotope ^{13}C (abundance 1.11%) was also discovered by a similar study of the band spectra of gaseous carbon compounds by Birge and King[11].

1.4 ISOTOPE EFFECTS

Since isotopy is of nuclear origin, isotope effects attributable directly to differences in the nuclear composition and other nuclear characteristics as spin may be termed *nuclear isotope* effects, distinguish to the same from chemical *isotope effects*. The latter arise indirectly from differences in atomic masses, as mobility of the atoms involved in diffusion, and of ions involved in electromigration, in zero-point energy affecting energy states relevant to spectral transitions, and in bonding energy controlling vapour pressure, etc. The isotope effects may be classified, thus:

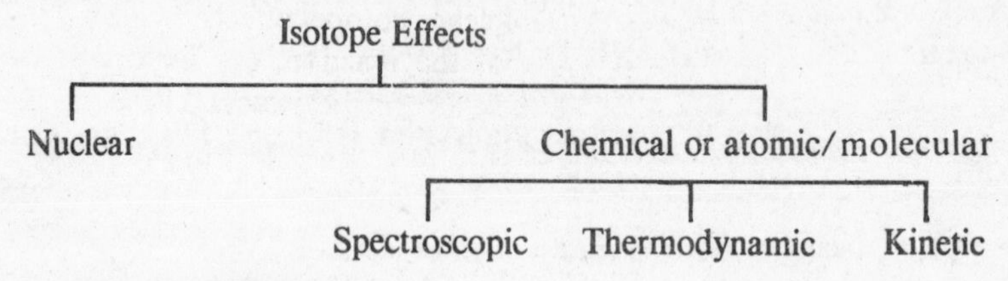

These are discussed below.

1.4.1 Nuclear Isotope Effects

Isotope effects in *nuclear reactions* are more profound, often ending in reactions of a different type. These effects arise directly from differences in atomic mass and nuclear characteristics, as even or odd nature of the number of protons and neutrons, their proximity to a magic number*, spin, parity, reaction cross section, etc. Values of neutron capture cross sections of selected elements listed in Table 1.4, show the enormity of differences in nuclear behaviour of isotopes. These are of particular relevance in the atomic age.

Table 1.4 Neutron Capture Cross Sections [12]

Nuclide	σ_c/barn**	Nuclide	σ_c/barn
^{1}H	0.332	$^{135}_{54}Xe_{81}$	2.65×10^{6}
^{2}H	5.2×10^{-4}	$^{136}_{54}Xe_{82}$	0.16
^{3}H	6×10^{-6}	$^{235}_{92}U_{143}$	582.8
^{6}Li	$\sigma(n,\alpha)$ 942	$^{238}_{92}U_{146}$	2.7
^{7}Li	0.045		

* A nuclear shell is complete when its proton or neutron number equals a magic number, i.e. 2, 8, 20, 50, 82, 126. A nucleus with completed, or closed shells, assumes greater stability.

** Unit area of cross section is 1 barn = 10^{-28} m^2.

The very low neutron capture cross section of ^{2}H, in contrast to ordinary hydrogen, makes heavy water an excellent reactor moderator. The very high cross section of ^{6}Li for the reaction ^{6}Li (n, α) T, has led to the commercial isolation of ^{6}Li from neutral lithium in which it occurs to 7.5 atom %. The exceptionally large difference in the neutron capture cross section between the xenon isotopes with 81 and 82 neutrons, (as also between $^{2}H_1$ and $^{3}H_2$) brings out the effect of the magic number, i.e. when a nucleus has already a magic number of neutrons its tendency to capture one more will be very feeble in contrast with its isotope having one neutron less than a magic number.

The most important difference in nuclear isotope effect is the fissionability of uranium isotopes. While ^{235}U can be fissioned by thermal neutrons (σ = 582.2 b), ^{238}U is virtually resistant to fission by thermal neutrons (σ = 2.7 b) (Table 1.4).

These considerations make clear as to why the isotopic separations on a commercial scale of ^{2}H (as heavy water), ^{6}Li and ^{235}U have assumed strategic importance in the present day, in spite of all isotopic separations, or even partial enrichments, being extremely difficult and expensive.

It may further be noted that nuclear isotope effects, which are more profound and alter the very nature of the reaction, are important over all mass ranges, unlike the thermodynamic or kinetic isotope effects. In fact, all nuclear properties including its very stability or otherwise, may vary greatly between isotopes of the same element. For instance, of the 12 isotopes of chlorine, of mass numbers from 32 to 43, only two are stable, namely ^{35}Cl and ^{37}Cl, and the rest are radioactive, decaying by either β^{+} activity (^{32}Cl, ^{33}Cl and ^{34}Cl), or by β^{-} activity (mass numbers 36, and 38 to 43), with half-lives varying from milliseconds, seconds and minutes, and in one case (^{36}Cl) 3×10^{5} years.

1.4.2 Chemical Isotope Effects

Most of the physico-chemical differences in the behaviour of isotopes and isotopic molecules are the result of either a *thermodynamic isotope effect*, involving a shift in the equilibrium in reactions between two isotopic reactants; or a *kinetic isotope effect* involving a difference in the rates of reactions with the two isotopic species. All these isotope effects, whether of the thermodynamic or kinetic type, represent only small changes in the magnitude of the yield or the rate, but no qualitative change in the nature of the reaction. Also, as mentioned earlier, these effects are pronounced only in the case of light elements and cease to be important as the ratio of the masses of the two isotopic species approaches unity. In other words, these isotopic effects become negligible with medium and heavy elements.

The chemical isotope effects occurring either as of thermodynamic or kinetic consequence, are considered in Chapter 3. But it would be useful first, to consider certain theoretical concepts relevant to these effects at this stage.

1.5 SOME QUANTUM-MECHANICAL CONCEPTS

The central problem of interest to the chemist is to know the molecular structure of substances, the location of the different atoms in the molecule, and the bond energies as reflected by the respective frequencies of vibration. Since these are more complicated in polyatomic molecules, a useful beginning is to understand the vibrations in a diatomic molecule which constitute a simple harmonic oscillator.

1.5.1 Energy of a Harmonic Oscillator

The *old* quantum theory for the energy E_v of a harmonic oscillator, had the value

$$E_v = v\, h\nu \tag{1.9}$$

where h is Planck's constant, ν the fundamental vibration frequency and v the vibration quantum number, 0,1,2,3... Thus, for the lowest quantum number, ($\nu = 0$), the oscillator energy would be zero, and all vibrations cease. This in its turn means that in the case of a diatomic molecule, the position, energy, and hence momentum, are each completely defined. However, this is not in accord with observation.

1.5.2 The Uncertainty Principle

A fundamental principle was suggested by Heisenberg, now known after him as *the Principle of Uncertainty*. It is based on the fact that it is impossible to know *exact* values of all the dynamical variables of a system at one and the same time. For instance, in determining the position and momentum of a particle at the same time, there must be some uncertainly Δx in position and Δp in momentum. Both Δx and Δp cannot be reduced to zero simultaneously. The uncertainty principle states that the product of the two errors is approximately of the order of Planck's constant, rather

$$\Delta p \cdot \Delta x = h / 2\pi \tag{1.10}$$

This situation is independent of the perfection of experimental techniques of measurement.

1.5.3 The Zero-Point Energy

The anomaly in regard to the harmonic oscillator due to these considerations is surmounted in the wave-mechanical treatment by adding a constant, $1/2\ h$, to the Planck expression. The energy of the oscillator then becomes:

$$E_v = (v + 1/2)\, h\nu \tag{1.11}$$

This means that even at the temperature of 0 K, and for the lowest vibration quantum number of zero, a *residual energy* of $(1/2)\ h\nu$ remains which is referred to as the *zero-point energy* and the uncertainty principle is accommodated.

The fundamental frequency of a diatomic molecule is given by:

$$\nu = \frac{1}{2\pi}\sqrt{\frac{k}{\mu}} \tag{1.12}$$

where k is a force constant, or the restoring force governing the oscillations between the atoms A and B of a diatomic molecule of reduced mass μ. Assuming that for a pair of isotopic molecules A_1B and A_2B, the force constant remains the same, we have for the ratio of the fundamental frequencies of the two molecules:

$$\nu_2 / \nu_1 = (\mu_1 / \mu_2)^{1/2} \tag{1.13}$$

This means that the shift in the vibration frequency is given by:

$$\frac{\nu_1 - \nu_2}{\nu_1} = 1 - \left(\frac{\mu_1}{\mu_2}\right)^{1/2} \tag{1.14}$$

Table 1.5 lists the values of the reduced masses (μ)*, fundamental frequencies (ν_0)**, zero-point energies (E_0) and the dissociation energies (D_0) of the hydrogen isotopic molecules.

Table 1.5 Characteristics of some isotopic diatomic molecules[13]

Molecule	μ*	$\omega_0/\mathrm{cm}^{-1}$**	$E_0/\mathrm{kJ\ mol}^{-1}$	$D_0/\mathrm{kJ\ mol}^{-1}$
H_2	1/2	4405.3	25.991	430.115
HD	2/3	3817.1	22.552	433.546
D_2	1	3118.8	18.456	437.646
HT	3/4	3608.4	21.225	—
DT	6/5	2845.6	16.861	—
T_2	3/2	2553.8	15.075	—
$H^{35}Cl$*	35/36	2989.0	17.723	427.186
$H^{37}Cl$*	37/38	2987.5	17.715	427.186

1.5.4 Anharmonicity in Molecular Vibrations

An ideal harmonic oscillator is one where the restoring force is propotional to the displacement r and the potential energy curve is parabolic (Fig. 1.2) the potential energy being proportional to r^2. Such a model cannot represent molecular vibrations correctly, except at low energies close to the bottom of the curve. Such a model also cannot account for the occurrence of molecular dissociation when the amplitude of vibration exceeds a certain value. In other words, some anharmonicity has to be envisaged for the vibrations, the

* Following the general practice, the spectroscopic frequencies are given in wave numbers (ω or $\bar{\nu}$) in units of cm^{-1}. The frequency ν per second, is related to it through the velocity of light in cm s^{-1}, thus

$$\nu = c\omega = c\bar{\nu}$$

** in units of mass of H

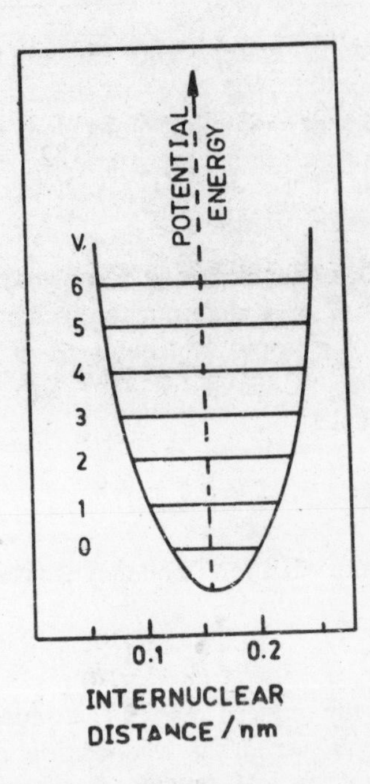

Fig. 1.2: **An ideal harmonic oscillator**—potential energy *vs* internuclear distance,. v—vibration quantum number

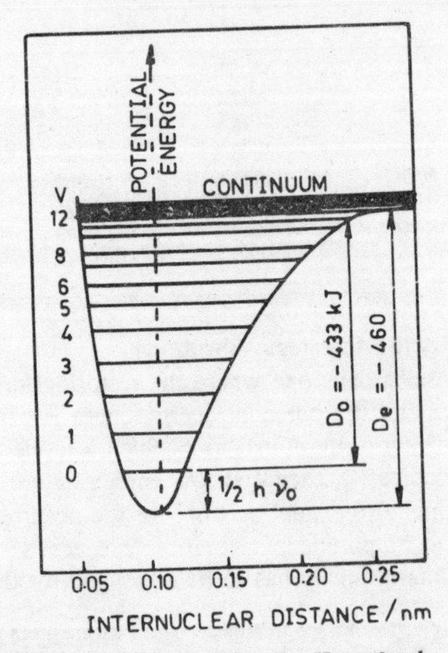

Fig. 1.3: **Potential energy curve for the H_2 molecules showing anharmonicity at high energies.** v—vibration quantum number

restoring force, given by $-(\partial E/\partial r)$ diminishing to zero at a sufficiently large value of r, as depicted in Fig. 1.3 for the hydrogen molecule. The energy levels corresponding to such an anharmonic potential energy curve may be expressed as a power series in $(\nu + 1/2)$, ν being the vibration quantum number.

$$E_v = h\nu \left[\left(v + \frac{1}{2}\right) - \alpha \left(v + \frac{1}{2}\right)^2 + \beta \left(v + \frac{1}{2}\right)^3 \right] \quad (1.15)$$

As may be seen from Fig. 1.3 the energy levels get closer with the increase in the quantum number, till they merge into a continuum. The figure also shows two values of the molecular dissociation energy; the *spectroscopic* value D_e, being the height of the continuum from the minimum of the curve and the *chemical* value D_0 being the height of the continuum above the ground state of the molecule, i.e. at $v = 0$, the difference being simply the zero-point energy:

$$D_e - D_0 = E_0 = \frac{1}{2} h\nu \quad (1.16)$$

References

1. J.J. Thomson, *Phil. Mag.*, 1907, *13*, 561.
2. F.W. Aston, *Mass Spectra and Isotopes*, Longmans Green Co., 1933.
3. F. Soddy, *Ann. Rep. Progress of Chemistry*, Chem. Soc., London, 1911, *99*, 72, *Chemistry of Radioelements*, Longmans Green & Co., 1914.
4. M. Haissinsky, *Nuclear Chemistry and its Applications*, Addison-Wesley Publishing Co. Inc., Reading, 1964.
5. A. K. Kimball, H. C. Urey and I. Kirshenbaum, *Bibliography on Heavy Hydrogen Compounds*, McGraw-Hill, Book Co., Inc., New York, 1949.
6. H.C. Urey, F.G. Brickwedde and G.M. Murphy, *Phys. Rev.* 1932, *40*, 1.
7. G. Wolf. *Isotopes in Biology*, Academic Press, New York, 1964.
8. I. Kirshenbaum, *The Physical Properties of Heavy Water*, McGraw-Hill Book Co., Inc., New York, 1954.
9. *Book of Data*, Nuffield Advanced Series, Penguin Books Ltd., Harmondsworth, 1972.
10. W. F. Giaque and H. I. Johnston, *Nature*, 1929, *123*, 318, 831.
11. R. T. Birge, and A. S. King, *Phys. Rev.*, 1929, *34*, 379.
12. C. M. Lederer and V. M. Shirley, *Table of Isotopes*, Wiley Interscience, New York, 1978.
13. A. I. Brodsky, *Isotope Chemistry*, Academy of Sciences, Moscow, 1957.

Chapter 2

STATISTICAL CONSIDERATIONS

Isotope effects, and some of the important methods of isotope separations, depend on the small differences in the average behaviour of the two isotopic species in their mixture. To understand these differences it is necessary to know how the individual species, with slightly differing specific characteristics, behave when present in a large assemblage of themselves. Expressing properties of bulk and of individuals comprising it, one in terms of the other, the distributions or partitions and their summations, are studied in statistical thermodynamics. Some basic concepts of this subject are outlined here, in so far as they are relevant to an understanding of isotope separations.

2.1 THE BOLTZMANN DISTRIBUTION

We shall be mainly considering systems which have a large but definite number of identical units and fixed total energy, the units non-interacting amongst themselves. Such a collection of units is called a *microcanonical ensemble*. The units may be particles, atoms, molecules, etc., each unit possessing a finite energy. The identical units may be, for certain purposes, distinguishable from one another, while for others they may not be so. The identical units in a crystal lattice are still distinguishable by virtue of their location at distinct sites. On the contrary, the molecules in a gas are entirely delocalized by incessant free translational motion and thus become indistinguishable.

Statistical thermodynamics envisages the assembly of total N units to be partitioned into N_1 units of permitted* energy ϵ_1, N_2 of energy ϵ_2 .... , each above the lowest or the ground state energy ϵ_o possessed by N_0 units, satisfying the requirements:

$$\text{total number of units} = N = \Sigma N_i \quad (2.1)$$

$$\text{total energy of the assembly} = E = \Sigma N_i \epsilon_i \quad (2.2)$$

$$\text{and average energy of a unit} = \bar{E} = E/N$$
$$= \Sigma N_i \epsilon_i / \Sigma N_i \quad (2.3)$$

*Allowed by the laws of quantum theory.

The actual partition of the energy into different groups at a given temperature is given by the *Boltzmann distribution* law.

$$N_i = N_o \exp(-\epsilon_i / kT) \qquad (2.4)$$

where k is the Boltzmann constant ($= 1.3806 \times 10^{-23}$ JK^{-1} $= 8.62 \times 10^{-5}$ eV K^{-1})*. The Boltzmann distribution law applies to all assemblies, whether the constituent units are distinguishable amongst themselves or not.

2.1.1 Degeneracy

Sometimes, more than one, say g_i, distinct states may have the same energy ϵ_i. This *multiplicity* confers a statistical weightage to the particular state and the total number of units in the state is now greater by the factor g_i, which is also referred to as the *degeneracy* of the state or level,

$$N_i = N_o g_i \exp(-\epsilon_i / kT) \qquad (2.5)$$

where the energy ϵ_i is possessed uniquely by a single state, the multiplicity is one and there is no degeneracy, i.e. $g_i = 1$.

So we finally have

$$N = \Sigma N_i = N_o \Sigma g_i e^{-\epsilon_i / kT} \qquad (2.6)$$

This gives for the number in the ground state

$$N_o = N \left[\Sigma g_i e^{-\epsilon_i / kT} \right]^{-1}$$

and allows Eq. 2.5 to be recast as

$$N_i = N \frac{g_i e^{-\epsilon_i / kT}}{\Sigma g_i e^{-\epsilon_i / kT}} \qquad (2.7)$$

Equation 2.7 represents the Boltzmann distribution law in its most general form.

2.2 THE PARTITION FUNCTION

The summation in Eq. 2.6 and in the denominator of Eq. 2.7, made *over all the energy states*, gives the probability that a unit (*e.g.* a molecule) may be present in the (permitted) state of energy ϵ_i. This summation is referred to as the *partition function*, z**.

$$z = \Sigma g_i e^{-\epsilon_i / kT} \qquad (2.8)$$

---

* This is same as the gas constant per *molecule* i.e. $k = R/L$, where R is the gas constant 8.314 3 J mol^{-1} K^{-1} and L is the Avogadro constant $= 6.022\,2 \times 10^{23}$ mol^{-1}

** z stands for *Zustandssumme* (sum-over-states in German).

This is an important function in chemical statistics as it determines the distribution of units among the different energy states in a system.

2.2.1 The Spread of the Partition Function

Omitting the degeneracy factors, partition function can be written as an infinite series, in energy terms:

$$z = \Sigma e^{-\epsilon_i / kT} = e^0 + e^{-\epsilon_1 / kT} + e^{-\epsilon_2 / kT} + \cdots \qquad (2.9)$$

The starting term of the series must be unity as each energy value, ϵ_1, ϵ_2,... ϵ_i is the *energy* in excess of the ground state energy. As only the terms for which $\epsilon \ll kT$ are significant, the series converges more rapidly with larger spacing between successive energy terms and lower temperature.

In an extreme case, as in electronic excitation, where the energy of the lowest excited state $\epsilon_1 \gg kT$ over the usual temperature range, there is no contribution to the partition function from any of the terms except from the ground state term. Hence in such a case, $z = 1$. However, if for several terms ϵ_1, ϵ_2... , the $\epsilon < kT$, then $z > 1$, increasingly so the more such terms. The *magnitude of z is a measure of the extent to which the units of a given assembly can spread* over the various permitted quantum states.

2.3 PARTITION FUNCTION OF MULTIPLE DEGREES OF FREEDOM : THE TOTAL PARTITION FUNCTION

In general, a unit (particle, molecule, etc.) may possess energy in several degrees of freedom as, translation (t), rotation (r), vibration (v) and electronic excitation (e). The total energy of the unit is resolvable into these modes:

$$\epsilon_{total} = \epsilon_t + \underbrace{\epsilon_r + \epsilon_v + \epsilon_e}_{\text{internal}} \qquad (2.10)$$

Of these, free translatory motion is *external* to the unit, displayed only by gases. It is absent in liquids and in crystals (solids). Energy in the other three degrees of freedom, viz., rotation, vibration and electronic excitation, are *internal* modes, being internal to the unit. These latter modes are displayed to different extents by crystals, liquids and polyatomic gases.

All partition functions being exponential terms, we have

$$z_{total} = z_t \cdot z_r \cdot z_v \cdot z_e \qquad (2.10\ a)$$

Of these, z_e being generally unity, may be dropped out. Each of these partition functions is related by precise equations with relevant basic properties of molecules as their masses, moments of inertia, symmetry characteristics, inter-nuclear distances, bond energies and vibration frequencies, rotation energies, etc., which can all be experimentally determined. It is this fact which gives a useful and concrete meaning to partition functions and lifts them from the realm of pure theory. This aspect is considered later (§2.5).

2.4 PARTITION FUNCTIONS AND THERMODYNAMIC PARAMETERS

The partition function is an extremely valuable concept as it permits the calculation of all the thermodynamic parameters of a macroscopic assembly purely from the results of spectroscopic measurements involving groups of individual atoms and molecules.

These relations are illustrated here by expressing the more important thermodynamic quantities as derivatives of partition functions.*

2.4.1 Total Internal Energy

The total internal energy (i.e. the energy due to all internal degrees of freedom) of a system is given by

$$E = \Sigma N_i \epsilon_i \qquad (2.11)$$

Replacing the value of N_i from Eq. 2.7,

$$E = N \frac{\Sigma g_i \epsilon_i e^{-\epsilon_i / kT}}{\Sigma g_i e^{-\epsilon_i / kT}} \qquad (2.12)$$

It may be noted that the summation in the numerator of Eq. 2.12 is simply kT^2 times the differential of the denominator with respect to temperature. Hence

$$E = N k T^2 \frac{\frac{a}{dT}\left[\Sigma g_i e^{-\epsilon_i / kT}\right]}{\Sigma g_i e^{-\epsilon_i / kT}} \qquad (2.13)$$

Since the term under both the summations is nothing but the partition function, z, the above expression for the total internal energy, simplifies to

$$E = N k T^2 \left[\frac{1}{z}\frac{dz}{dT}\right]_V^{**} = N k T^2 \left[\frac{d \ln z}{dT}\right]_V$$

$$= k T^2 \left[\frac{d \ln Z}{dT}\right]_V \qquad (2.14)$$

since $N \ln z = \ln z^N = \ln Z$,

---

* For a deeper study, any work on statistical thermodynamics may be consulted *e.g.*, L.K. Nash, *Elementary Statistical Thermodynamics*, Addison-Wesley, Reading, Massachusetts, 1968. Some more are listed in the *Bibliography*.

** To avoid change in energy due to volume change, accompanying a change in temperature, it is necessary to stipulate that the differentiation with respect to temperature is at constant volume.

where Z is referred to as the *grand or molar partition function,* related to z, the molecular partition function, by the relation

$$Z = z^N \text{ or } z^L \tag{2.15}$$

2.4.2 Molar Heat

Since molar heat at constant volume is by definition:

$$C_v = \left[\frac{dE}{dT}\right]_V = \frac{d}{dT}\left[KT^2\frac{d\ln Z}{dT}\right]$$

$$C_v = \frac{k}{T^2}\left[\frac{d^2\ln Z}{d(1/T)^2}\right] \tag{2.16}$$

2.4.3 Enthalpy

Since by definition $H \equiv E + PV$, by replacing the values for one mole i.e. $PV = RT$ and $Nk = R$ into Eq. 2.13, we arrive at an expression for enthalpy in terms of the partition function,

$$H = RT^2\left[\frac{d\ln Z}{dT}\right]_V + RT \tag{2.17}$$

2.4.4. Entropy

For a change occurring at constant volume, the change in entropy is by definition,

$$dS \equiv q_{rev}/T = C_v\, dT/T \tag{2.18}$$

Replacing C_v by its value in terms of the partition function Eq. 2.16 and by carrying out certain additional operations*, one arrives at

$$S = k\ln Z + kT\left[\frac{d\ln Z}{dT}\right]_V + S_0 \tag{2.19}$$

where S_0 is a constant of integration, which can be identified as the entropy of the system at $T = 0$, for as $T \to 0$, all units of the system tend to settle down in the ground state, making $Z \to 1$. Further in Eq. 2.19, the only terms independent of temperature are $k \ln Z$ and S_0. Hence,

$$\text{at } T = 0,$$
$$S_0 = k\ln Z = 0 \tag{2.20}$$

S_0 being zero, we arrive at

$$S = k\ln Z + kT\left[\frac{d\ln Z}{dT}\right]_V$$

The last term in the above being equal to E/T by virtue of Eq. 2.14, we end

* *See* L.K. Nash, *Ref.* 1.

up as the final expression for entropy

$$S = E/T + k \ln Z \tag{2.21}$$

Incidentally, Eq. 2.19 provides the statistical mechanical basis for the *Third Law of Thermodynamics*, which states that at the temperature of zero *K*, the entropy of perfectly crystalline substances is zero.

2.4.5 Helmholtz Free Energy

The Helmholtz free energy is defined as

$$A \equiv E - TS \tag{2.22}$$

A simple substitution of the partition function value of *S* from Eq. 2.21 leads to

$$A = -kT \ln Z \tag{2.23}$$

2.4.6 Gibbs Free Energy

Lastly, as molar Gibbs free energy is related to the molar Helmholtz free energy by the relation

$$G \equiv \bar{A} + P\bar{V} = \bar{A} + RT,$$

it follows immediately from Eq. 2.23

$$G = -RT \ln\left(\frac{z}{L}\right) \tag{2.24}$$

The functions z, or Z, relevant to this, are both total partition functions.

2.5 PARTITION FUNCTIONS AND MOLECULAR PROPERTIES

It is useful to note that the partition functions (z or Z) appear only as $\ln z$, or $\ln Z$, in the expressions for *all* the thermodynamic quantities. This permits a conversion of the Z_{total} from a product into a sum of separate terms, as shown below, for the internal energy as an example:

$$E = kT^2 \left[\frac{d \ln Z_{total}}{dT}\right]_V \tag{2.14}$$

$$= k\,T^2 \frac{d}{dT} (\ln Z_t \times \ln Z_r \times \ln Z_v \ldots)$$

$$= k\,T^2 \left[\frac{d \ln Z_t}{dT}\right]_V + k\,T^2 \left[\frac{d \ln Z_r}{dT}\right]_V + k\,T^2 \left[\frac{d \ln Z_v}{dT}\right]_V$$

Thus, to find the total internal energy, it is only necessary to know the separate contributions of the different partition functions and add them up. Generally, these individual partition functions of translation, rotation and vibration are related to the physical characteristics and the spectral data of the molecules concerned.

We will now consider the partition functions of translation, rotation and vibration. These relations are briefly presented here for simple types of molecules, without going through their derivations, which are discussed in detail in text books on statistical thermodynamics[1-5]

2.5.1 Energy of Translation

While the classical equation for the kinetic energy of a particle is $(1/2)\,mv^2$, the wave-mechanical expression for the permitted energy of translatory motion, ϵ_t of a particle of mass m is

$$\epsilon_{t(x)} = \frac{n_x^2 h^2}{8\,mx^2} \tag{2.25}$$

where h is the Planck constant, x the distance the particle is free to move along the direction parallel to the X axis and n_x is an integer 0, 1, 2, ... (a quantum number). The general expression for free motion in space along the X, Y and Z directions is, then,

$$\epsilon_t = \frac{h^2}{8m}\left[\frac{n_x^2}{x^2} + \frac{n_y^2}{y^2} + \frac{n_z^2}{z^2}\right] \tag{2.26}$$

The presence of h^2 in the numerator of Eq. 2.26 makes the energy levels lie so closely packed (spacing between the energy levels being of the order of 10^{-40} joule), they may be considered to form a continuum, without any serious error.

On the basis of the equipartition of energy of $(1/2)\,kT$ *per degree of freedom*, the total kinetic energy for the three degrees of freedom due to translatory motion is : $(3/2)\,kT$ per molecule or $(3/2)\,RT$ per mole.

The partition function for one degree of freedom for translatory motion, say along the X direction, is given by:

$$z_{t(x)} = \Sigma e^{-\epsilon_{t(x)}/kT}$$

$$= \Sigma e^{-n_x^2 h^2/8\,mx^2\,kT} \tag{2.27}$$

Because the energy levels form a continuum, as pointed out, the summation in Eq. 2.27 may very well be replaced by an integral

$$z_{t(x)} = \int_0^{\infty} \exp(-\,n_x^2 h^2\,/\,8\,mx^2 kT)\,dn_x$$

which yields[2]

$$z_{t(x)} = \frac{\sqrt{(2\pi\, m\, k\, T)}}{h}\, x \qquad (2.28)$$

And so the complete partition function for the three degrees of freedom of translational motion, is

$$z_t = z_{t(x)} \cdot z_{t(y)} \cdot z_{t(z)} = \left[\frac{2\pi\, m\, k\, T}{h^2}\right]^{3/2} V \qquad (2.29)$$

2.5.2 Energy of Rotation

While in the case of monoatomic gases, the only kind of motion possible is of translation; in molecules containing two or more atoms, rotational and vibrational motions, besides translation, become possible. The classical equation for the energy of rotation is $(1/2)\, I\, \omega^2$, where I is the moment of inertia and ω is the angular velocity of rotation. For a diatomic and linear polyatomic molecules (as O = C = O), there are two degrees of freedom of rotational motion along the two axes perpendicular to the line joining the atoms. The wave-mechanical expression for the permitted rotational energy of such (linear) molecules, considered as rigid rotators (i.e. assuming the absence of atom to atom vibration)*, is

$$\epsilon_{r(J)} = \frac{J\,(J+1)\, h^2}{8\pi^2\, I} \qquad (2.30)$$

where J is the rotational quantum number 0,1,2,... Except for light diatomic molecules as H_2, HD and D_2, the moment of inertia is high. This fact and the presence of h^2 makes the energy levels closely packed to permit them to be considered a continuum. The energy spacing between the rotational levels is of the order of 10^{-23} joule. On the basis of equipartition of energy, the energy for the two degrees of freedom of rotation is kT per molecule or RT per mole.

Finally one arives at

$$z_r = \frac{8\pi^2\, I\, k\, T}{\sigma\, h^2} \qquad (2.31)$$

for the rotational partition function where σ is a *symmetry number* which is the number of times the molecule presents an identical or indistinguishable position during one full rotation of 2π along the axis of symmetry.

2.5.3 Energy of Vibration

Where the moleule has two or more atoms, there can be vibratory or oscillatory motions between every pair of atoms forming a bond. Though atomic vibrations must include some anharmonicity for reasons explained in Sec. 1.5.4, the vibration levels for the present purpose may be assumed to

---

* The case of non-linear molecule is not considered as they involve several moments of inertia.

correspond to simple harmonic oscillators. This is permissible as the vibrational contributions are small at ordinary temperatures. Thus, the energy of an oscillator is

$$\epsilon_v = (v + 1/2)\, h\nu \qquad (1.11)$$

where ν is the fundamental frequency of vibration given by Eq. 1.12 and v is a vibrational quantum number* , 0,1,2,... The quantity (1/2) $h\nu$ is the zero-point energy, persisting even at the temperature 0 K, as described earlier (Sec. 1.5.3). In other words, the lowest vibrational ground state energy is not zero (unlike in translation and rotation) but a finite amount equal to $h\nu/2$. Thus the vibrational levels will be evenly spaced at $1/2\, h\nu$, $3/2\, h\nu$, $5/2\, h\nu$ . . . . .

The partition function for each degree of vibrational freedom is

$$z_v = \sum_v \exp\{-(v + 1/2)\, h\nu/kT\} \qquad (2.32)$$

Drawing the constant zero-point energy factor outside the summation,

$$z_v = \exp(-h\nu/2\,kT) \cdot \sum_v \exp(-vh\,\nu/kT) \qquad (2.33)$$

As the term under summation in Eq. 2.33 includes all values of v = 0, 1, 2, 3, ... the series is

$$1 + e^{-x} + e^{-2x} + e^{-3x} + \ldots$$

whose sum is given by $1/(1-e^{-x})$. Hence,

$$z_v = \exp(-h\nu/2\,kT)\,[1-\exp(-vh\nu/kT)^{-1}] \qquad (2.34)$$

Sometimes, to have a common zero for all modes of motion, translation, rotation and vibration, the zero energy level is taken as the one corresponding to *separated* atoms at rest. In this case, the energy of dissociation of the molecule ϵ_D is to be added to each vibrational level, and the *new* value is $z_v \exp(\epsilon_D/kT)$.

2.6 PARTITION FUNCTIONS AND EQUILIBRIUM CONSTANT

The equation relating Gibbs molar free energy to total partition function,

$$G = -RT \ln\left(\frac{z}{L}\right) \qquad (2.24)$$

is extremely important in chemical thermodynamics, as it enables the

* The vibrational quantum number v should be distinguished from the ν used as an abbreviation for vibrational term in ϵ_v.

calculation of the equilibrium constant for a reaction, i.e. how far the reaction would go at a given temperature, *from purely experimentally determinable quantities*. These are the total partition functions of the reacting substances. We shall illustrate this in the case of a general equilibrium reaction involving only ideal gaseous reactants and products.

The molar free energy of n moles of an ideal gas in the standard state (i.e. at a pressure of one atmosphére), in accord with Eq. 2.24 is

$$\Delta G^o = n\bar{G}^o = - n\,RT \ln \left(\frac{z^o}{L}\right)$$

This holds good for each reactant and each product gas in their mixture under equilibrium at a given temperature, i.e. we can express the standard free energy of any component species G_i^o in terms of its total partition function $z^0{}_i$.

The relation between the net change in free energy ($\Delta G^o = \Sigma G^0{}_{\text{products}} - \Sigma G^0{}_{\text{reactant}}$) and the equilibrium constant K_p is through the van Hoff isotherm

$$- \Delta G^0 = RT \ln K_p \tag{2.35}$$

Applied to a general reaction in equilibrium

$$aA + bB + ... \rightleftharpoons mM + nN + ...$$

Eq. 2.35 takes the form

$$RT \ln K_p = - \Delta G^o = RT \ln \left[\left\{ \frac{(z_M^o/L)^m\,(z_N^o/L)^n \ldots}{(z_A^o/L)^a\,(z_B^o/L)^b \ldots} \right\} e^{-\Delta E_d / RT} \right] \tag{2.36}$$

The quantity in the large (outer) brackets is readily identifiable as the equilibrium constant, made up of two factors. One consists of the product of the total partition functions of the reactants (denominator) and the product of the total partition functions of the resultants (numerator), and the other of $e^{(-\Delta E_d / RT)}$. The former represents the change in free energy of the reaction between the reactants and the products, each component being assumed to be in the *atomic ground state*.

Actually, the equilibrium constant is to represent the *normal molecular* state of the components; hence, the second correcting term involved is in the conversion of atomic $\rightleftharpoons$ molecular states. The *energy input* in splitting *a* molecules of *A*, *b* molecules of *B*, ... into their atoms in the ground states is $(a\ \epsilon_{dA} + b\ \epsilon_{dB} + ...)$, while the *energy output* as the product atoms in the ground state recombine to form m molecules of M, n molecules of N, is $-(m\ \epsilon_{dM} + n\ \epsilon_{dN} + ...)$.

The *net* difference of these two operations is

$$\Delta \epsilon_d = -[\, m\ \epsilon_{dM} + n\ \epsilon_{dN} + \cdots - a\ \epsilon_{dA} - b\ \epsilon_{dB} - \cdots\,]^* \tag{2.37}$$

* $\Delta \epsilon_d$ may also be looked upon as the energy of the reaction at the temperature of 0 K.

The corresponding value for a moles of A reacting with b moles of B to form m moles of M and n moles of N is $\Delta E_d = L \Delta \epsilon_d$; hence, the correcting factor in Eq. 2.37.

Focussing on the main result, we learn that the equilibrium constant of a reaction can be calculated in terms of the total partition functions of the component species participating in the equilibrium, and these functions in their turn can be obtained from purely physical spectroscopic data. The importance of this cannot be over emphasised.

Expressed in terms of relevant partition functions, the equilibrium constant acquires a simple statistical interpretation as the ratio of the sum of the probabilities, that the system can be found at equilibrium in one of the energy levels of the product components, to the sum of the probabilities that it be found in one of the energy levels of the reactant components. *In simple words, the equilibrium constant indicates the extent to which a reversible reaction would go in the given direction at a given temperature.*

References

1. L.K. Nash, *Elements of Statistical Thermodynamics*, Addison-Wesley Publishing Co., Reading, Mass., 1968.
2. T.L. Hill, *Introduction to Statistical Thermodynamics,* Addison Wesley Publishing Co., Mass. 1960.
3. O.K. Rice, *Statistical Mechanics, Thermodynamics and Kinetics*, W.H. Freeman & Co., San Fransisco., 1967.
4. J.E. Mayer and M. Mayer, *Statistical Mechanics*, Wiley, New York, 1940.
5. G.S. Rushbrooke, *Introduction to Statistical Mechanics*, Oxford University Press, London, 1949.

Chapter 3

CHEMICAL ISOTOPE EFFECTS

3.1 REVERSIBLE AND IRREVERSIBLE PROCESSES

Any purely physical or physico-chemical phenomenon may be symbolically represented by

$$\text{State I (initial)} \longrightarrow \text{State II (final)}$$

The transformation may be *either* one of *reversible equilibrium process*

$$\text{State I} \rightleftharpoons \text{State II}$$

the changes in the two directions proceeding with equal rates and hence independent of time, *or* it may be an *irreversible kinetic* (rate) process state I going over to state II with a certain rate and hence time-dependent, and there is no equilibrium. There can be isotope effects in both the equilibrium and the kinetic processes. We will now consider the two separately.

3.1.1 Thermodynamic or Equilibrium Isotope Effect

Let us consider a system under equilibrium as

$$AB + C \rightleftharpoons ABC$$

Though the equilibrium considered here is a typical chemical one, the treatment holds good for all types of equilibria, whether it be chemical, or purely physical as

$$\text{liquid} \rightleftharpoons \text{vapour system.}$$

If A consists of two isotopic atoms A_1 and A_2 , there would be two isotopic molecules A_1B and A_2B reacting with C and in equilibrium with two isotopic product molecules A_1BC and A_2BC involving two equilibria with different constants K_1 and K_2, mainly due to the difference in the zero-point energies of A_1B and A_2B

$$A_1B + C \rightleftharpoons A_1BC \ ... \ (K_1) \qquad \text{(i)}$$

$$A_2B + C \rightleftharpoons A_2BC \ ... \ (K_2) \qquad \text{(ii)}$$

Let us put x and $(1-x)$ for the mole fractions of A_1B and A_2B in the species AB, and y and $(1-y)$ for the mole fractions of A_1BC and A_2BC in the species ABC. Further let the total concentrations of the species be C_{AB}, C_C and C_{ABC}, the subscript distinguishing the species. We then have

$$K_1 = \frac{[A_1BC]}{[A_1B]\,[C]} = \frac{y\,C_{ABC}}{x\,C_{AB} \cdot C_C}$$

$$K_2 = \frac{[A_2BC]}{[A_2B]\,[C]} = \frac{(1-y)\,C_{ABC}}{(1-x)C_{AB} \cdot C_C}$$

Replacing K_1/K_2 by K we arrive at

$$\frac{y}{1-y} = K\,\frac{x}{1-x} \qquad (3.1)$$

$$\text{or} \quad K = \frac{y}{x} \cdot \frac{1-x}{1-y}$$

Here $(K-1)$ is the *thermodynamic* or *equilibrium isotope effect.* Where only two simple isotopic molecules are concerned, the ratio of their total partition functions Z_1/Z_2 directly gives the net equilibrium constant i.e.

$$K = Z_1 \,/\, Z_2 \qquad (3.2)$$

By reversing the sense of (ii) and adding the same to (i), we have

$$A_1B + A_2BC \rightleftharpoons A_2B + A_1BC \qquad \text{(iii)}$$

which is characterized by the same equilibrium constant $K = K_1/K_2$. Reactions of this type are referred to as *isotope exchange reactions.* The product ABC enriched in one of the isotopes can be easily separated, being chemically different from AB, which is enriched at the same time in the other isotope.

If A_1 and A_2 were totally identical with the same zero-point energy, the total partition functions and the chemical potentials of the different isotopic molecules would have been the same and $K_1 = K_2$ and $K = 1$. The larger the deviation of K from unity, the larger the equilibrium isotope effect, being the same as $(K-1)$. The effect is the largest between the isotopes of hydrogen as revealed by the values of K at 25°C for the following exchange reactions[1-3]

$$H_2 + D_2 \rightleftharpoons 2\,HD \qquad K = 3.28 \text{ (experiment)} = 3.33 \text{ (theory)}$$

$$HT + H_2O \rightleftharpoons H_2 + HTO \qquad K = 6.26 \text{ (experiment)} = 6.19 \text{ (theory)}$$

As an example we cite the exchange of the nitrogen isotopes between ammonia gas and ammonium ions in solution,

$$^{15}NH_{3(g)} + {}^{14}NH_4^+(aq) \rightleftharpoons {}^{14}NH_{3(g)} + {}^{15}NH_4^+(aq)$$

the equilibrium constant has the value[1-3]

$$\frac{[^{14}NH_{3(g)}]\ [^{15}NH^+_{4(aq)}]}{[^{15}NH_{3(g)}]\ [^{14}NH^+_{4(aq)}]} = K = 1.034 \text{ (experiment)} = 1.035 \text{ (theory)}$$

This shows the decrease of the isotope effect with increase of the mass. Though $(K-1)$ is not large, it is adequate to permit the separation of pure ^{15}N (natural abundance = 0.36 per cent) by this exchange reaction, the technical details of which are described in a later section (Sec. 7.3.6). Values for the equilibrium isotope effect have been determined for a very large number of exchange reactions involving the isotopes of H, C, N, O, S, Cl, Br, I and some other elements[1,4].

3.2 THE KINETIC ISOTOPE EFFECT

In addition to the equilibrium effect considered above, two isotopic molecules A_1B and A_2B may react with C with *different rates* R_1 and R_2 involving *rate constants* k_1 and k_2*

$$A_1B + C \rightarrow A_1C + B \quad \ldots \quad R_1, k_1$$

$$A_2B + C \rightarrow A_2C + B \quad \ldots \quad R_2, k_2$$

If Δn_1 and Δn_2 be the amounts of A_1C and A_2C formed in a small interval of time Δt, and adopting the same notation for the respective mole fractions as under § 3.1, we have

$$y = \frac{\Delta n_1}{\Delta n_1 + \Delta n_2} \quad \text{and} \quad (1-y) = \frac{\Delta n_2}{\Delta n_1 + \Delta n_2}$$

---

* According to the law of mass action, the *rate* of the reaction is given by

$$R = -\,d[AB]\,/\,dt = k\,[AB]\,[C]$$

which yields

$$\frac{y}{(1-y)} = \frac{\Delta n_1}{\Delta n_2} = \frac{K_1 [A_1B][C] \Delta t}{K_2 [A_2B][C] \Delta t} = \frac{k_1 x}{k_2 (1-x)} \tag{3.3}$$

If we replace the *ratio* of the two rate constants by k we arrive at

$$\frac{y}{(1-y)} = k \frac{x}{(1-x)} \tag{3.4}$$

This is the *kinetic isotope effect*, formally similar to the equilibrium isotope effect (Eq. 3.1)*. A separation of the product from the reactant results in the enrichment of the isotopes.

The kinetic isotope effect, as all other isotope effects, is maximum for the isotopes of hydrogen, being 60 and 18 at 25°C in k_H/k_T and k_H/k_D respectively. The value falls to 1.3 in k_{10B}/k_{11B} and ceases to be important in other elements.

3.2.1 Theory of the Kinetic Isotope Effect

As the mean thermal velocity of a gas molecule varies inversely as the square root of its mass**, isotopic molecules will have a small difference in their mean velocities, which implies a lower collision frequency***. This, together with the difference in the zero-point energies of isotopic reactant molecules (§ 1.5.3), and the difference in the energy of activation needed for a reaction (see below), results in the kinetic isotope effect.

The theory of the kinetic isotope effect is an extension of Eyring's classical theory[5] of absolute reaction rates which postulates the formation of an *activated complex* $[A....B...C]^{\ddagger}$ between the reactant molecules AB and C, as an intermediate in a *transition state* which eventually breaks into product molecules AC + B. The activated complex is invariably in a higher energy state ϵ_a over the ground state of the reactant molecules and only a fraction ($e^{-\epsilon_a/kT}$) of the latter possesses the energy of activation ϵ_a and these only can form the activated complex. A thermodynamic equilibrium is envisaged between the *reactant molecules* and the *complex* as shown below

$$AB + C \underset{}{\overset{K\ddagger}{\rightleftharpoons}} [A..B..C]^{\ddagger} \longrightarrow AC + B$$

---

* The mole fractions x and y will recur in the same form in the expression for the isotope separation factor as well (§ 4.1).

** The mean molecular velocity is given by $C = (8kT/\pi m)^{1/2}$

*** The collision frequency (in unit volume in unit time), is given by $f = \sigma n^2 \bar{c} \sqrt{2}$, where n is the number of molecules in unit volume, $\bar{c}$ the mean molecular velocity, and σ is the collision cross section, given by πd^2, where d is molecular diameter.

where K‡, the equilibrium constant, is given by

$$K^{\ddagger} = \frac{[A..B..C]^{\ddagger}}{[AB][C]} \tag{3.5}$$

It is thermodynamically related to the change in the free energy $(\Delta G^o)^{\ddagger}$ and the relevant partition functions by the relation:

$$RT \ln K^{\ddagger} = -\Delta G^{o\ddagger}$$

$$= RT \ln \left[\frac{(z_t^o / L)^{\ddagger}_{A..B..C}}{(z_t^o / L)_{AB} (z_t^o / L)_C} e^{-\Delta E_d / RT} \right] \tag{3.6}$$

where z_t^o is the total partition function of the species indicated by the subscript, and ΔE_d is the correction factor involved in the conversions of atomic-molecular states of reactants and products. The rate of the overall reaction of transformation of reactant molecules into the products is given by*

$$-\frac{d[AB]}{dT} = k' = \frac{kT}{h} K^{\ddagger} \tag{3.7}$$

where kT/h represents the frequency of the activated complex breaking into the products.

The kinetic isotope effect, or the difference in the rate constants between two isotopic reacting molecules on this theory, is simply the difference in the equilibrium constants $K_1^{\ddagger}$ and $K_2^{\ddagger}$ and the latter are defined by the partition functions as in Eq. 2.37. Though the total partition function for a polyatomic molecule is a lengthy and complicated expression involving several rotational and vibrational constants characteristics of the molecule, most of these fortunately cancel out—being the same for a pair of *isotopic* molecules. Thus, for two isotopic molecules A_1B and A_2B in their reaction with C.

$$A_1B + C \underset{}{\overset{K_1^{\ddagger}}{\rightleftharpoons}} [A_1BC]^{\ddagger} \overset{k_1}{\rightleftharpoons} A_1C + B$$

$$A_2B + C \underset{}{\overset{K_2^{\ddagger}}{\rightleftharpoons}} [A_2BC]^{\ddagger} \overset{k_2}{\rightleftharpoons} A_2C + B$$

---

* The symbol for the constant of the overall reaction rate is primed (k') to distinguish the same from that of the Boltzmann constant (k) when both occur in the same equation.

we have by Eq. 3.7

$$\frac{k_1}{k_2} = \frac{K_1^{\ddagger}}{K_2^{\ddagger}} = \frac{[z]^{\ddagger}_{A1BC}}{[z]^{\ddagger}_{A2BC}} \times e^{-(\Delta E_{D1} - \Delta E_{D2})} \qquad (3.8)$$

Bigeleisen, who has made important contributions to the subject, showed that the ratio of the rates is determined by the symmetry numbers ($\sigma,\sigma^{\ddagger}$), the function of zero-point energies (ϵ_{01} and ϵ_{02}) and the activation energies (E_{a1} and E_{a2}), for the two isotopic reactant molecules. The final expression reads[6]

$$\frac{k_1}{k_2} = \frac{\sigma_1\,\sigma_2^{\ddagger}}{\sigma_2\,\sigma_1^{\ddagger}} \left(\frac{\mu_2}{\mu_1}\right)^{1/2} \left[1+f(\epsilon_0) - f(\epsilon_0)^{\ddagger}\right] \qquad (3.9)$$

The lowest vibration frequency and hence the zero-point energy for the heavier isotopic molecule (symbol with subscript 2) is lower, including its energy of activation. However, normally $\Delta E_a < \Delta\epsilon_0$ (*vide* Fig. 3.1).

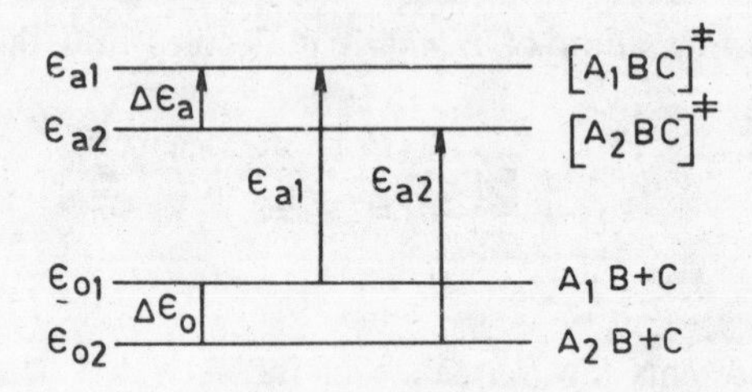

Fig.3.1: **Energy levels of isotopic reactant molecules,** $A_1B + C$ **and** $A_2B + C$; $[A_1...B...C]^{\ddagger}$ **and** $[A_2...B...C]^{\ddagger}$ **transition complexes**

It needs to be clarified that the reduced mass and the frequency of vibration involved are for that pair of atoms in the molecule *the vibration between which becomes rigid and ends in translation* in the direction of the reaction coordinate, resulting in the rupture of that particular bond. This is the primary step in the activated complex leading to products.

For instance, in the reactions

$$CH_3I + HI \rightarrow CH_4 + I_2$$

and

$$CH_3I + DI \rightarrow CH_3D + I_2$$

for which k_H/k_D is 1.42 at 270° C[7], the activated complex is $[H_3C..I..I..H]^{\ddagger}$ and $[H_3C..I..I..D]^{\ddagger}$, respectively. The isotope effect is in the change of a vibrational motion into translation between the pair of atoms $[...I..H]^{\ddagger}$ and $[I...D]^{\ddagger}$. The parameters μ, ϵ_0 and E_a of Eq. 3.9 refer to this pair of atoms in the reacting molecule. Similarly, in the decarboxylation of malonic acid:

$$H_2{}^{12}C\begin{cases} {}^{12}COOH \\ {}^{13}COOH \end{cases} \rightarrow \begin{cases} {}^{12}CH_3{}^{13}COOH + {}^{12}CO_2 & (k_{12}) \\ {}^{12}CH_3{}^{12}COOH + {}^{13}CO_2 & (k_{13}) \end{cases}$$

the crucial bond is $[..{}^{12}C...{}^{12}C..]^{\ddagger}$ or the $[..{}^{12}C...{}^{13}C..]^{\ddagger}$ whose rupture is involved in the kinetic isotope effect amounting to $k_{12c}/k_{13c} = 1.028$ at 138 °C[8].

References

1. M. Haissinsky, *Nuclear Chemistry and Its Applications* (Addison-Wesley Publishing Co. Inc., Reading, Mass, 1964).
2. H.C. Urey, J.R. Huffmann, H.G. Thode and M. Fox, *J. Chem. Phys.*, 1937, *5*, 856.
3. H.G. Thode and H.C. Urey, *ibid*, 1939, *7*, 34.
4. D.R. Stranks, and G.M. Harries, *J. Am. Chem. Soc.*, 1953, *75*, 2015.
5. H. Eyring, *J. Chem. Phys.*, 1935, *3*, 107 and *Chem. Revs.*, 1935, *17*, 65.
6. J. Bigeleisen, *J.Chem. Phys.*, 1949, *17*, 344, 425, 675; 1953,*21*, 1333; *J. Phys. Chem.* 1952, *56*, 823.
7. T.W. Newton, *J. Chem. Phys.*, 1950, *18*, 797
8. P.E. Yankwich, A.L. Promislow and R.F. Nystrom, *J. Am. Chem. Soc.*, 1954, *76*, 5893.

Chapter 4

ISOTOPE ENRICHMENT

4.1 COSMOS OF ISOTOPIC HOMOGENEITY

The results of the previous sections show the feebleness of chemical isotope effects, except in the case of hydrogen. This implies that any significant isotopic enrichment would be a difficult process, which is reflected in the near constancy of the isotopic composition of all terrestial matter. Except for small variations in the *D/H* ratio around the mean value of 0.0148, in the composition of waters of different sources, and in the isotopic ratios of matter of meteoretic or extra-terrestrial origin, the cosmos we live in is isotopically uniform.

Life is possible without any need for isotopically altered matter, except in the case of a few elements as D, ^{13}C, ^{15}N and, relatively recently ^{57}Fe*, and these too were needed in minute amounts for research. These demands were met by a few Isotope Pools maintained in laboratories.

4.2 NEEDS OF THE ATOMIC AGE

But the atomic age has changed all this. Today vast amounts, measured in tonnes, of highly enriched, sometimes absolutely pure isotopes of certain elements are needed. These include ^{2}H, ^{3}H, ^{6}Li, ^{7}Li, ^{10}B, ^{233}U, ^{235}U and some others, each on account of its certain special nuclear property. Continuous work on perfecting several sophisticated techniques, each specially suited for these isotopes, has been in progress since the Second War, with excellent results, though at phenomenal costs.

4.3 ISOTOPE SEPARATION AND ENRICHMENT FACTORS

4.3.1 The Separation Factor

The abundance of a particular isotope in an element or in a mixture of

* The demand for the 2.14% abundant ^{57}Fe, never felt in the past, suddenly rose since the discovery of the Mössbauer effect (the recoilless nuclear resonance absorption) in 1958.

isotopic molecules is given by

$$R = \frac{x_0}{1-x_0} \tag{4.1}$$

where x_0 is the mole fraction of that isotope, and $1-x_0$ the mole fraction of all the other isotopes. When the mixture is subjected to a unit operation of isotope separation by any method, there result two fractions—one slightly enriched and the other correspondingly depleted in respect of the particular isotope. If y and x are the mole fractions of the isotope of interest in the enriched and depleted fractions, we arrive at the ratios:

$$R_{(enriched)} = \frac{y}{1-y}$$

and

$$R_{(depleted)} = \frac{x}{1-x} \tag{4.2}$$

and the single operation or *elementary separation* factor (α) is defined as the ratio

$$\alpha = \frac{y}{1-y} \Big/ \frac{x}{1-x} \tag{4.3}$$

More often the separation factor, following a single operation, is taken simply as the ratio f'/f where f' and f are the abundances of the isotope in the *enriched* and *initial* samples. This is permissible as long as α is only slightly in excess of unity. When the same operation is repeated successively N times on a sample to achieve higher enrichment, the *overall or global* separation factor is given by $A = \alpha^N$

4.3.2 The Enrichment Factor

It is conventional to choose for reference an isotope that would lead to $\alpha >$ 1, such that the *enrichment factor*

$$\epsilon = (\alpha - 1) \tag{4.4}$$

stays positive. It may be noted that in an isotopic exchange equilibrium, as described under § 3.1.1, the separation factor as defined by Eq. 3.1 also represents the equilibrium isotope effect:

$$\alpha = \frac{K_1}{K_2} = K \tag{4.5}$$

as both are represented by the ratio of isotopic abundances. Later, it would be shown to be applicable to irreversible processes as well.

4.3.3 The Production Separation Factor and the Fractionation Factor

Sometimes it becomes necessary for theoretical calculations to include the *production separation* and the *fractionation factors.* The former (β) correlates the isotopic composition of the product (enriched or impoverished) with that of the initial mixture, being defined in terms of the corresponding mole fractions.

$$\beta = \frac{y}{1-y} \bigg/ \frac{x_0}{1-x_0} \tag{4.6}$$

which is formally similar to the definition of α (Eq. 4.3).

Most often, in actual isotope separation processes, β is nearly the same as the elementary separation factor α, the more nearly so when the actual mass m of the separated product is smaller, relative to the initial mass M of the mixture treated. In other words, where the composition of the *residual* material after processing differs only slightly from the untreated initial substance, $\beta \simeq \alpha$.

The relation between the two separation factors is through θ the *fractionation factor* whose value lies $0 < \theta < 1$, the precise relation being

$$\beta - 1 = \frac{(\alpha-1)(1-\theta)}{1+\theta(\alpha-1)(1-y)} \tag{4.7}$$

as θ tends to 0, β tends to equal α. If in a special case where the treatment leads to nearly equal fractionation of M (the initial mass) into product and residue, i.e. $\theta = 1/2$ Eq. 4.7 simplifies to

$$(\beta - 1) \simeq 0.5\,(\alpha - 1) \tag{4.8}$$

This follows the assumption that $(\alpha - 1)$, in the denominator of Eq. 4.7, being very small, the entire denominator is sensibly equal to unity.

4.3.4 The Enhancement Factor

Another mode of expressing the isotopic enrichment adopted by physicists employing electromagnetic separators, is the *enhancement factor Y;* defined in terms of percentage compositions, C and $(100-C)$ as the percentage compositions of the isotope of interest, and of the composition of the other isotope(s).

The enhancement factor is defined as

$$\gamma = \frac{C_f}{100-C_f} \bigg/ \frac{C_i}{100-C_i} \qquad 4.8\,(a)$$

where the subscripts i and f stand for the initial and the separated final samples. Obviously, this is the same as the more commonly used separation factor α defined in terms of mole fractions x and y (Eq. 4.3).

4.4 OPERATION IN A CASCADE

As the enrichment resulting in a single operation is very feeble (except in an electromagnetic separator), the system is repeatedly subjected to the same operation in a countercurrent fashion. A certain number of units are joined in series such that the processed material of the *i*th stage becomes the starting material of the $(i + 1)$th stage.

At each stage, the enrichment goes on increasing progressively, upto a point. The initial material enters at some middle stage and the two fractions, one enriched and the other depleted in respect of the selected isotope, move countercurrently along the cascade.

4.4.1 The Functioning of a Cascade

A unit stage (*n*th) of a cascade is shown in Fig. 4.1(a). It is assumed that equilibrium is attained between the phases at the barrier. The total flux x of initial material entering the unit is resolved in two fractions : y_P the enriched product leaves at the top and x_R the depleted reject leaves at the bottom. The complete cascade consists of several such stages in a series-parallel arrangement. Stages above, as $(n + 1)$, $(n + 2)$, ... constitute enrichment stages, while those below as $(n - 1)$, $(n - 2)$, ... constitute depletion stages, (Fig.4.1 (*b*)). The upper and lower limits of enrichment and depletion and

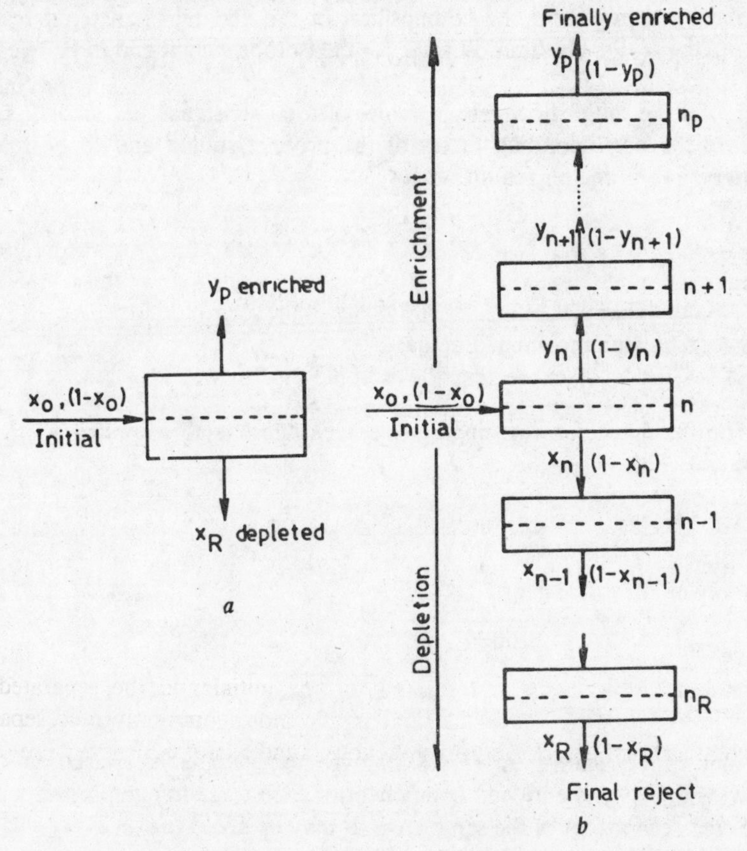

Fig. 4.1: **Isotope separation by gaseous diffusion:**a—unit stage; b—multistage cascade; ----diffusion barrier.

hence the total number of stages in a *series* are determined by the degree of enrichment desired and the value of the initial material affordably rejected respectively. The number of units in *parallel* to any stage only multiplies the yield of that stage.

For convenience and clarity, we will now repeat the system of symbols used, all *x*s and *y*s being mole fractions of the species as indicated below (Fig.4.1 (*b*)).

x_o and $1-x_o$ represent the composition of the initial substance for the isotope sought and the other isotope(s) to be rejected.

y_n and $1-y_n$ represent the composition of the fraction leaving stage *n* and reaching stage $n+1$ for the isotope sought and the other(s) to be rejected.

x_n and $1-x_n$ represent the composition of the fraction leaving stage *n* and descending to stage $n-1$ for the isotope sought and to be rejected.

y_P and $1-y_P$ represent the composition of the end product collecting at the topmost stage for the isotope sought and to be rejected.

x_R and $1-x_R$ represent the composition of the end reject material reaching the bottommost stage for the isotope sought and to be rejected.

If J_o is the initial flow rate in moles of total substance per second, and J_P and J_R the corresponding fluxes of the product sought and to be rejected, conservation of matter requires that

$$J_o = J_P + J_R \tag{4.9}$$

Conservation applied to the isotope sought requires

$$J_o\, x_0 = J_P Y_P + J_R X_R \tag{4.10}$$

We further have for the single stage separation factor, applied to the *n*th stage (Eq. 4.3),

$$\frac{y_n}{1-y_n} = \alpha \frac{x_n}{1-x_n}$$

On solving for y_n we get

$$y_n = \frac{\alpha x_n}{1+(\alpha-1)\, x_n} \tag{4.11}$$

A plot of y_n against x_n (Fig.4.2) gives the progressive variation of the composition of the enriched component of each stage in equilibrium with the depleted component of the same stage. It may be noted that at every stage except at the extreme ends of $x = 0$ and $x = 1$ which are irrelevant for any mixture of isotopes.

A cascade may be operated under condition of (*a*) *total reflux* resulting in no net yield or (*b*) *steady flow* of initial substance and continuous withdrawal of the two end products enriched and depleted. Working under total reflux is of no practical interest as there is no yield, yet the results are of importance as they set limiting conditions of relevance in designing a steady flow cascade. The two modes of cascade functioning are briefly described below.

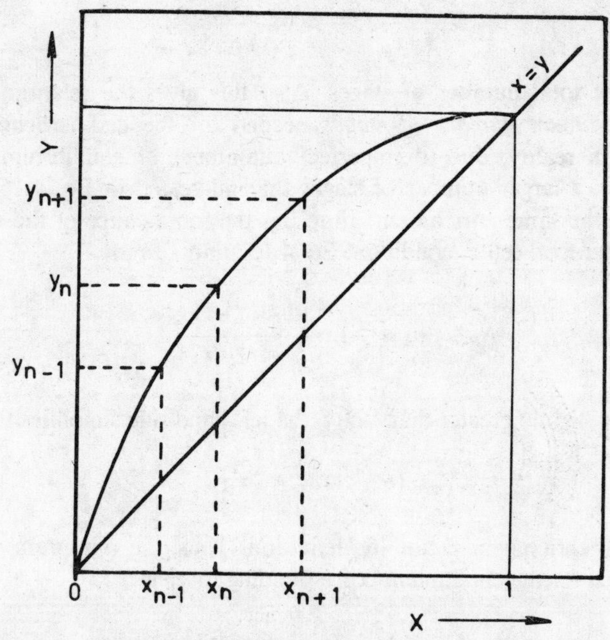

Fig. 4.2: **Isotope equilibrium compositions at different stages:** *y*s are enriched fractions moving forward and *x*s are depleted fractions moving backward

4.4.2 Cascade Under Total Reflux

Under conditions of total reflux the system works in a closed cycle with no net yield of either fraction. However, it is assumed that there is complete equilibrium across the barrier of each stage, i.e. no further change in isotopic compositions occurs with the passage of time. The flux in the upward direction must equal that in the countercurrent downward direction, i.e. $\overrightarrow{J} = \overleftarrow{J}$. Applied to the fluxes at the *n*th stage

$$\overrightarrow{J} = ky_n = kx_{n+1} = \overleftarrow{J} \tag{4.12}$$

where k is flux constant in moles/s.

Hence,

$$y_n = x_{n+1} \tag{4.13}$$

This last permits the recasting of the expression in the single stage separation factor Eq. 4.3 as

$$\frac{x_n + 1}{1 - x_{n+1}} = \alpha \frac{x_n}{1 - x_n} \tag{4.14}$$

An extension of this concept leads to the relation between the end products represented by mole fractions y_p and x_R and the global separation factor α^N

$$\frac{y_p}{1 - y_p} = \alpha^N \frac{x_R}{1 - x_R} \tag{4.15}$$

where N is the total number of stages. Also this gives the relation for the limiting or *minimum number of stages* needed for the end enrichment y_P sought after. In reality, due to imperfect attainment of equilibrium in the different stages, a larger number of stages than indicated in Eq. 4.15 would be needed for the same enrichment. Here lies the significance of the analysis of cascade under total reflux condition. Eq. 4.15 simplifies to

$$N_{min} \cdot \ln \alpha = \ln \frac{y_P/(1-y_P)}{x_R/(1-x_R)}$$

Since α is only slightly greater than unity, the left hand side simplifies to

$$N_{min}\,(\alpha - 1) = N_{min}\,\epsilon$$

where ϵ is the enrichment factor by definition. Thus, the minimum number of stages needed for the enrichment corresponding to y_P is

$$N_{min} = \frac{1}{\epsilon} \cdot \ln \frac{y_P/(1-y_P)}{x_R/(1-x_R)} \tag{4.16}$$

4.4.3 Functioning of a Cascade under Steady Flow

Next we consider the functioning of the cascade under conditions of steady flow i.e. the initial substance is continuously fed into the cascade at some stage at a steady rate and the enriched and rejected samples withdrawn at higher and lower stages respectively also at steady rates. Evidently, under these conditions the thermodynamic equilibrium is perturbed at every stage and the global enrichment is less than under total reflux, which in other words means, the number of plateaux or stages needed is much higher than under total reflux for a given enrichment. Evidently, lower the yield extracted, i.e. the lower the flow rates (J_o, J_P and J_R) less the deviation from equilibrium in each stage and the cascade functioning approaches more closely the condition of total reflux.

It is just this factor which is of supreme importance, namely the optimization of the flow rates. As cost considerations also enter, the calculations become involved. The principal parameter is the ratio of the flux of the final product to that of the input (J_P/J_O), and this is to be optimized. However, increasing the ratio beyond a critical value results in zero enrichment. The optimum condition is found to be

$$\left(\frac{J_p}{J_o}\right)_{max} \simeq \frac{Y_p - X_n}{X_n} \frac{1}{\alpha - 1} = \frac{Y_p - X_n}{\epsilon X_n} \quad (4.17)$$

This is the condition to meet minimum reflux or optimum yield. The essential point to note from Eq. 4.17 is that *the input flux at every stage has to vary with the stage number*, i.e. increase with the latter.

4.4.4 The Ideal Cascade

As indicated in Fig. 4.1 (*a*) three parameters are involved at each stage : x, the composition of the sample entering before the barrier of the stage, y, the composition of the enriched product leaving at the top to below the barrier of the next higher stage, and x', the composition of the depleted residue returning to below the barrier of the stage immediately below. These are depicted in Fig. 4.3. The material fed below the barrier of the $(n+1)$th stage

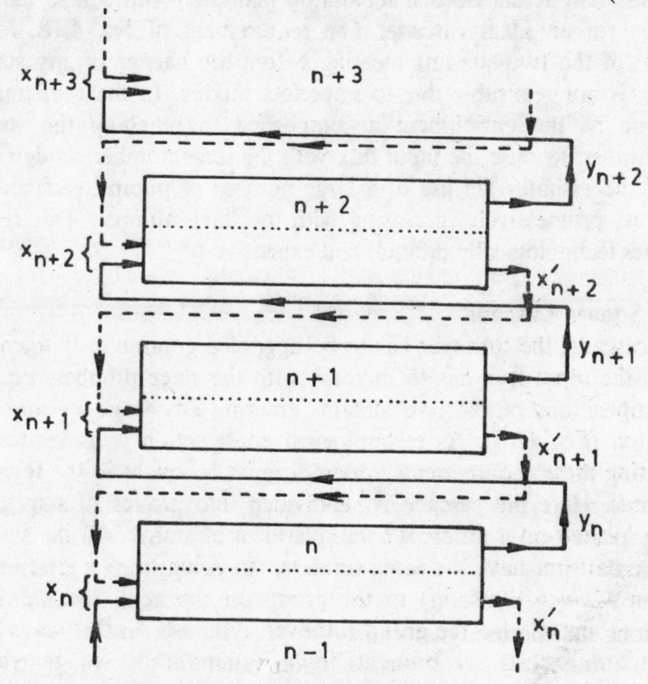

Fig. 4.3: Principle of the ideal cascade

is x_{n+1} and this is made up of two streams (1) an enriched fraction y_n coming from the stage just below it (the nth stage) and (2) a depleted fraction x_{n+2} rejected by the next higher stage $(n + 2)$.

$$x_{n+1} = y_n + x'_{n+2}$$

In an *ideal cascade*, the composition of the fractions before each barrier is the same, or very nearly the same, i.e.

$$x_{n+1} = y_n = x'_{n+2} \tag{4.18}$$

Defined this way, it can be shown that the minimum number of stages for an ideal cascade for a given enrichment y_P is

$$N_{min} = \frac{1}{\beta-1} \ln \frac{y_P / (1-y_P)}{x_R / (1-x_R)} \; * \tag{4.19}$$

where $\beta-1 = (\alpha-1)/2$ or $\epsilon/2$ (Eq. 4.8).

In an ideal cascade, all stages operate under identical conditions and hence have the same stage separation factor α_N. Further, there are no losses due to inefficient mixing.

A comparison of Eq. 4.19 with Eq. 4.16 shows that the number of stages needed in an ideal cascade is twice the number for a cascade functioning under total reflux.

In the case of an actual isotope separation plant, it is difficult to maintain the conditions of an ideal cascade. The requirement of Eq. 4.18, for the compositions of the two streams meeting before the barrier of any stage to be the same, is not realizable due to imperfect mixing. In the remixing that follows, some of the enrichment accomplished in reaching the stage is destroyed. Further, to raise the input flux with the stage number, as demanded by Eq. 4.17, necessitates the use of a large number of pumps, each rated to provide a flux, progressively increasing with the stage number. This requirement becomes technologically difficult and expensive.**

4.4.5 The Square Cascade

In an ideal cascade, the turnover has to be upgraded continuously from stage to stage, i.e. the input flux has to increase with the stage number, Eq. 4.17, and the compositions of the two streams entering any stage y_N and x'_{N+2} have to match (Eq. 4.18). As technological construction considerations go against meeting these requirements, a compromise is sought in the form of a *square cascade*. Here the cascade is subdivided into groups of stages, each group being located on a different level, platform or storey. All the stages in a group or a platform have the same turnover. In going from a group at one level (i.e. on a given platform) to the group on the next, the number of stages in the group, as also the group turnover, diminish. In this way, design and manufacturing costs are brought down pyramidically. In general, the

* See Sec. 4.3.3 for the definition of the product separation factor β and its relation to α the single stage separation factor.

** To effect a ten-fold enrichment of ^{235}U from its abundance of 0.72% in natural uranium by the process of gaseous diffusion of UF_6, for which α is 1.0014, a cascade of 1650 stages is needed (see § 10.2.2).

minimum turnover (moles/s) for a stage of any group in a square cascade is about half the turnover in an ideal cascade.

In a square cascade, the mole fraction of the enriched output sample is not a simple or unique function of the stage number, as in an ideal cascade*. The turnover at the bottom of the cascade, i.e. for stages of low level groups, approaches the value of an ideal cascade, but as the group level rises about four-fold, the situation approaches that of total reflux of no yield.[1-2]

4.5 METHODS OF ISOTOPE SEPARATION

Having considered the different ways of expressing the results of isotope enrichment and the general modes of operating a process in a cascade of series-parallel units, we shall now present the more important methods of isotope separations. Each method will be described, first by discussing the principles involved and in the end indicating some examples of its application.

It will be found convenient to group the isotopes needed in enriched forms, into three classes :

First, those isotopes needed in minute amounts for research and for certain special purposes as, ^{13}C, ^{15}N, ^{18}O, ^{57}Fe and ^{198}Hg, and some others. The methods of their separation or enrichment will be considered under the general methods in the next four chapters (5-8).

Second, those isotopes needed for atomic energy purposes in large amounts, in tonnes in some cases, and in a state of high degree of enrichment bordering on near isotopic purity. This class includes ^{2}H, ^{3}H, ^{6}Li, ^{7}Li, ^{10}B, ^{235}U and ^{239}Pu (Chapters 9, 10).

Third, an increasingly large number of radioactive isotopes needed for peaceful uses in research, medicine and industry as ^{14}C, ^{22}Na, ^{24}Na, ^{35}S, ^{59}Fe, ^{58}Co, ^{60}Co, ^{63}Ni, ^{99}Mo, ^{99}Tc, ^{126}I, ^{131}I and many others (Chapter 11).

4.5.1 Classification of the Methods of Isotope Separation

The more important methods of isotope separation can be classified as under:

(1) *Electromagnetic Methods*

These methods are unique in many ways. They lead to the complete separation of all the isotopes present in the sample in one operation, though the yields are poor.

(2) *Physical Methods*

These methods depend, either directly on the difference in the masses of the isotopic atoms/molecules, and/or on the mean difference in the velocities of their thermal motions. Some of these are irreversible statistical processes. They include :

(i) Gaseous diffusion (ii) Nozzle jet separation, (iii) Thermal diffusion, and (iv) Ultracentrifugation which is, however, a reversible statistical process.

---

* For details see H. London[1] or K. Cohen[2].

(3) *Chemical Methods*

These methods avail of small differences in physico-chemical characteristics of isotopic atoms and molecules such as differences in their structure, bond energies, etc. The chemical methods include both reversible statistical processes as : (i) Isotope exchange reactions under chemical equilibrium, (ii) Fractional distillation, (iii) Solvent extraction, (iv) Chromatographic methods and (v) Irreversible statistical process like molecular distillation.

(4) *Electrochemical Methods*

These include (i) Electrolysis—an irreversible statistical process depending on differences in the rates of exchange between the ion and electrolyte species for the two isotopic ions/molecules, and (ii) Ionic migration methods, based on differences in the specific mobilities of the isotopic ions; which are also irreversible statistical processes.

(5) *Photochemical Methods*

These methods depend on specific differences in the bond energies of isotopic molecules leading to well defined differences in their absorption spectra. These are specific irreversible processes. These include the application of lasers.

(6) *Biological and Microbial Methods*

They depend on differences in the behaviour of certain microbes to specific isotopes.

References

1. H. London, *Separation of Isotopes* (George Newnes Ltd., London, 1961)
2. K. Cohen. *The Theory of Isotope Separation as Applied to Large Scale Production of* ^{235}U. (National Nuclear Energy Series, Div. III, Vol. 1B, McGraw-Hill Book Co., 1951).

Chapter 5

ELECTROMAGNETIC SEPARATION

As stated in the beginning, the early experiments of Sir J. J. Thomson[1] (1911-13) on the analysis of positive ion beams from a discharge tube in crossed electric and magnetic fields, led to the construction of the first mass spectrometer by Aston, which had been continually improved upon since then. As the need arose for separated isotopes, at first in micro and milligram amounts for research, and suddenly in the early 1940s for larger amounts, the mass spectrometers were converted to product collectors, which thus became isotope separators.

5.1 MASS SPECTROMETRY

5.1.1 The Early Mass Spectrometers

The classical work of Thomson was followed by the illustrious researches of Aston[2] (1919) who constructed a greatly improved mass spectrograph in which an electrostatic field (E) first bends the positive ion beam (I) into a velocity or energy spectrum, of which a narrow ribbon of approximately one energy is selected by a diaphragm (D). This latter then enters a strong magnetic field (B), acting perpendicularly to (E) and so oriented as to reverse the initial deviation by more than twice the angle, with the result that all particles of a given e/M are brought to focus at one spot on a photographic plate (F) (Fig. 5.1). With successive improvements, Aston[3] obtained a mass

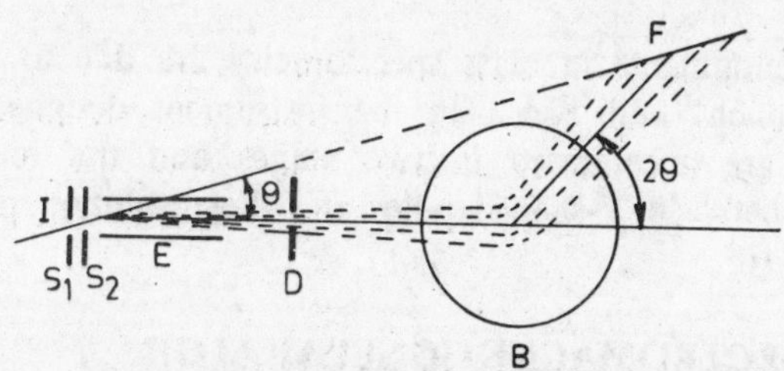

Fig. 5.1: **Aston's mass spectrograph:** I—positive ion beam defined by slits S_1 and S_2; E—electrostatic field which bends I by θ D—diaphragm selecting a beam of nearly one energy; B—magnetic field which reverses the deviation due to E by 2θ; F—focus spot on photographic plate for particles of a given e/m

resolution power $M/\Delta M$ of 10^4 and was able to show that 71 of the chemical elements examined by him consisted of more than one isotope.

5.1.2 Improved Mass Spectrometers

The next significant improvement in mass spectrometry was due to Dempster[4] who used a positive ion beam accelerated to a potential (V) (about 2.5 kV). A narrow ribbon of the accelerated ion beam, defined by slits, entered a strong homogeneous magnetic field (B) which bent the trajectory into a semi-circle of radius (r). Finally, a second set of slits permitted canalyzing the ions traversing a particular curvature, to reach an electrometer which measured its intensity. Ions of a given e/M follow a trajectory of radius (r) related to B and E by

$$\frac{B^2 r^2}{2V} = \frac{M}{e} \tag{5.1}$$

Hence,

$$r = \frac{1}{B}\left(\frac{2MV}{e}\right)^{1/2}$$

The mass resolution, $\Delta M/M$ is directly related to variations in the electric field and magnetic induction,

$$\frac{\Delta M}{M} = \frac{2\,\Delta B}{B} = \frac{\Delta V}{V}$$

If the two isotopes to be separated have masses close to 250* and if the fluctuation of the image (focus of product collection) is not to exceed 10%, at the distance between the collectors, we must make sure that the field fluctuations do not exceed the limits

$$\frac{\Delta V}{V} \simeq \frac{1}{2500} \quad \text{and} \quad \frac{\Delta B}{B} \simeq \frac{1}{5000}$$

Further developments in mass spectrometry are due to Bainbridge and Jordan[5], Mattauch[6] and Nier[7]. In the instrument designed by Nier, the positive ions are accelerated in two stages and the magnetic field is homogeneous between V-shaped poles and the resolution power $M/\Delta M$ is of the order of 10^6.

5.2 THE ELECTROMAGNETIC SEPARATOR

Though based on the same central principle as the mass spectrometer, the machine to serve as an isotope separator has to be designed with necessary

* As in the case of uranium isotopes.

modifications. The mass spectrometer aims at a high resolution power to enable the accurate determination of atomic masses to within 10^{-7} or 10^{-8} u, but as the resolution goes up, the *yield* of the isotope dwindles to insignificant amounts. Nier's early mass spectrometer (1940) yielded only a few micrograms of ^{235}U separated from natural uranium*.

The fall in the yield is of no great consequence in a mass spectrometer as the signals can be amplified electronically, whereas in an isotope separator the yield is very important.

The yield m (in micrograms) of an isotope of atomic mass M and abundance f is given by

$$m = 0.037\,3\, f\, M\, I\, t^{**} \tag{5.2}$$

where I is the ion current in microamperes and t is time in hours. The numerical constant includes the reciprocal of the Faraday constant and a mean (assumed) value of the overall efficiency of the separator. Eq. 5.2 shows that for the time of collection of the product to be within reasonable limits, the ion current has to be at least some 10^6 times greater than in a mass spectrometer. Also for a given current, the mass collected (m) is larger higher the isotopic mass M, which is another unique feature of the electromagnetic method, unlike in most other methods where an inverse relation holds good.

5.2.1 Essential Components of an Electromagnetic Separator

The basic components of an electromagnetic separator are the same as in a mass spectrometer, except for the presence of the collector. The major components consist of :

(*a*) An ion source, with the associated systems of extraction and acceleration electrodes.

(*b*) A chamber under high vacuum of 10^{-5} – 10^{-6} torr and at a high potential with associated heavy duty pumps with a pumping out speed of the order of 20 000 l/s,

(*c*) An analysing magnetic field which resolves the heterogeneous ion beam into a velocity or energy spectrum, and

(*d*) Collectors to receive the resolved isotopes.

Each of the above involves intricate technological precision designing, the details of which can be found in standard works on the subject[8,9]. Only some general facts are presented here.

---

* It is ironical to note that it was studies with this microgram trace amounts that led to the discovery that it is this rarer isotope of uranium (^{235}U) which is responsible for the fission of uranium by slow neutrons. This finding led to spectacular developments as the atom bomb, necessitating the construction of huge electromagnetic separators rated to yield kilogram amounts of ^{235}U.

* Eq. 5.2 is same as the classical Faraday's law of electrolysis which states that the mass deposited is proportional to the quantity of electricity passed ($I \cdot t$).

5.2.2 The Ion Source

The ion source plays an important role as it governs the magnitude of the ion current which in turn determines the isotope output in the collector at the end. An ion source is a fairly complex assembly of the following components : a hot cathode, a furnace to volatilize the charge material, baffles, a magnetic field, slits, an ionization chamber, and extraction and acceleration electrodes of (usually) graphite.

A heated cathode furnishes electrons which are accelerated to 100-300 V energy before reaching the ionization chamber which contains the gas whose isotopes are to be separated at a pressure of 10^{-3} mm Hg. If the substance is not gaseous, the solid element (if volatile) or a suitable charge material, most often a chloride or other halide, is volatilized in a small furnace at the appropriate temperature* from whence the vapour enters the ionization chamber. The electron current may be 0.5 to 5 A. A magnetic field imparts to the electrons a helicoidal path in the ionization chamber, thereby lengthening their trajectory in the vapour medium and hence increasing the ionization collisions with the atoms/molecules of the vapour (Fig. 5.2). The magnetic field also helps in deflecting a part of the positive ions towards the cathode filament which helps in neutralizing some of the negative space charge there, which in turn results in greater electron emission. An electrode at the same potential as the filament at the other end of the chamber (the anticathode) serves to reflect back the electrons and this results in a greater ion yield.

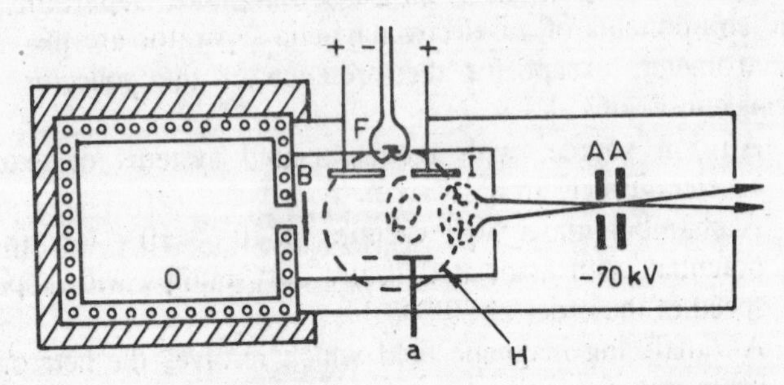

Fig. 5.2: **The Nier type ion source:** O—oven where the material is volatilized; B—baffle; I—ionization chamber, F—hot filament providing electrons; H—magnetic field; ++300 volts positive to F; a—anticathode at same negative potential as F; AA—accelerating field (−70 kV).

5.2.3 The Ionization Processes

The conditions in the ion chamber correspond nearly to those of a glow discharge and the ionization processes occurring are mainly due to collisions between energetic electrons and the atoms/molecules of the vapour, the main reactions being :

---

* If the volatilization temperature be above 1200 °C, the charge material is bombarded by 30 kV electrons.

$$M + e^- \rightarrow M^+ + 2\,e^- \qquad \text{primary ionization}$$

(secondary)

$$M + e^- (\text{sec}) \rightarrow M^* + e^- \qquad \text{excitation}$$

$$\left.\begin{array}{rl} M^* + e^- \rightarrow & M^+ + 2\,e^- \\ M^* + h\,\nu\,(\text{glow}) \rightarrow & M^+ + e^- \end{array}\right\} \text{secondary ionization}$$

A part of the charge material yielding ions, other than the one for which the separator is tuned (Eq. 5.1), is invariably lost*.

An ideal ion source should be able to ionize nearly all elements in vapour form with high efficiency, defined as the ratio of the number of ions to the number of atoms/molecules introduced into the chamber. The source should also yield high density monoenergetic ion beam at the exit slit in the wall of the chamber which is penetrated by the accelerating field due to about 70 kV. Figure 5.2 shows a typical Nier ion source.

5.2.4 The Product Collector

The designing of the collector of an electromagnetic separator is associated with its own problems which become evident on noting the high level of energy dissipation in the small space of the collector. A beam of 100 mA accelerated to 40 kV dissipates a power of 4 kW over a small space, the gap available being around 15 mm and 5 mm in the collectors of ^{235}U and ^{239}Pu respectively. This necessitates the use of special ceramic refractory materials in its construction and the provision of efficient local cooling. Secondly, the large momentum which the incoming ions bring in, specially in the case of ions of heavy elements, results in nearly 50 per cent loss by the rebounding of arriving ions and their escape out of the collector with or without charge neutralization. In addition, there is a partial re-evaporation/sputtering of the isotope atoms already deposited. To contain these losses to within 20 per cent, the collector is usually given the shape of a closed pocket to trap all the incoming particles, whose cooled surface is protected from direct impact of incoming particles by shaping it appropriately (Fig. 5.3).

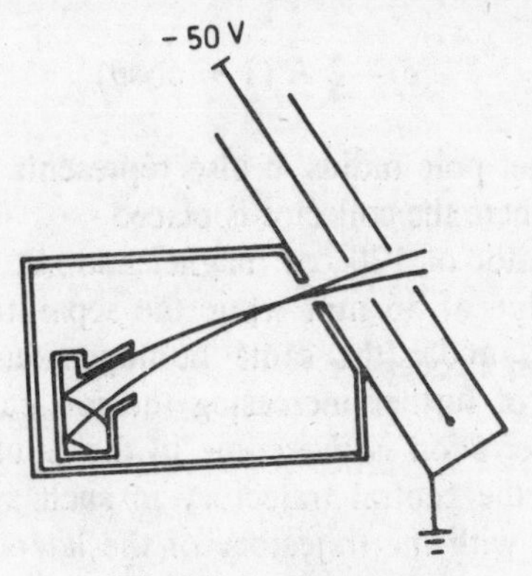

Fig. 5.3: **Isotope collector**

* For instance, with UCl_4 as the charge material only 60-70 per cent appears as U^+ ions and the rest as UCl^+, UCl_2^+, UCl_3^+, UCl_4^+, Cl^+, Cl_2^+ and U^{++} which are all lost.

5.2.5 The Calutron

Recognizing the wartime importance of separating the ^{235}U isotope (abundance 0.72 per cent) from natural uranium, the U.S. government launched in 1942-43 a well guarded gigantic project for this with the cooperation of a team of international experts under the direction of E.O. Lawrence*. Eq. 5.2 shows that raising the yield from μg to mg amounts per hour requires a correspondingly larger beam current, mA in place of μA. For this, a widening of the slits would be necessary, but that would result in poorer focussing or imperfect mass resolution. Besides, the condensation of uranium vapour over different parts of the machine would lead to increased space charges and alteration of the applied field intensity. This would necessitate frequent cleaning which is time consuming and interrupts operations.

The strategy adopted by Lawrence's team in solving the problem was to draw ion currents of higher intensity by widening the slits and tracking the dispersing beam to focus by employing a progressively inhomogeneous magnetic field towards the centre, guided by intricate calculations in regard to the precise shaping of the profile of the magnet pole pieces. A separator so designed is referred to as the CALUTRON**. Any serious deviation from Eq. 5.1 in regard to the accelerating voltage or the magnetic excitation current would result in an overlapping of the separated isotope beams and in displacing the same away from the mouth of the collector placed at a precise position.

The electromagnet of a calutron usually has an air gap of 30-60 cm between the pole pieces and a radius of 120 cm for the pole face. The vacuum chamber containing the source and the collector are situated between the pole pieces in one plane. Let us suppose a monoergic ion beam while starting from an exit point of the source diverges by an angle θ initially. The divergence reaches a maximum when a quarter circle of the trajectory in the magnetic field is completed and convergence sets in thereafter tending to near merger again on completing a semicircle of the path, not quite but with a short dispersion δ (Fig. 5.4), given by

$$\delta = 2\,R\,(1 - \cos\theta) \tag{5.3}$$

where R is the magnet pole radius. δ also represents the width of the beam landing at the point where the collector is placed.

Applied to a separator of 120 cm magnet and slit widened to make $\theta = 10°$ gives for δ a value of 36 mm while the separation between ^{235}U and ^{238}U is only 15 mm under the same homogeneous magnetic field. This shows the nonutility of further increasing the ion current by widening the slit. However, this aberration is overcome in the calutron by enhancing the magnetic field along the central trajectory in such a way as to modify its curvature to coincide with the trajectory of the lateral or flank components

* The renowned inventor of the cyclotron.

** Derived from *Cal*ifornia *U*niversity cyclo*tron*.

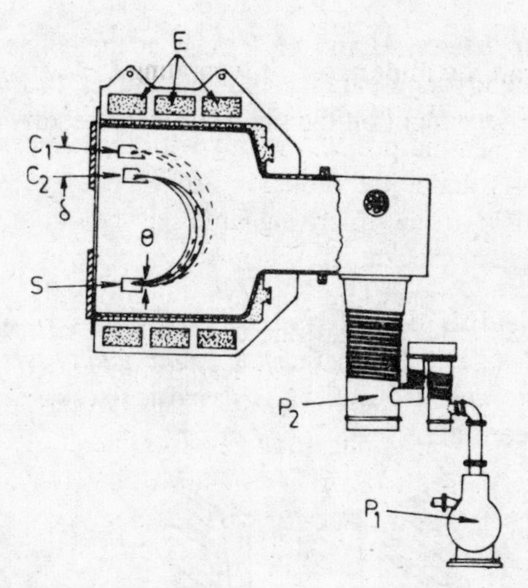

Fig. 5.4: **Principle of the calutron**: S—ion source with widened slits. angle θ; C_1, C_2—Collectors; δ—displacement; P_1: primary pump; P_2: oil diffusion pump; S: ion source; E: magnet coils (Reprinted with permission from *La Séparation des Isotopes* by M. Chemla and J. Périé. p.47 © Presses Universitaires de France, 1974)

subjected to a relatively weaker field. In other words, by such a calculated boosting of the magnetic field on the central trajectory over the outer ones, the final beam width is reduced to a value less than the separation between the two isotopes of uranium. Thus, a typical calutron drawing ion currents of centiamperes can yield centigram amounts of ^{235}U per hour [10,11]. Within two years of the start of the project, an enormous battery of nearly 68 calutrons was assembled at Oak Ridge (Tennessee, USA) which yielded kilograms of ^{235}U*. This must be conceded as an extraordinary achievement of American technology, consideration of costs aside.

In 1957 it was decided to set up a reserve stock of isotopes of all elements, an *Isotopes Pool*, to meet the research requirements of all countries. A certain number of the calutrons at Oak Ridge are set aside for this, specially those which can treat many (about six) different elements simultaneously.

5.2.6 Inhomogeneous Field Separator

In the calutron just described, the slit is widened to obtain a much higher beam current and the magnetic field is enhanced for the central trajectory over the value for the flank trajectories so that the former is bent more and its curvature increased and made to coincide with those of the latter. Apart from the calutron, separators have been constructed with inhomogeneous magnetic analysers with cylindrical symmetry[8]. This is done to obtain increased *dispersion* (δ) with the same magnet, the dispersion being the distance which

* Within another two years, (in July—August 1946) three atom bombs, (two of ^{235}U and one of ^{239}Pu), were exploded.

separates neighbour masses, M and $M + \Delta M$ at the collector. Here the pole pieces are made slightly non-parallel and this makes the field non-unitorm, the field being now dependent on the radius of the trajectory (Fig. 5.5a)

$$B_{(r)} = B_0 \left(\frac{r_0}{r}\right)^n \tag{5.4}$$

where B_0 is the field at central trajectory of radius r_0 and n is the *field inhomogeneity index*. Eq. 5.4 defines a *cylindrically symmetrical field*, of which the usual homogeneous field is a particular case with $n = 0$ and hence $B_{(r)} = B_0 =$ constant.

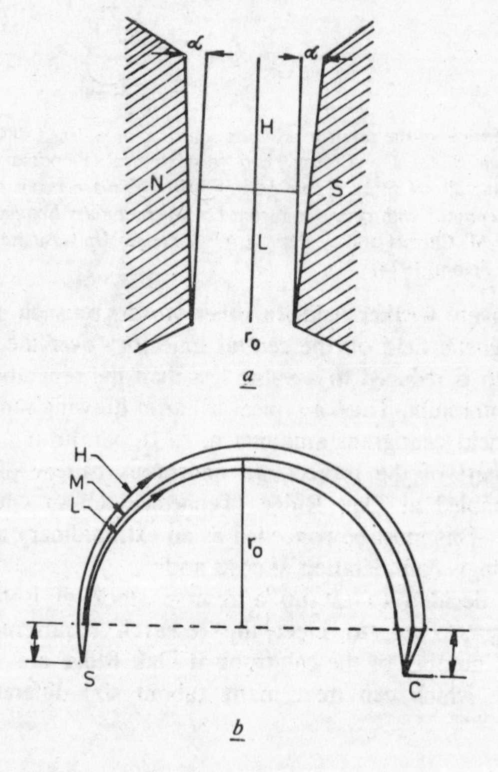

Fig. 5.5: **Inhomogenous field analyser:** (a) magnet with non-parallel pole pieces; (b) The inhomogenous field $B(r) = B(r_o)$ $(r_o/r)^{1/2}$; S—ion source: C—collector; H. M. L—trajectories of heavy, medium and light mass ions
(From *Separation of Isotopes* by H. London, p. 461 © George Newnes Ltd., 1961: *reprinted by permission of William Heinemann Ltd*)

The larger dispersion available with such an analyser becomes apparent on considering Fig. 5.5 (b) where the source S and collector C are shown outside the inhomogeneous magnet. The trajectories of the light and heavy isotopes which are deflected more and less, are respectively shown by L and H while the medium mass isotope, if present, would follow the middle trajectory M, their radii being

$$r_L < r_M < r_H$$

In accord with Eq. 5.4, the less deflected heavier isotope is subjected to a weaker field B_H while the more deflected lighter isotope is subjected to a stronger field B_L. These effects add up and reinforce each other leading to a greater dispersion than if a field of the same strength had acted on both isotopes. It can be shown that the dispersion for $n = 1/2$ is twice than that in $n = 0$ (homogeneous field), for the same magnet. The choice of a value of 1/2 for n leads to radial and axial focussing. That is : a point object gives a point image for $n = 1/2$, while it would be a line in a homogeneous field ($n = 0$).

The first separator of this kind was developed in Moscow with a 130 cm magnet leading to a total deflection of 225°. This has been used for separating ^{206}Pb and ^{207}Pb, as well as for the separation of certain transuranic elements[8].

5.2.7 Double Stage Separators

Two types of two stage separators have been developed. These are:

(a) *A magnetic analyser followed by an electrostatic analyser*

The ions are first analysed in a 60° magnetic field of 50 cm radius and focussed on an intermediate diaphragm. The beam selected by the latter enters an electrostatic analyser consisting of a cylindrical condenser in the form of a 105° sector of 50 cm radius (Fig. 5.6). Such a magnetic +

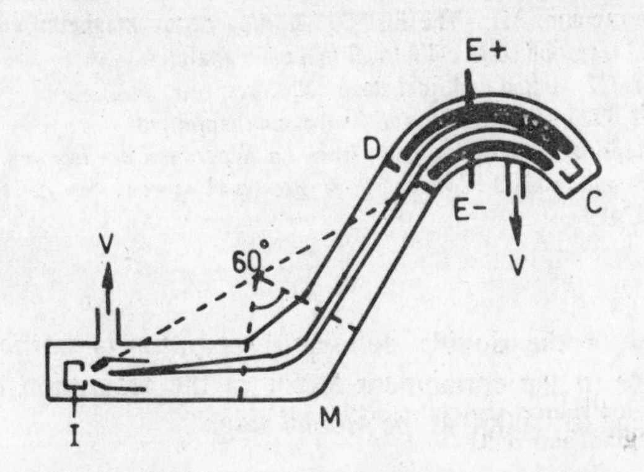

Fig. 5.6: **Double-stage separator with electrostatic analyser:** I—ion source; M—magnet, 60°; D—diaphragm; E , E —cylindrical electrodes; C—collector; V—to vacuum
(From *Separation of Isotopes* by H. London, p.465 © George Newnes Ltd., 1961: *reprinted by permission of William Heinemann Ltd*)

electrostatic stage separator has been in use in Saclay (France) for the separation of ^{235}U. This has the drawback of *space charge* effects in the second stage.

(b) *A homogeneous magnetic analyser followed by an inhomogeneous magnetic analyser*

A separator of this kind was developed by Bernas[11] at Orsay (France) by adding to the first stage magnetic separator, the same as in the Saclay machine, an inhomogeneous magnetic field of cylindrical symmetry.

The first magnetic analyser of radius 50 cm bends the beam by 60° and when it arrives at the first collector, a diaphragm selects a part of the band to enter the second (inhomogeneous) magnetic field of radius 40 cm which bends the same by 191°.4. The final product collects in C_2 (Fig. 5.7). The

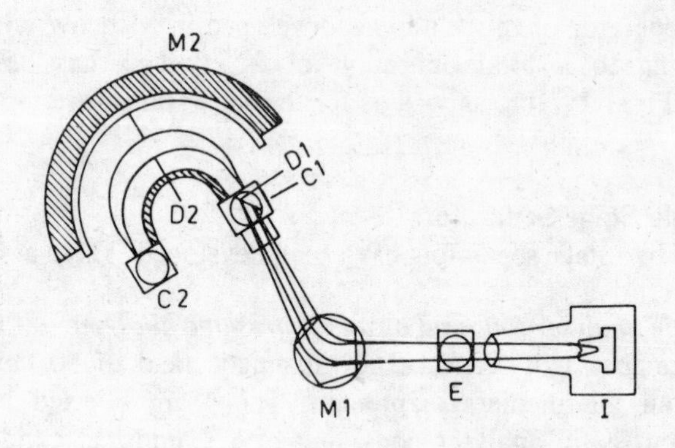

Fig. 5.7: **Double-stage separator with magnetic analyser:** I—ion source: E—to vacuum; M1—The first 60° homogeneous magnetic analyser; M2—second stage cylindrical magnetic analyser: C1, C2—initial and final stage collectors; D1, D2—intermediate and final stage diaphragms. (Reprinted with permission from *La Séparation des Isotopes* by M. Chemla and J.Périé, p.51 © Presses Universitaires de France, 1974)

overall efficiency of the double deflexion separator is borne out by the enormous increase in the enrichment factor in the separation of ^{235}U from 200 at the first stage to 14 000 at the second stage[12,13].

5.2.8 Other Special Separators

Several other separators of varying novel designs as the resonance separator and the isotron, have been experimented upon in the period 1942-44 when almost any method proposed for the separation of ^{235}U was considered seriously in America and work initiated irrespective of the costs involved. Most of these did not get beyond the laboratory research stage.

5.2.9 Separators for Radioactive Isotopes

An electromagnetic separator using a 90° homogeneous magnet bending the ion beam into a curvature of radius 130 cm has been constructed at Harwell (U.K.) for the separation of ^{239}Pu from ^{240}Pu and some other transuranic elements. The separator has been christened HERMES. As it is to handle intensely radioactive materials all the operations are remote controlled and the entire unit is in a sealed room wherein a small negative pressure is maintained which prevents the leak of activity. It has an enhancement factor (γ)* of 250 in the separation of ^{239}Pu from ^{240}Pu.

References

1. J. J. Thomson, *Rays of Positive Electricity (1913), Phil. Mag.*, 1907, *13*, 561.
2. F. W. Aston, *Mass Spectra and Isotopes* (Longman Green and Co, 1933).
3. F. W. Aston, *Proc. Roy. Soc.*, 1927, *115*, 487.
4. A. Dempster, *Phys. Rev.*, 1918, *11*, 316; 1922, *20*, 631.
5. K. T. Bainbridge and E. B. Jordan, *Phys. Rev.* 1936, *50*, 282.
6. J. Mattauch, *Phys. Rev.*, 1940, *57*, 1155.
7. A. O. Nier, *Rev. Sci. Instruments*, 1940, *11*, 212.
8. H. London, *Separation of Isotopes* (George Newnes Ltd, London, 1961).
9. J. Koch, *Electromagnetic Isotope Separators and Applications of Electromagnetically Enriched Isotopes* (North Holland Publishing Co., 1958).
10. H. D. Smyth, *Atomic Energy for Military Purposes* (Princeton University, Princeton, N. J. 1945), *Revs. Mod. Phys.*, 1945, *17*, 351.
11. R. H. Bernas, *J. Phys. Radium*, 1953, *14*, 34.
12. J. R. Richardson, *Electromagnetic Separation of Isotopes in Commercial Quantities*, TID 5 217 (Eds. R. K. Wakerling and A. Guthries, 1949).
13. M. Chemla, and J. Périé, *La Séparation des Isotopes* (Presses Universitaires de France, 1974).

* See § 4.3.4.

Chapter 6

PHYSICAL METHODS OF ISOTOPE SEPARATION

Physical methods depend either directly on the masses of the isotopic atoms or on the difference in the mean molecular velocities of the isotopic molecules concerned.

The following methods are considered in this chapter.

1. Gaseous diffusion across a membrane,
2. Diffusion across an inert gas,
3. Nozzle separator,
4. Thermal diffusion, and
5. Ultracentrifugation.

All these except ultracentrifugation are irreversible statistical proceses.

6.1 GASEOUS DIFFUSION ACROSS A MEMBRANE

The escape of a gas through an orifice or through the pores of a membrane is referred to as diffusion, though it is really a process of *effusion.* In as far back as 1846, the escape rate had been shown by Thomas Graham to be inversely proportional to the square root of the density and hence the molecular weight of the gas. This is in accord with the kinetic theory expression for the mean molecular velocity:

$$\bar{c} = (8kT/\pi M)^{1/2} \tag{6.1}$$

6.1.1 Theory of the Diffusion Process

The flow of a gas through a membrane has partly the character of a gas escaping through an orifice where the gas viscosity (η) is involved in accord with Poiseuille's relation and partly of a gas flowing through a capillary as a molecular flow related to $M^{-1/2}$ in accord with Knudsen's relation. The overall result is a sum of the two, expressed by

$$N = \frac{a}{\sqrt{M}}(P_F - P_B) + \frac{b}{\eta}(P_F^2 + P_B^2) \quad (6.2)$$

where P_F and P_B are the high and low pressures, before and after the barrier respectively. The mean free path of the gas molecules (λ) relating to the pore or capillary radius is an important parameter which determines the relative contributions of the two terms. The mean free path is dependent on the nature of the gas, its molecular mass M, its viscosity η and on pressure P, and temperature T, by

$$\lambda = \frac{16\eta}{5P}\left(\frac{RT}{2\pi M}\right)^{1/2} \quad (6.3)$$

where λ is much greater than the diameter and the length of the pore, molecular flow dominates as, under this condition, the gas molecules can freely traverse the barrier before suffering a collision with other gas molecules.

Suppose a gas consisting of two isotopic molecules of masses M_1 and M_2 and of partial pressures P_1 and P_2, is placed in a vessel initially to one side of a porous membrane, whose pore dimensions are smaller than the mean free path of the gas molecules. Assuming that the back pressure on the other side of the membrane is negligible, the ratio of the fluxes J_1 and J_2 of the two isotopic molecules (i.e. the number of their molecules crossing unit area of the barrier in unit time) is given by

$$\frac{J_1}{J_2} = \frac{P_1}{P_2}\left[\frac{M_2}{M_1}\right]^{1/2} = \left[\frac{P_1 D_1}{P_2 D_2}\right] \quad (6.4)$$

where D_1 and D_2 are the diffusion rates of the two species.

If x be the mole fraction of the isotope of interest in the initial sample and y in the sample which has got across the barrier, x and $(1 - x)$ would represent the initial partial pressures before diffusion and y and $(1 - y)$ would represent the fluxes of the two isotopes in the sample which have diffused. Eq. 6.4 can then be written in terms of the mole fractions,

$$\frac{y}{1-y} = \frac{x}{1-x} \cdot \frac{D_1}{D_2} = \frac{x}{1-x}\sqrt{\frac{M_2}{M_1}} \quad (6.5)$$

Hence the separation factor is given by

$$\alpha = \sqrt{\frac{M_2}{M_1}} = \frac{D_1}{D_2} \quad (6.6)$$

i.e. the separation factor is the inverse ratio of the square root of the isotopic masses.

When $(M_2 - M_1) \ll M_1$

$$\alpha = 1 + \frac{\Delta M}{2M} \quad (6.7)$$

where $2\overline{M} = M_1 + M_2$ and hence the enrichment factor is given by

$$\epsilon \simeq \frac{\Delta M}{2\overline{M}} \tag{6.8}$$

This value of α represents the *maximum*; in actual practice the value is lower due to imperfect mixing, back diffusion and other factors. To multiply the effect, the process is worked in a cascade of a large number of stages, which brings in the problem of operating an equal number of specially designed pumps (Fig. 6.1 *a* and *b*).

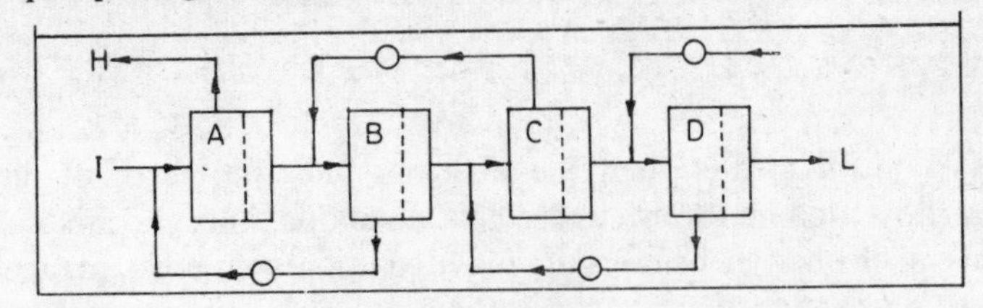

Fig. 6.1(a): **Isotope separation by diffusion across a membrane:** A, B, C, D... different stages; I—initial sample enters; H—heavy fraction leaves; L—light fraction leaves; O: pump; ---- porous barrier.

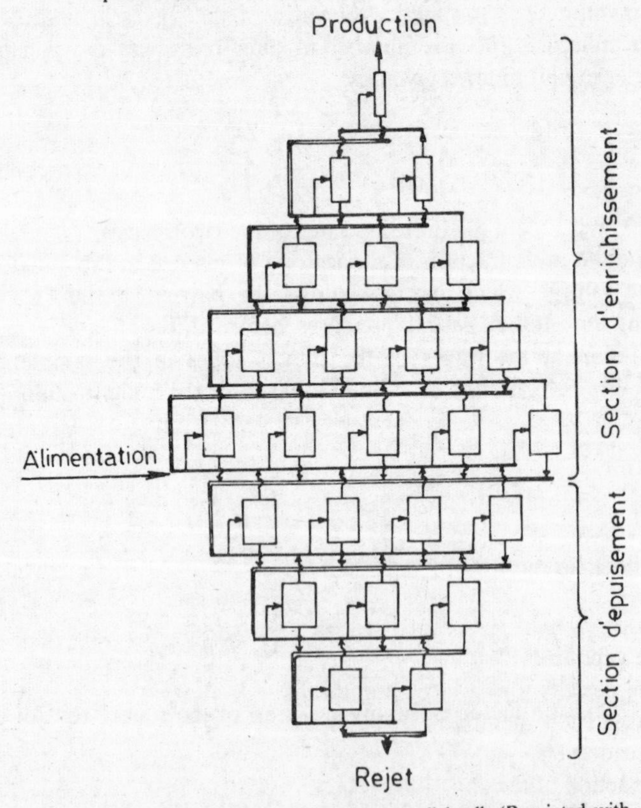

Fig. 6.1(b): **Pyramidal cascade of series**—parallel cells (Reprinted with permission from *La Séparation des Isotopes* by M. Chemla and J. Périé, p. 93 © Presses Universitaires de France, 1974).

6.1.2 Applications of Diffusion

Apart from electromagnetic separation, gaseous diffusion was the first method used in the separation of isotopes. Aston[1(a)], in 1913, partially separated the neon isotopes ^{20}Ne and ^{22}Ne by diffusing the gas across a pipe clay tube repeatedly. Later, in 1921 Harkins[2] succeeded in enriching the ^{35}Cl and ^{37}Cl isotopes by diffusing HCl gas also across a pipe clay tube. Using a cascade of 50 stages, Hertz[3,4] obtained practically prue ^{22}Ne from natural neon in which the abundance of the isotope is 9.2 per cent. Hydrogen enriched in ^{2}H and methane enriched in ^{13}C were obtained by Wooldridge and Jenkins[5].

A new dimension was brought into the development of gaseous diffusion as a technique for the separation of isotopes when the same began to be applied to uranium since the discovery of nuclear fission in 1939, more particularly in the period 1941-44. Nearly 99 per cent pure ^{235}U had been obtained in kilogram quantities to meet the needs of the atomic age by the process of diffusion of the vapour of uranium hexafluoride (b.p. 56° C) employing cascades each of several thousand stages. The number of stages needed is fantastically large because of the very low single stage separation factor of 1.0043, being the square root of the ratio of the masses of $^{238}UF_6/^{235}UF_6$. Solutions to the problems of designing and operating the same number of very special pumps, the selection of the proper ceramic or fritted membrane material with pore size of 0.015-0.02 μm, having the necessary mechanical strength and inertness towards the corrosive action of UF_6, and its decomposition into fluorine compounds, at the temperature involved remain closely guarded secrets. The large scale separation of ^{235}U isotopes is considered in more detail in Chapter 10.

6.1.3 Limitations of the Diffusion Process

However advantageous the method may be in the separation or enrichment of isotopes on a small scale, at least five major factors disfavour the process of diffusion for adoption to the separation of isotopes on an industrial scale. These are:

(*a*) The technical and engineering difficulties in fabricating the special membrane material of a total area of several hundred square metres, and having extremely fine and regular pores of size less than the mean free path of the gas molecules,

(*b*) Imparting sufficient mechanical strength to the membrane to withstand pressure difference at the temperature involved,

(*c*) Ensuring that the selected material is chemically resistant to the action of the vapour of the substance at the temperature involved,

(*d*) The designing of a large number of special pumps needed to pull the gas across each membrane and for recycling the same when operating in a cascade. The pump components too have to be chemically resistant to the action of the vapours.

(*e*) Lastly, the vast number of stages and the associated pumping, cooling and heating systems have to be all, as one unit, under one roof which makes the process unwieldy, extremely expensive and, above all, strategically vulnerable to enemy action.

6.2 DIFFUSION THROUGH ANOTHER GAS

Separation of a gaseous mixture, say of isotopes, by diffusion through another (a third) inert gas or vapour was first demonstrated by Hertz[6] who used mercury vapour of a diffusion pump for the separation of the neon isotopes. Maier[7] carried out an extensive experimental study of the process of separating hydrogen from methane in a mixture of the two by this process. The separation is governed by the difference in the interdiffusion coefficients of the components for diffusion into the third gas, also referred to as the *separating agent*. There are two ways of adopting this technique: (1) *mass diffusion* and (2) *sweep diffusion*. These are described below.

6.2.1 Mass Diffusion

In the mass diffusion, a porous filter is still used to complete the separation resulting in the interdiffusion of the components (isotopic species) into the separating agent, the third gas or vapour. Maier[7] as well as Benedict and Boas[8] studied this process in detail. The unit employed by Maier is outlined in Fig 6.2. A diffusion screen in the form of a tube coaxial with the main

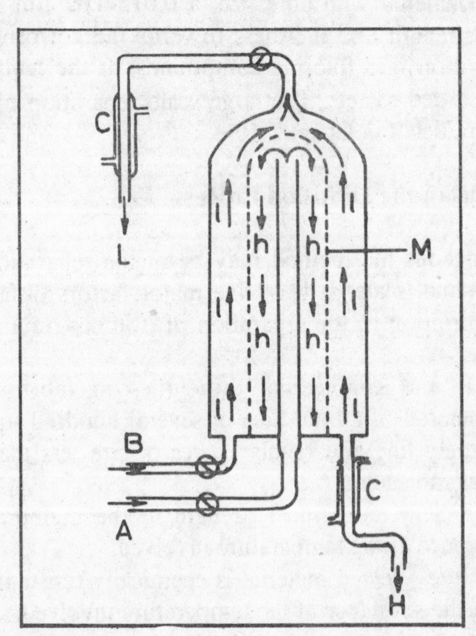

Fig. 6.2: **Maier's mass diffusion,** A:process gas enters; B:separating agent gas enters; C:condensers, Q:valve; H:heavy stream; L:light stream; M:mass diffusion screen.
(From *Separation of Isotopes* by H. London, p. 346 © George Newnes Ltd., 1961 : reprinted by permission of William Heinemann Ltd.)

unit divides the latter into two zones. The gas whose components (isotopes) are to be separated, referred to as the *process gas*, enters through a central coaxial tube of narrow diameter which reaches nearly upto the top of the cylinder, from whence the gas flows downward inside the diffusion screen. The separator-sweep vapour, which is often steam, enters at the bottom of the outer zone and flows upward over the outer surface of the screen. The inner zone being connected to a water cooled condenser ensures a pressure gradient for the steam across the screen. While passing through the screen (from the outer to the inner zone), the steam mixes with the process gas along all the pores of the screen. In this process, more of the heavier isotope having a lower diffusion coefficient is dragged down with the steam and the two flow out of the system along an outlet at the bottom close to the axis of the unit, whence the steam is condensed and the heavy isotope collected. A greater fraction of the lighter isotope, having a higher diffusion coefficient on the other hand, escapes the steam and crosses the screen into the outer zone whence it escapes from an outlet at the top. Separate valves control the rates of flow of the incoming and outgoing gases

Maier's results show that the size of pores of the separating screen ought to be between 0.08 and 0.4 mm for a screen of alundum or asbestos, or 0.01 mm for a metal foil screen.

6.2.2 Sweep Diffusion

There is no porous barrier in the technique of sweep diffusion developed by

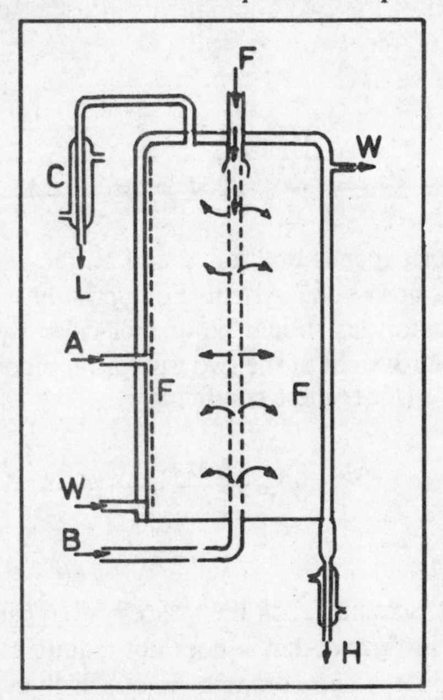

Fig. 6.3: **Principle of sweep diffusion;** A: process gas enters; B: separating agent gas enters; C: condenser; F: cold liquid feed for film; W: cooling water; L: light fraction out; H: heavy fraction + waste out.

Cichelli, Weatherford and Bowman[9]. The separating gas or vapour is admitted into a porous feed tube along the axis of the tube whose walls are cooled by flowing water. The process gas is admitted in the outer region at about the middle height (Fig. 6.3). The separating agent condenses along the walls of the tube and flows down and this cold steam drags with it more of the heavier component, while more of the lighter component escapes at the top.

6.2.3 Theory of Mass and Sweep Diffusion

We present here the relevant conclusions of the theory applicable to both mass and sweep diffusion techniques of separating the isotopes of a gas by a separator vapour. Details of this theory involving the flow of three gaseous components are to be found in the original paper of Cichelli, *et al*[9]. In the separation of the components of a binary mixture of two isotopic species of masses M_1 and M_2 by the separator gas of mass M_s, the theory gets simplified under the condition that $(M_2 - M_1) \ll M_1$. A quantity of importance is the *separability* γ defined by Benedict and Boas [8] as

$$\gamma = \frac{\Delta D}{\overline{D}} \tag{6.9}$$

where ΔD is the difference between the inter-diffusion coefficients D_{s1} and D_{s2} for the diffusion of the two components into the separator gas, and $\overline{D}$ is the average value of the two $= 1/2\,(D_{sl} + D_{s2})$. Under the condition where $\Delta M \ll M_1$; we have

$$\gamma = \frac{\Delta D}{\overline{D}} = \frac{\Delta M}{2M} \times \frac{M_s}{M_s + \Delta M} \tag{6.10}$$

The separation factor depends both on γ and on the partial pressure gradient of the separator gas across the system. For optimum separation, it is considered that the separator gas should be of molecular weight M_s nearly equal to the mean molecular weight of the two isotopic components, $M_s \simeq \frac{1}{2}(M_1 + M_2)$. Under these conditions

$$\gamma = \frac{\Delta M}{4\overline{M}} = \frac{\epsilon}{2} \tag{6.11}$$

6.2.4 Limitations

The one important advantage of the process of separation of isotopes by diffusion across a third gas is that it does not require a fine-pored membrane and the battery of pumps and compressors of a high compression ratio, both of which are essential for the process of gaseous diffusion across a membrane. This advantage, however, is offset to a large extent by the higher power requirement which is computed to be about 50 times more[10]. In addition, a great amount of energy is spent in vapourizing the condensed separator agent in every cycle. However, for separations on a small scale the method of separation by diffusion across an inert gas is more convenient.

6.3 THE NOZZLE SEPARATOR

The nozzle separator is another application of the diffusion process, developed by Becker, Bier and Burghoff[11] on the same basic principles. The gaseous mixture consisting of two isotopic molecules, the process gas, is fed under high pressure p_B at a flow rate of F moles per second to a convergent slit type nozzle C (Fig. 6.4). from whence the gas mixture emerges in the form

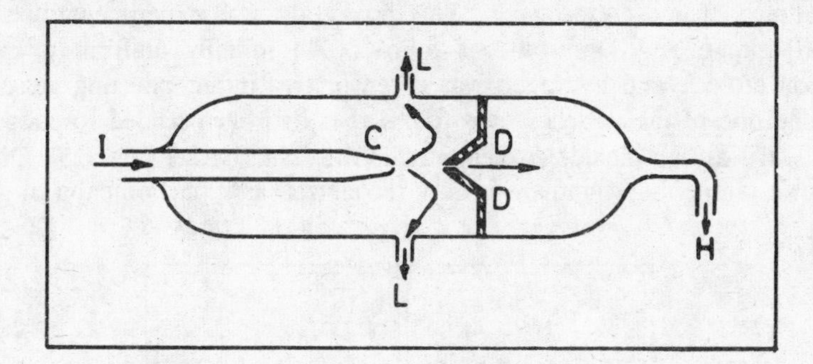

Fig. 6.4: **The nozzle separator**—I:initial sample enters; C:convergent slit nozzle; D:diaphragm; H:heavy fraction leaves; L:light fraction leaves.

of an expanding jet. A diaphragm D divides the jet into two fractions: the first, enriched in the lighter isotope is diverted towards the periphery, and the second, enriched in the heavier isotope, concentrates in the core fraction of the jet. The two fractions are pumped off at pressures p_L and p_H respectively.

If y and x be the mole fractions of the lighter isotope in the peripheral and core fractions respectively, the separation factor α is given by

$$\frac{y}{1-y} = \alpha \frac{x}{1-x} \tag{6.12}$$

The magnitude of α is of the same order as in gaseous diffusion.

According to Boltzmann's equation for diffusion flow in a binary gas mixture there exist for a freely expanding jet, concentration, pressure and temperature gradients leading to three corresponding diffusions. However, of these the pressure diffusion generally dominates[12].

In the early stages, the nozzle separator appeared to be of interest in the separation of uranium isotopes. Becker and Schutte[13] obtained considerable data on the separation of ^{235}U by driving UF_6 vapour into a ceramic nozzle of 0.085 mm width fixed 0.11 mm before a 0.29 mm wide diaphragm. For an optimum pressure of 20 mm Hg for the UF_6 vapour entering the nozzle, the enrichment factor was about 4×10^{-3}. To arrive at a significant separation, it is necessary to multiply the effect by employing a series of nozzles joined in a cascade. The major drawback of the nozzle method is the need for high suction volume compressors due to the low operating pressure. Details of the separation of the uranium isotopes by this method are presented in § 10.2.3

6.4 THERMAL DIFFUSION

Theoretical studies of Enskog and Chapman[12] showed as long back as 1917 that a gaseous mixture when subjected to a temperature gradient should differentiate into a lighter and a heavier component, concentrating at the hot and the cold ends respectively. This prediction was experimentally verified by Chapman and Dootson[14] as follows. An initially uniform mixture of carbon dioxide and hydrogen was taken in two interconnecting vessels. On heating one of the vessels it was found that hydrogen tended to concentrate in it while carbon dioxide concentrated in the colder vessel (Fig. 6.5). Diffusion under a temperature gradient is now recognized as a phenomenon of general

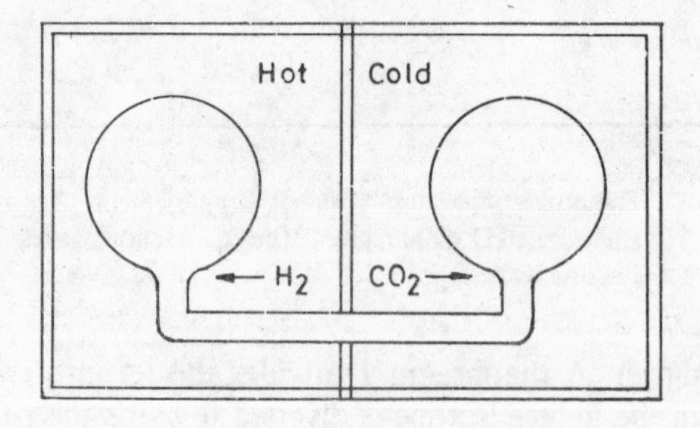

Fig. 6.5: **Principle of thermal diffusion**

occurrence. The long known Soret effect (1879) wherein the heavier and lighter components of a liquid mixture separate if one part of the system is heated with respect to the other, is considered as one of the forms of thermal diffusion. The Soret effect and the still earlier observation of Ludwig (1856) on the partial separation of constituents of a liquid mixture under a temperature gradient, remained undeveloped for want of an adequate theory of the liquid state at that time.

As diffusion proceeds under a temperature gradient, a concentration gradient sets in which tends to mix back the separated constituents. The temperature and concentration diffusions proceeding in reverse directions end up in a steady state where the overall separation can only be of a feeble magnitude. A significant development resulted when Clusius and Dickel[15] of Switzerland demonstrated that the temperature gradient also generates convection currents and that the separations are cumulative under the joint action of temperature-induced diffusion and thermal siphoning. This combination of effects is termed *thermal diffusion*, which is now recognized as an efficient technique for separating isotopes.

6.4.1 Theory of Thermal Diffusion

Suppose a gaseous mixture of two isotopic molecules is admitted into a Clusius-Dickel column which is simply a narrow tube along the axis of which is stretched a wire or a coaxial tube of a narrower diameter, heated to a high temperature (T′) with respect to the outer wall (T) (Fig. 6.6). Under the action of the temperature gradient the lighter isotope moves towards the

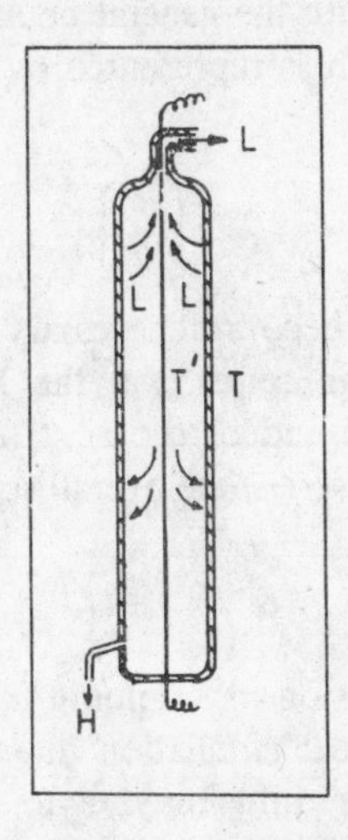

Fig. 6.6: **A Clusius-Dickel column of thermal diffusion**—T′: temperature of wire ~ 500₁ °C; T: temperature of outer tube ~ 50₁ °C; H: heavy isotope stream; L: light isotope stream.

hot wire and the heavier isotope towards the cold walls of the outer tube. At the same time, thermal convection currents lift the heated lighter isotope molecules upwards and sink the colder heavier isotope molecules downwards. These inward-upward movement of the lighter constituent and the outward-downward movement of the heavier constituent prevent remixing due to concentration gradients to a significant extent. Under these conditions, the column functions as a cascade of a large number of stages or equilibrium plates, leading to two well separated fractions; the one enriched in the lighter isotope escaping at the top and closer to the inner tube (or wire), and the other enriched in heavier isotope collecting at the bottom closer to the outer tube. A single 4-6 m column of 1 cm diameter with a temperature difference of about 500 °C was found to function as the equivalent of a few thousand theoretical plates *.

---

* For the significance of a theoretical plate, see § 7.1.2.

6.4.2 Thermal Diffusion Factor

On the attainment of a steady state, which is always slow*, the *thermal diffusion factor* A_T is given by

$$\frac{y}{1-y} = \frac{x}{1-x}\left[\frac{T'}{T}\right]^{AT} \tag{6.13}$$

Where T and T' are the temperatures of the wall and the wire and y and x, as usual, are the mole factions of the lighter isotope of interest in the top and bottom fractions. In accord with the general notation used by us hitherto, the single stage separation factor α is represented by the factor bracketed in Eq. 6.13, *viz.***

$$\alpha = \left[\frac{T'}{T}\right]^{AT} \tag{6.14}$$

Since α is usually small, it becomes necessary to work in a cascade by combining several columns in series, *i.e.* the hot end output of the *i*th column to be fed to the cold end of the next $(i+1)$th column. Where n columns are thus combined in series, the overall separation factor is

$$\alpha^n = \left[\frac{T'}{T}\right]^{nA_T} \tag{6.15}$$

The theory of thermal diffusion in a column involving radial diffusion due to temperature gradient, vertical circulation due to convection currents and both in opposition to reverse diffusion due to concentration gradients—is indeed complicated.

Attempts have been made to obtain a theoretical expression for the thermal diffusion factor in terms of molecular interaction constants which envisage a short range repulsion and a long range attraction. In this context, the model of Lennard-Jones, involving two variables, λ and μ for the interaction energy, W, appears to be the simplest.

$$W(r) = \lambda r^{-12} - \mu r^{-6} \text{***} \tag{6.16}$$

Two parameters of importance are (i) ϵ the minimum energy in the W vs r curve and (ii) the distance r_m where the minimum occurs. Fig. 6.7 shows the variation of interaction energy with distance for H_2 molecules at an ϵ / k i.e. at a temperature of 3.3 K. The maximum attraction of 2.5 meV occurs at $r_m = 0.33$ nm (Lennard-Jones' model). A quantity, *reduced temperature* $T^* = kT/\epsilon$, (or $T(\epsilon/k)^{-1}$), is also envisaged, the value

* It may take days and weeks for the steady state to be reached. This depends on several factors such as the column dimensions, the thermal conductivity of the gas, etc.

** Some of the original workers have used q for the single stage separation factor and α_T for the new quantity, the thermal diffusion factor[10]. To avoid confusion, we are consistently retaining the symbol α for the separation factor and a different symbol A_T for the new quantity, Eq. 6.14 relates the two quantities.

*** This is also known as the (12,6) model.

of ϵ/k for a gas being about 0.75 times its critical temperature or 1.4 times its boiling point[16].

Eq. 6.16 is then written in the form

$$W = \epsilon \left[\left(\frac{r_m}{r}\right)^{12} - 2 \left(\frac{r_m}{r}\right)^{6} \right] \tag{6.17}$$

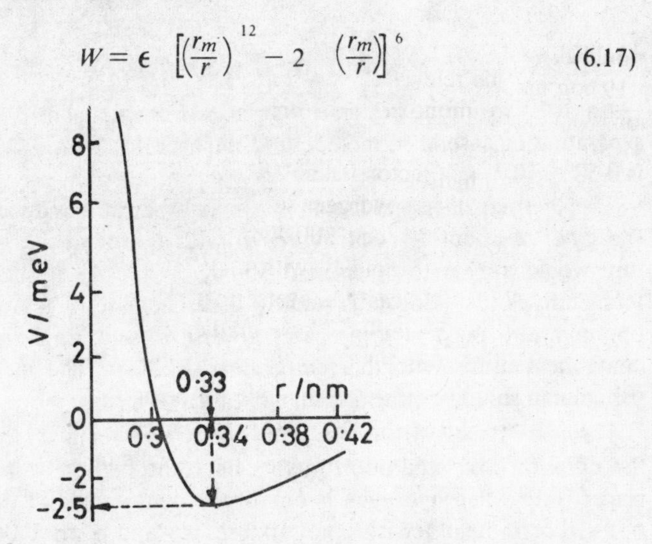

Fig. 6.7: **Interaction energy (W) *vs* distance (r) for hydrogen:** Model of Lennard - Jones.
(From *Separation of Isotopes* by H. London, p. 303 © George Newnes Ltd., 1961:*reprinted by permission of William Heinemann Ltd*)

A more refined model with three variables, known as the *exp-six* model has also been proposed[17]. Though in general, for a binary mixture of two gases A and B, three interaction forces are involved as f_{A-A}, f_{B-B} and f_{A-B}, with as may ϵ and r_m values, the situation simplifies where A and B are isotopic molecules, the three forces being nearly identical.

At the origin lies the *mass ratio* of the isotopic molecules $(M_2 - M_1)/(M_1 + M_2)$ which is small for most mixtures of isotopic molecules. According to Clark Jones[18], the thermal diffusion factor is given by

$$A_T = \theta \, \frac{M_2 - M_1}{M_1 + M_2} \tag{6.18}$$

where θ is a combination of ratios of collision integrals A^* B^* and C^* given by

$$\theta = \frac{15\,(6C^* - 5)\,(2A^* + 5)}{2A^*\,(16A^* - 12B^* + 55)} \qquad 6.18\,(a)$$

These ratios all being the same for molecules which are assumed to behave as rigid elastic spheres, θ becomes 105/118 (= 0.89) and under these conditions the thermal diffusion factor Eq. 6.18 approaches the mass ratio since θ is close to unity.

Where polyatomic molecules are involved with significant rotational contribution to the total kinetic energy, a second term involving the moments of inertia I_1 and I_2 of the two isotopic molecules is to be added to obtain A_T

$$A_T = \theta \frac{M_2 - M_1}{M_1 + M_2} + \phi \frac{I_2 - I_1}{I_1 + I_2} \tag{6.19}$$

Eq. 6.19 would be relevant where ΔM is small but ΔI is large, as in the separation of carbon monoxide and nitrogen for which ΔM is zero but $d\ I$ is 0.58×10^{-47} kg/m^2.

The thermal diffusion factor is a maximum at a reduced temperature T^* $(=kT/\epsilon)$ of about 10. For hydrogen with ϵ around 2.5 meV, this temperature would appear to be around 300 K. While A_T is slightly less at higher temperatures, its value falls rapidly with decrease of temperature below the optimum and may, in some cases, reverse in sign; implying that the heavier component diffuses up the temperature gradient. The mean temperature in the column should not be less than $5\ \epsilon/k$.

The mass ratio $[(M_2 - M_1)/(M_1 + M_2)]$ being the major determinant, the diffusion and separation factors have the highest values of 0.4 and 1.3 respectively when hydrogen is one of the constituents of the mixture and $T' = 2T$. For most other isotopic mixtures A_T and α are 0.04 and 1.03 or less for the same ratio of T'/T.

6.4.3 **Innovations**

Clusius not only pioneered work on thermal diffusion but contributed extensively to the development of the subject over two decades. Besides effecting the separation of the isotopes of a large number of elements in a state of high purity (*vide infra*), Clusius and coworkers introduced several innovations of which we shall describe here two: (*a*) swing separation, and (*b*) use of an auxiliary gas.

(a) *Swing Separation*

The equipment of earlier design consisted simply of a series of vertical tubes whose top ends were located in a heated enclosure while the bottom ends were maintained at a lower temperature. All the tubes were joined in series *via* a pair of capillaries which connected the hot top end of one tube with the cold bottom end of the next one so that convectional circulation was maintained in all the tubes *via* these capillaries. This, however, resulted in the hold-up of considerable amount of gas in the dead space of the capillaries, and in the heavy consumption of energy, close on 20 per cent of the total energy, only in maintaining convectional circulation.

Clusius and Huber[19] overcame this drawback to a considerable extent by using a single capillary, in place of two, and of a narrower diameter (~ 2mm) reducing thereby the dead space gas hold up and by the introduction of a reciprocatory motion throughout the gas mass from the first to the last tube. This last was effected by inserting two reservoirs R_1 and R_2 at each end (Fig. 6.8) which were alternately heated and cooled which made the gas swing to and fro periodically across all the tubes. This device ensured adequate mixing of the gas at the extreme ends of each column and in the joining capillaries.

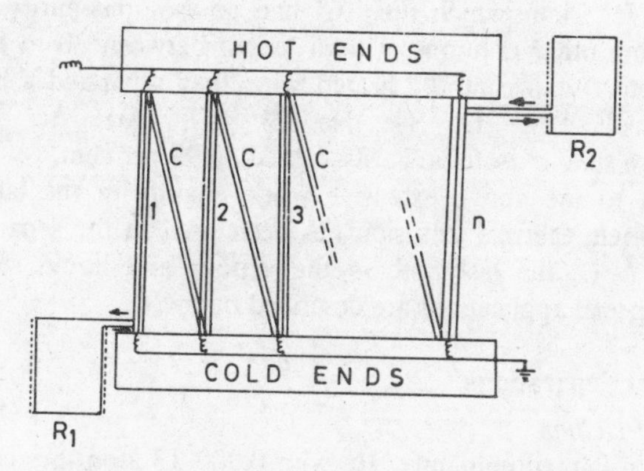

Fig. 6.8: **A cascade of Clusius-Dickel columns with swing operation—** 1,2,...n columns in a series cascade; C:capillary joining hot end 1 to cold end 2, ...; R_1 R_2: Two gas reservoirs each alternately heated and cooled to induce reciprocatory gas motion.

(b) *Use of an Auxiliary Gas*

Another drawback of the technique of isotope separation by the thermal diffusion method is the incompleteness of the separation *i.e.* the residual presence of the *other* isotope in the output samples. Theoretical treatment of the subject of diffusion in mixed gases by Chapman and Cowling[12] had revealed an important result that the relative proportion of two components diffusing in the presence of an added foreign gas, would be the same as in its absence. Clusius[20,21] put to good use this theoretically anticipated proposition when he modified the technique by admitting an auxiliary gas as a third component of molecular weight close to the mean of the isotopic components under separation. It was experimentally shown that with increasing amounts of the auxiliary gas, the two isotopic components are progressively pushed apart from the middle part of the column till they are finally separated from each other nearly completely at the top and bottom ends of the column. The two separated isotopes no doubt would be mixed with the third gas but this latter being nonisotopic can be easily removed by chemical means. This innovation of admitting an auxiliary gas proved a great success in the separation of the isotopes of neon and argon each of which consists of three isotopes of which the middle mass isotope serves as the auxiliary constituent. These are considered under § 6.4.5.

6.4.4 **Applications**

The earliest application of the process of thermal diffusion to isotope separation was due to Clusius and Dickel themselves and to their group. Using a 2.6 m long column, they first succeeded in enriching neon in respect of ^{22}Ne from its natural content of 9.2 per cent to 31 per cent[15]. This was followed by the use of multiple columns of pyrex glass in series of a total length of 36 m and a platinum alloy wire along the axis heated to 650-750° C for the separation of $H^{37}Cl$ in which the ^{37}Cl had an isotopic purity of 99.4 per cent. Employing other columns of total length between 30 to 80 m and of varying diameters with centrally heated wire, they separated a large number of isotopes as: ^{13}C, ^{15}N, ^{18}O, ^{21}Ne, ^{22}Ne, ^{35}Cl, ^{37}Cl, ^{36}Ar, ^{38}Ar, ^{84}Kr, ^{86}Kr and ^{136}Xe, each in a state of isotopic purity exceeding 99 per cent*.

In addition to the above extensive work, mainly in the laboratories of Clusius in Zurich, thermal diffusion has been used in the separation of ^{2}H, ^{3}H, ^{3}He and ^{235}U, the last both in the vapour and liquid states of UF_6. Some of the special applications are described below.

6.4.5 Special Separations

(i) Isotopes of Helium

Natural helium is predominantly ^{4}He with 0.000 13 atom per cent of ^{3}He. It is the second lightest element and has the lowest boiling point of 4.21 K under one atmosphere pressure. Liquid helium boiling under reduced pressure has provided low temperatures well below 1 K needed for the study of special properties of materials. Helium (^{4}He) has no solid-liquid-vapour triple point, instead it has two liquid phases, liquid He-I and liquid He-II which are in equilibrium with vapour at the triple point 2.71 K and 38 mm pressure. Of these, the liquid He-II has many extraordinary properties displayed by no other liquid. For instance, its viscosity falls to nearly zero as the temperature is lowered, making it a *superfluid* which cannot be contained in any vessel as it creeps up and seeps through the finest pores and escapes. The formation of such a material hinders and limits the use of liquid helium in obtaining much lower temperature well below 0.01 K. On the contrary, the rarer isotope ^{3}He on liquefaction yields a normal liquid with no superfluidity**, hence the need for isolating this isotope for meeting the requirements of ultra-low temperature research.

The earliest attempts to obtain pure ^{3}He were made in Harwell using the thermal diffusion process. Bowring and Davies [23] used three columns each of 4.5 m length of which the first two were of the coaxial tubes with the inner of stainless steel of diameter 7.94 mm and the outer of brass of 30 mm diameter, maintained at 680 and 15 °C respectively. The third column was also of brass but of 13.1 mm diameter surrounding a nichrome wire of diameter 0.71 mm stretched along the axis and heated to 645° C. Special seals were designed to withstand a pressure of a few atmospheres. With this equipment, they succeeded in isolating 100 cm^3 of pure ^{3}He at the end of one year, and in the process 10^5 litres of natural helium had to be treated which indicates the prohibitively expensive nature of the project.

---

* For detailed references, see H. London [10], or Grew and Ibbs [22].

** The superfluidity of ^{4}He is due to its being a *boson i.e.* a nuclide with an even number of nucleons in the nucleus following the Bose-Einstein statistics. The nuclide ^{3}He differs from this in being a *fermion* with an odd number of nucleons and following the Fermi-Dirac statistics. For details any book on Nuclear Physics may be consulted.

An alternative and probably less expensive process was launched in the Mond Laboratory using the nuclear reaction.

$$^{6}\mathrm{Li}\ (n, \alpha)\ T$$

which needs, to start with, the isotope ^{6}Li, itself to be separated from natural lithium at great expense*. The above nuclear reaction is mainly carried out to obtain tritium in large amounts needed for studies on thermonuclear reactions. Tritium, the product of the reaction, by β^{-} decay with a half-life of 12.33 years, yields ^{3}He as the final stable product.

$$^{3}\mathrm{H} \rightarrow {}^{3}\mathrm{He} + \beta^{-} + \nu$$

Thus, the overall product of these two reactions on long standing is a gaseous mixture of ^{3}H, ^{3}He and ^{4}He from which the ^{3}He is to be separated. For this the mixture is passed through palladium metal which absorbs all the tritium and the residue, consisting of only the two isotopes of helium, is fed into a specially designed thermal diffusion column which separates them. The output at the hot end of the column is 99.999 96 per cent pure ^{3}He containing no more than 10^{-10} per cent of tritium impurity [24].

In fact the separation of the helium isotopes should present no problem. The mixture has only to be liquified and placed in a vessel of pore size of the order of 10 nm which would be impervious to all normal liquids. The superfluid will leak out fast and leave behind ^{3}He in an isotopically pure state, an excellent example of the Maxwellian devil[24]!

(ii) *Isotopes of Uranium*

No less than 2100 columns were set up in multiple cascades of series-parallel arrangement, capable of treating 50 g of uranium per day. The limited object was to obtain sufficiently enriched samples suitable for use in the calutron. Thermal diffusion in liquid uranium hexafluoride had also been investigated. However, the yields being incommensurate with the energy input, the project was abandoned in preference to that of gaseous diffusion. We shall revert to this in a later chapter.

(iii) *Triisotopic Systems*

The thermal diffusion column functions best in extracting only one isotope, either the lightest or the heaviest, allowing all other isotopes of intermediate masses in a polyisotopic system to accumulate in the column. It is exactly in such a case that the idea of Clusius of admitting auxiliary gases of different masses, correctly chosen, and which could be later easily separated by simple chemical means, proved successful, as pointed out earlier (§ 6.4.3 (b)). It is only the thermal diffusion column which has the facility of admitting and removing the foreign gas without altering the relative proportion of the isotopic components. We shall illustrate this with respect to the separation of

* The separation of ^{6}Li is considered in Chapter 9, devoted to isotopes of special importance in the atomic age.

the three isotopes of neon and of argon successfully effected by Clusius and coworkers[19].

(a) *The Neon Isotopes*
Natural neon is a mixture of 90.5 atom per cent of ^{20}Ne, 0.27 of ^{21}Ne and 9.2 of ^{22}Ne. Beginning with 60 litres of the natural gas, Clusius obtained, following an initial separation, a sample enriched in ^{21}Ne to the extent of 6.6 per cent, mostly concentrated over the middle segment (6 to 16 m) of the column. To this was added methane deuterated to different extents as, $^{18}(CD_2H_2)$, $^{19}(CD_3H)$ and $^{20}(CD_4)$. Such an admixture led to all the ^{22}Ne diffusing towards the bottom of the column whence it was drawn out in 99 per cent purity. The residue in the column now was only ^{21}Ne and ^{20}Ne. At this stage $^{20}(CD_4)$ was withdrawn from the column and replaced by normal methane $^{16}(CH_4)$. This resulted in the diffusion of ^{21}Ne down the column whence it was collected in a purity of 99.6 per cent, leaving in the column only the lightest isotope ^{20}Ne free from the other isotopes. Thus all the three isotopes of neon were separated in a high state of purity exceeding 99 per cent by the use of auxiliary gases of intermediate molecular masses. The overall separation factor was computed to be 96 700[10,25].

(b) *The Argon Isotopes*
Natural argon consists of three isotopes ^{36}Ar 0.34, ^{38}Ar 0.06 and ^{40}Ar 99.6 atom per cent. The larger mass difference of two units between the isotopes makes their separation relatively easier than in the case of neon. Most of the heaviest isotope is eliminated in the first diffusion process, the residue being greatly enriched in the lighter isotopes, thus : $^{36}Ar = 74$, $^{38}Ar = 14.5$ and $^{40}Ar = 11$ per cent. By successively admitting the different isotopic molecules of hydrogen chloride gas, Clusius[25] obtained the three isotopes of argon each in a nearly pure state. While $D^{35}Cl$ gets in between ^{36}Ar and ^{38}Ar, the species $D^{37}Cl$ gets in between ^{38}Ar and ^{40}Ar. At the same time the species $H^{35}Cl$ and $H^{37}Cl$ effect the dilution of ^{36}Ar and ^{38}Ar respectively, the separations being easier in the expanded state. The net result is a redistribution of the argon isotopes and the auxiliary hydrogen chloride gases as pictured schematically in Fig. 6.9.

6.4.6 Evaluation

The positive aspects of the thermal diffusion as a process of isotope separation are (i) the extreme simplicity of the equipment involved consisting mainly of coaxial tubes, wires, valves and special seals, and (ii) large overall separation factors the process is associated with, leading to high purity separations. As against these advantages of simplicity and performance efficiency, the process proves to be very expensive when account is taken of the heavy input of energy needed to maintain a high temperature gradient across the inner and outer tubes over prolonged periods. The time needed for the attainment of equilibrium is of the order weeks or even months. The period may be shortened slightly by maintaining a slow flow rate in open tubes, but the separation factor goes down compared to operating with closed tubes.

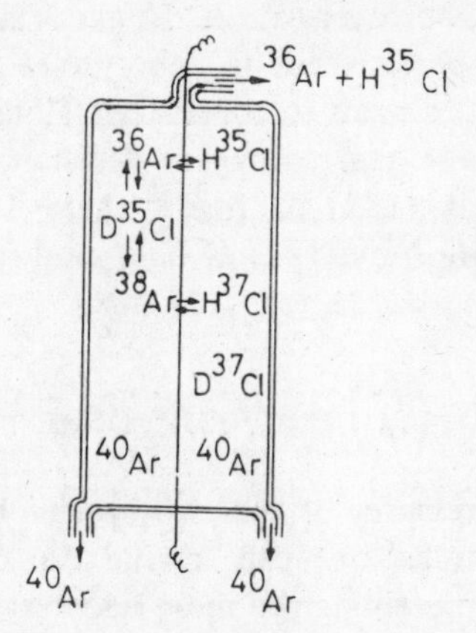

Fig. 6.9: **Separation of isotopes of argon**—a triisotopic system. by an auxiliary gas in a Clusius column.

6.5 ULTRACENTRIFUGATION

Soon after the discovery of isotopy, Lindemann and Aston[25] tried to effect a separation of gaseous isotopic molecules in the high gravitational field of a centrifuge, but without success. The attempts of Joly and Poole[26] and of Mulliken[27] in the early twenties too met with no better success. It was only after the development of the vacuum centrifuge some sixteen years later, that Beams and coworkers[28] succeeded in centrifuging carbon tetrachloride vapour to obtain a sample partially enriched in ^{37}Cl. This was followed by Humphrey's[29] work in altering the nearly 1:1 ratio of ^{79}Br to ^{81}Br to a value some ten per cent higher by centrifuging the vapour of ethyl bromide. About the same time, Martin and Kuhn[30] presented a theory of the thermally controlled countercurrent gas centrifuge. However, it was Cohen[31] who developed a consolidated theory of the ultracentrifuge, specially applicable to the separation of the isotopes of uranium, the essential features of which are considered below.

6.5.1 Cohen's Theory

Isotope separation in a gravitational field is a reversible statistical process based on the pressure gradient of the gas in the field governed by the classical barometric expression

$$P_h = P_o \exp\,(-\,Mg\,h/RT) \qquad (6.20)$$

where P_o and P_h are the pressures at the ground level and at height h, M the the molecular weight of the gas and g the acceleration due to gravity. With isotopic gases of masses M_1 and M_2, the heavier species M_2 tends to concentrate at the ground level and the lighter species M_1 at the top level. We have for the two species:

$$\left(\frac{P_1}{P_2}\right)_h = \left(\frac{P_1}{P_2}\right)_o \exp\left[-(M_1 - M_2)g\,h/RT\right] \qquad (6.21)$$

In a centrifugal field, however, where the gas is subjected to spinning at high speeds, the heavier species tends to concentrate at the periphery (radius r), while the lighter species tends to concentrate along the axis ($r = 0$). Also the energy per mole of the gas rotating at a distance r from the axis with an angular velocity ω is $(1/2)\ M\ (\omega r)^2$ where ωr is the equivalent linear velocity. Replacing $Mg\ h$ by $(1/2)\ M\ (\omega r)^2$ and reversing the order, Eq. 6.21 becomes

$$\left(\frac{P_1}{P_2}\right)_o = \left(\frac{P_1}{P_2}\right)_r \exp[-(M_1 - M_2)(\omega\ r)^2/2RT] \quad (6.22)$$

Lastly, the partial pressures of the two species being proportional to their respective mole fractions, one can rewrite Eq. 6.22 in terms of the usual quantities y and x representing the mole fractions at the axis ($r = 0$) and the periphery (r) respectively, thus:

$$\frac{y}{1-y} = \frac{x}{1-x}\ e^{-(M_2 - M_1)(w\ r)^2/2RT} \quad (6.23)$$

whence the separation factor comes to be

$$\alpha = \exp\left[\Delta M(\omega r)^2/2RT\right] \quad (6.24)$$

where $\Delta M = M_2 - M_1$

Finally, in view of α being only slightly in excess of unity*, the same can be taken as

$$\alpha = 1 + \Delta M(\omega r)^2/2RT \quad (6.25)$$

The *overall optimum separation factor* A_{opt} is a function of the single stage factor α_o and the dimensions of the centrifuge, length L and radius r_a, given by

$$\ln A_{opt} = (L/2r_a) \ln \alpha_0 \quad (6.26)$$

The quantity $L/2r_a$ represents *multiplication factor K.*

The Separative power δu is defined as

$$\delta u = \frac{\pi \rho D L}{2}\left[\frac{\Delta M(\omega r_a)^2}{2RT}\right]^2 \quad (6.27)$$

where ρ is the density, D the diffusion coefficient and ωr_a is the peripheral velocity.

The most important feature of separation by the centrifuge is that α *depends simply on* ΔM, unlike in gaseous diffusion where it depends on $\sqrt{M_2/M_1}$, or on $\Delta M/(M_1 + M_2)$ as in thermal diffusion. Hence, the method is applicable to the separation of isotopes of heavy elements with the same efficiency as with light elements. Also, modern ultracentrifuges spinning

---

* For $\Delta M = 1$ and a peripheral velocity $\omega r = 300$ m/s, the separation factor α at room temperature ($T = 300$ K), comes to be 1.018 3

at 90,000 revolutions per minute generate accelerations equivalent to 10^5 times earth's gravity.

6.5.2 The Ultracentrifuge

The gas centrifuge for isotope separation has been developed in three stages. The earliest was of the simple evaporation type, designed by Mullikan[27], with a single flow corresponding to the elementary separation factor.

In his early work, Beams[28] used this type with a vertical rotation axis of the rotor which was suspended from a thin, highly elastic tubular shaft. Such a unit was sufficient for small quantities of enriched samples needed for research but was not suited for large scale separations. The difficulty was to design a shaft which allows sufficient flow of the gas and also is elastic enough to provide necessary mechanical disconnection of the rotor from the bearings at the ends of the rotating system. To meet these opposing demands, a double coaxial tube shaft was designed in which the rotor was suspended from the inner tube while the outer tube was divided into two parts connected by thin walled highly flexible bellows.

Later came the centrifuge with *cocurrent* flow type in which there were two flows in the same direction, one closer to the axis and the other closer to the outer wall. Lastly, the centrifuge with *countercurrent* flow was developed, in which the stream closer to the axis flowed down and the other closer to the wall moved upwards. Under such a countercurrent flow, the separation factor gets multiplied. In the separation of the isotopes of xenon, for instance, a multiplication factor of 4.5 was attained[28a].

A typical countercurrent centrifuge consists of an aluminium cylindrical rotor about 1 m high and 30 cm diameter, resting on a pointed pivot which minimizes friction. Tempered steel, or alloys of aluminium, or of titanium, or of fibre glass have also been found suitable as a rotor material. As the shape of the rotor was found to significantly affect the performance, various shapes, besides the cylindrical, as the disc and tubular pieces radiating to the axis were evaluated. In place of thc normal straight vertical, the performance of curved vertical shafts too had been studied.

The separation power was found to be proportional to the length of the centrifuge and to the *fourth power* of the peripheral velocity i.e. $(\omega r)^4$. The maximum velocity is, however, limited by the tensile strength of the rotor material. For safe performance, the tensile strength has to be greater than $\rho\omega^2r^2$ where ρ is the density of the material. The maximum permissible peripheral velocity of a thin-walled ring-shaped rotor made of fibre glass is stated to be 527 m/s, while the corresponding values for tempered steel, aluminium and titanium alloys are 318, 366 and 420 m/s respectively[10].

Elastic shock absorbers are fitted to minimize effects of vibrations due to very high spinning speeds. The rotor is generally driven by an induction motor and the stator consists of external winding fitted just below. The entire unit is placed inside a rigid enclosure casing under vacuum to eliminate frictional heating effects. An important result of the inner cylinder revolving at such high speeds of 50 000 to 100 000 revolutions per minute is that it *functions as a molecular pump* generating high vacuum. The trapped air is pushed out to the outside. Thus the centrifuge functions as a *self-contained unit*, not requiring another pump.

In the rotor designed by Martin and Kuhn[30], the same is subdivided into several chambers and the process gas to be separated is led through the chambers in a countercurrent fashion. This has the effect of operating as a multiple stage cascade. Under the centrifugal action, the heavier species is pushed outwards while the lighter species concentrates towards the axis. It was, however, subsequently realized that even when not subdivided into successive chambers, the rotor displays the multiple stage effect. A typical centrifuge is shown in Fig. 6.10. An internal streaming of the gas parallel to the axis along the directions indicated by the arrows, produces a *cascade effect* whereby in a single unit the separation factor is multiplied some 20-50 times. This effect is limited only by the magnitude of back diffusion.

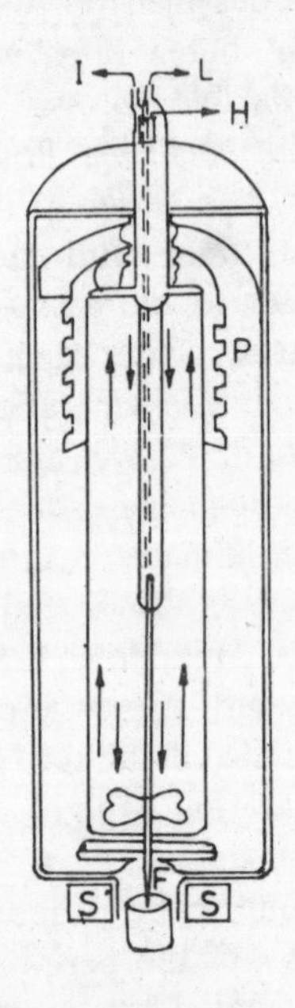

Fig. 6.10: **Ultracentrifuge with counter-current (Zippe type)**—I:initial fraction; H:heavy fraction; L:light fraction; P:molecular pump; S:Stator; F:needle.
Reprinted with permission from *La Séparation des Isotopes* by M. Chemla and J. Périé, p. 144 © Presses Universitaires de France, 1974.

The process of centrifugation generates a pressure gradient in the form of a depression along the axis and a compression close to the outerwall, which is adequate to maintain the above mentioned streaming parallel to the axis without the need of an external pump. At the equivalent of each stage, the lighter isotope is lifted to the next upper stage and the heavier isotope pushed down to the stage below.

6.5.3 The Hydrogen Stabilization Effect

Frequently, the isotope enrichment in a gas centrifuge falls short of the theoretical value. A principal cause of this is the spontaneous setting up of thermoconvection currents by the radial flow of the gas leading to expansions along the axis and compressions at the walls. If the resulting heat transfer to the surroundings does not keep pace, a temperature gradient results bringing in convection currents which tend to homogenize the gas composition. One way of overcoming this anti-separation tendency is to heavily admix hydrogen to the extent of about 90 per cent of the sample gas. On account of its lightness and very high thermal conductivity,* the temperature gradient gets nullified to a very large extent. This technique is referred to as *the hydrogen stabilization effect*[32] and leads to an improvement of the enrichment factor.

In addition, the inclusion of a heating element close to the rotor also leads to a larger enrichment factor. The heating element may simply be a 200-watt bifilar wound filament resting inside a quartz tube some 4 cm long and about 1 cm in diameter fitted to the rotor at about its middle height so that it revolves with the rotor. This set up is referred to as the *thermally controlled centrifugation.* Such a role of the inclusion of a heating element is well brought out in the separation of the xenon isotopes[33] with hydrogen stabilization.

6.5.4 Applications

The gas centrifuge has been widely used for obtaining small amounts of enriched samples of isotopes of Cl, Br, Se, Ge, Xe and a few other elements, mostly to meet the needs of research. Curiously, in some cases, notably in selenium and germanium, the actual enrichment realized was some 2 to 2.5 times *greater* than the theoretical expectation in accord with Eq. 6.6. Such a positive effect is considered the result of a temperature rise due largely to adiabatic gas circulation[34]. It is, however, not clear why the same is not observed as a general effect in the centrifugation of isotopic mixtures of all elements.

The importance of the centrifuge as an efficient device for effecting isotope separations suddenly rose when it was found adaptable to uranium isotopes. Today, the thermally controlled countercurrent ultracentrifuge has become a major competitor in the production of kilogram quantities of ^{235}U, starting with uranium partly enriched in the isotope by other techniques. We shall revert to this in a later chapter devoted to the separation of uranium isotopes.

6.5.5 Evaluation

Following are the positive counts commending the adoption of the ultracentrifuge for isotope separations :

---

* Amongst gases, hydrogen has the highest thermal conductivity. On an arbitrary scale if the thermal conductivity of H_2 is put equal to 100, helium would have a value of 78 and the values of all other gases would be well below 20. That is why hydrogen is preferred as the carrier gas in gas chromatography using a thermal conductivity cell as the detector.

(i) The most important feature of the centrifuge is that the separation factor depends on the mass difference of the two isotopes ΔM (= $M_2 - M_1$), unlike in all other methods available. This makes the instrument function in the separation of isotopes of heavy elements with the same efficiency as in the case of light elements, hence its importance to the separation of uranium isotopes.

(ii) The single process (elementary) separation factor with the centrifuge is around 50 times greater than in gaseous diffusion. This means far fewer stages in series would be needed with the centrifuge to arrive at a given enrichment. This advantage is, however, partly offset by a lower yield per stage compared to the process of gaseous diffusion. This only means a larger number of centrifuges have to be operated in parallel to multiply the net yield.

(iii) Another important feature of the centrifuge is the avoidance of a large number of cumbersome pumps indispensable in a diffusion plant. The centrifuge is thus an elegant and compact unit, all the components needed to boost the yield may be located close to each other in a small space, or the units may be scattered far apart if considered necessary. The situation is totally different in the case of a diffusion plant, where the thousands of stages in series have all to be contiguously located under one roof, which makes the complex not only extremely spread out but very vulnerable to enemy action.

6.5.6 Limitations

The most formidable factor against the deployment of the ultracentrifuge for the large scale separation of isotopes, specially of uranium, is in regard to the very stringent specifications which the materials of the rotor and the cylinder walls have to satisfy. These materials have to possess exceptionally high mechanical (tensile) strength to withstand the terrific centrifugal force generated, the equivalent of some $10^4 - 10^5$ times earth's gravity. Eq. 6.6 demands a high peripheral velocity to be developed to obtain a large separation factor and the former is limited by the tensile strength of the rotor.

Further, special devices are to be designed for introducing the feed gas at one point and withdrawing the enriched products from other points while the system is revolving at enormous speeds as in the ultracentrifuge which is operating in vacuum.

Lastly, the rotor, the cylinder and all the materials coming into contact with the gas under centrifugation have to be resistant to the strong corrosive effects of the gas. This is a particularly severe limitation in the choice of materials where uranium hexafluoride is the system. Earlier, these obstacles held up the adoption of the centrifuge for the separation of uranium isotopes. However, as a result of subsequent studies on the properties of a large variety of special materials needed for space projects, the preparation of materials suitable for the ultracentrifuge became possible and the same is now probably the best device available for the separation of ^{235}U.

References

1. F. W. Aston, *Mass spectra and Isotopes*, Longmans Green & Co.1933

1.(a) J. G. Aston, *Phil. Mag.*, 1920, *39*, 449.

2. W. D. Harkins, (*vide* S. Glasstone, *Source Book of Atomic Energy*, D. van Nostrand Co., Inc., New York).

3. G. L. Hertz, *Z. Phys.*, 1932, *79*, 108.

4. G. L. Hertz,*Naturwissen*, 1933, *21*, 884.

5. D. E. Wooldridge and F. A. Jenkins, *Phys. Rev.*, 1936, *49*, 404.

6. G. L. Hertz, *Z. Phys.*, 1934, *91*, 810.

7. C. G. Maier, *Bull. U.S. Bur. Min.* No. 431 (1940).

8. M. Benedict and A. Boass, *Chem. Engng. Progr.*, 1951, *47*, *2*, 51; *3*, 111.

9. M. D. Cichelli, *et al.*, *ibid*, 1951, *47*, *2*, 63; *3*, 123.

10. H. London, *Separation of Isotopes*, G. Newnes. Ltd., London, (1961).

11. E. B. Becker, K. Bier and H. Burghoff, *Z. Naturf.*, 1955, *10a*, 565.

12. S. Chapman and T. G. Cowling, *The Mathematical Theory of Non-uniform Gases*, Cambridge University Press, London, II Ed. (1951).

13. E. W. Becker and R.S. Schutte, *Z. Naturf.*, 1960, *15a*, 336.

14. S. Chapman and F. W. Dootson, *Phil. Mag.*, 1917, *33*, 248.

15. K. Clusius and G. Dickel, *Naturwiss.*, 1938, *26*, 546; 1939, *27*, 148, 487. *Z. Phys. Chem.*, 1939, *B 44*, 397, 451.

16. J. O. Hirschfelder, C. F. Curtiss and R. B. Bird, *Molecular Theory of Gases and Liquids*, John Wiley and Sons, New York, (1954).

17. E. A. Mason, *J. Chem. Phys.*, 1954, *22*, 169; 1955, *23*, 49; 1957, *27*, 75, 782.

18. R. Clark Jones, *Phys. Rev.*, 1940, *58*, 111; 1941, *59*, 1019.

19. K. Clusius and M. Huber, *Z. Naturf.*, 1955, *10a*, 230.

20. K. Clusius, *Helv. Phys. Acta*, 1949, *22*, 473.

21. K. Clusius, and E. Schumacher, *Helv. Chim. Acta*, 1953, *36*, 949; 969.

22. K. E. Grew and T. L. Ibbs, *Thermal Diffusion in Gases*, Cambridge University Press (1952).

23. R. W. Bowring and R. H. Davies, *Atomic Energy Research Establishment Report*, 1957, 1958, 2058 (2).

24. M. Chemla et L. Pe'rie', *La Se'paration des Isotopes*, Presses Universitaires, Paris. 1974.

25. F. A. Lindemann and F. W. Aston, *Phil. Mag*, 1919, *37*, 350.

26. J. Joly and J. H. J. Poole, *ibid*, 1920, *39*, 372.

27. R. S. Mulliken, *J. Amer. Chem. Soc.*, 1922, *44*, 1033.

28. J. W. Beams and C. Skarstrom, *Phys. Rev.*, 1939,*56*, 266.

28 (a) G. Hertz and E. Nann, *Z. Elektrochem*, 1954, *58*, 612.

29. R. F. Humphreys, *Phys. Rev.*, 1936, *56*, 684.

30. H. Martin and W. Kuhn, *Z. Phys. Chem.*,, 1940, *189*, 219 (Abs. A)

31. K. Cohen, *The Theory of Isotope Separation as Applied to the Large Scale Production of* ^{235}U. McGraw-Hill *Book* Co., New York (1951).

32. W. Groth and P. Harteck, *Z. Elektrochem.*, 1950, *54*, 129.

33. V. Faltings and F. Seehober, *ibid*, 1953, *57*, 445.

34. V. Faltings, W. Groth and P. Harteck, *Naturwissen*, 1950, *37*, 490.

Chapter 7

EQUILIBRIUM EXCHANGE PROCESSES

7.1 PHYSICAL EQUILIBRIUM EXCHANGE

An equilibrium exchange may be: (*a*) either between two phases holding the same substance, or (*b*) between two different substances in the same or different phases. These two types are sometimes distinguished as (*a*) physical and (*b*) chemical equilibria. All equilibrium processes are reversible and need much less energy for their sustenance, unlike in the case of irreversible processes considered hitherto*. For this reason it is more convenient and economical to scale up an equilibrium exchange process. Further, the separation factors of an equilibrium process can be much larger than the limiting value given by the $M^{-1/2}$ law for most irreversible processes. However, here too the separation factors fall below expected limits for isotopes of elements beyond about chlorine. Also as the temperature rises, the separation factor tends to unity meaning zero enrichment.

We shall first consider some physical equilibrium processes as (1) fractional distillation and (2) solvent extraction.

7.1.1 Fractional Distillation

It should be possible to bring about a separation of isotopic molecules by fractional distillation using the small difference in their boiling points. This would be a reversible statistical process. As the difference in the boiling points of a pair of isotopic molecules would generally be only of the order of a degree, the distillation process will have to be repeated very many times. It would be more convenient, however, to use a rectifying or fractionating column, of which several forms are possible (*vide* § 7.1.4). In these refluxing columns an upward stream of vapour is intimately contacted with a downward flowing stream of liquid. The refluxing consists in boiling the liquid phase at the bottom and condensing the vapour at the top. A continuous exchange of molecules takes place between the two phases, the more volatile component passing relatively more into the vapour phase and the less volatile in the

* Ultracentrifugation is, however, considered a reversible statistical process.

liquid phase.

$$\text{liquid} \rightleftharpoons \text{vapour}$$

7.1.2 Theory of Fractional Distillation

The existence of a difference in the vapour pressures of isotopic molecules is due to a difference in the bond vibration energies of the polyatomic molecules in the liquid state as anticipated on quantum theory*. This will be considered further in the next section.

If x and y are respectively the mole fractions of the more volatile component in the liquid phase and in the vapour phase at any plate, we have

$$\frac{y}{1-y} = \frac{p_1}{p_2} \times \frac{x}{1-x} \tag{7.1}$$

where p_1 and p_2 are the vapour pressures of the more volatile and the less volatile species at the temperature of equilibrium. Evidently, the ratio p_1/p_2 is the elementary separation factor:

$$\frac{y}{1-y} = \alpha \frac{x}{1-x} \tag{7.2}$$

However, if the vapour and the liquid on a given plate are not in complete equilibrium, the concentration of the more volatile constituent in the vapour would be less than equivalent to the mole fraction y, and we have

$$\frac{y}{1-y} = \beta \alpha \frac{x}{1-x} \tag{7.3 a}$$

where $\beta < 1$ is the efficiency of the plate. A *theoretical plate* corresponds to perfect liquid-vapour equilibrium, with $\beta = 1$. The height of a theoretical plate (HETP) is defined as the distance between two points A and B along the column, so chosen that the ratio of the concentrations of the two substances (say the two isotopic molecules), indicated by subscripts 1 and 2, *in the same phase* obeys the relation, under condition of total reflux:

$$\left(\frac{C_1}{C_2}\right)_A \Big/ \left(\frac{C_1}{C_2}\right)_B = K = \alpha = 1 + \epsilon \tag{7.3 b}$$

This definition of HETP implies that the two streams leaving a given cross section of the column in opposite directions are in equilibrium with each other. Hence, the number of plates that a column of length L is equivalent to is

$$n = L/\text{HETP} \tag{7.3 c}$$

Efficient columns and column packing materials have been developed (§ 7.1.4) which confer the equivalence of several thousand theoretical plates to columns of about 10 m height, the overall separation factor of n plates being α^n. It may be noted that for appreciable separation to be possible, α

* Classical theory envisages no such effect.

should not be less than 1.01.

7.1.3 Vapour Pressures of Isotopic Molecules

There are certain modes of vibrations and vibration-rotations present only in the *condensed* phase and absent in the vapour phase. The energies associated with these modes differ for polyatomic isotopic molecules and give rise to a small difference in their vapour pressures, mainly due to the higher zero-point energy of the lighter isotopic molecule accompanied by a lower latent heat of vaporization. This difference in vapour pressures is to be put to use in separating isotopes by fractional distillation where possible. Table 7.1 presents relevant data for selected molecules to illustrate the isotope effects in the vapour pressures of isotopic molecules.

The observed variations of vapour pressures of isotopic molecules cannot be explained on a single theory. Broadly two types of isotope effects seem to emerge out of the data, (i) a *normal* isotope effect and (ii) an *abnormal* effect, distinguishable by the following characteristics.

(i) *The normal isotope effect*

Here the lighter isotopic molecule has the higher vapour pressure and hence the lower boiling point compared to the heavier species, i.e. $p_1/p_2 > 1$. Also in this case, $\ln p_1/p_2$ decreases with rise of temperature following a T^{-2} relation. Bigeleisen[1,2] had arrived at an empirical relation for the variation of the vapour pressure ratio with temperature of the form :

$$\ln \alpha = \ln \frac{p_1}{p_2} = \frac{A}{24} \left[\frac{h}{2\pi kT} \right]^2 \left[\frac{1}{m_1} - \frac{1}{m_2} \right] \quad (7.4)$$

where m_1 and m_2 are the masses of the *isotopic atoms* (not molecules). Generally, monatomic molecules and homonuclear diatomic molecules display the normal vapour pressure isotope effect (*vide* Table 7.1).

(ii) *The abnormal isotope effect*

Contrary to the above, here it is the heavier isotopic molecule which has the higher vapour pressure and hence a lower boiling point, i.e., $p_1/p_2 < 1$. Also, in this case $\ln p_2/p_1$ falls of with T^{-1}. Some polyatomic isotopic molecules display the abnormal isotope effect.

London[1] had advanced the view that the substitution of the central atom of a polyatomic molecule by a heavier isotopic atom results in abnormal isotopic effect ($p_1 < p_2$), as in this case the internal vibrational frequencies in the liquid phase are lower than in the vapour phase. On the contrary, the substitution of a lighter atom in the peripheral part of the molecule by a heavier isotopic atom leads to the normal isotopic effect ($p_1 > p_2$). This view of London is exemplified by the isotopic molecules of carbon tetrachloride, *viz.*,

$$p^{12}C\,Cl_4 \,/\, p^{13}CCl_4 < 1 \text{ (abnormal)}$$

$$pC^{35}Cl_4 \,/\, pC^{37}Cl_4 < 1 \text{ (normal)}$$

However, it is not possible to generalize the above observation, for a totally reverse situation prevails in the case of the isotopic molecules of methane resulting from central and peripheral substitutions, *viz.*

$$p^{12}CH_4 / p^{13}CH_4 = 1.003 \text{ (at } T = 105 \text{ K) (normal)}$$

and

$$pCH_4 / pCH_3D = 0.997 \text{ (at } T = 100 \text{ K) (abnormal)}$$

Again, contrary to *both* the above cases, are the isotopic molecules of ammonia which display only the normal effect for substitutions whether in the centre or in the periphery, only the magnitude of the effect in the latter case is much larger (*vide* Table 7.1).

Table 7.1 Ratios of Vapour Pressures of Selected Isotopic Molecules[1],*

Isotopic pair	Temp./K	$p_1/p_2 = \alpha$	Ref.
(a) *Monatomic*			
^{20}Ne - ^{22}Ne	24.7	1.044 5	10
^{36}Ar - ^{40}Ar	83.8	1.006	11
(b) *Diatomic-homonuclear*			
H_2 - HD	20.4	1.73	16
N_2 - $^{14}N^{15}N$	64.8	1.006 2	9*a*
O_2 - $^{16}O^{18}O$	69.5	1.010 0	9
(c) *Diatomic-heteronuclear*			
^{12}CO - ^{13}CO	68.3	1.010 9	9
$C^{16}O$ - $C^{18}O$	69.0	1.007 9	9
(d) *Polyatomic*			
H_2O - HDO	313.1	1.059	7
	373.1	1.026	
NH_3 - ND_3	202.0	1.227	16*a*
	235.7	1.139	
$^{14}NH_3$ - $^{15}NH_3$	200	1.005	
$^{10}BF_3$ - $^{11}BF_3$	173.1	0.993 5*	1
$^{10}BCl_3$ - $^{11}BCl_3$	211.4	1.000 0	
	286.0	0.997 0*	16*b*
CH_4 - CH_3D	110.0	0.996 5*	16*c*
$^{12}CH_4$ - $^{13}CH_4$	104.8	1.003 5	9
C_6H_6 - C_6D_6	330	0.973*	8

* Abnormal isotake effect

It had also been suggested that both the normal effect, variation of $\ln p_1/p_2$ with T^{-2}, and the abnormal effect, variation of $\ln p_2/p_1$ with T^{-1} may be co-current to different extents. As the two effects are of opposite sign, there may be a temperature where the two just annul each other meaning that $p_1 = p_2$ or that no isotope enrichment is possible at that critical temperature. This is the case with $^{10}BCl_3$ - $^{11}BCl_3$ at 2.114 K, the normal effect prevails at temperatures below this value and the abnormal effect above this.

* (Data from *Separation of Isotopes*, by H. London, Table 4.1, p.45, © George Newnes Ltd, 1961, by permission of William Heinemann Ltd)

From the instances cited, it is clear that the diverse isotope effects observed in vapour pressures of liquids cannot be simply explained. Multiple factors, as the differences in zero-point energy and internal vibrational frequencies in the liquid phase, molecular symmetry, structure and polarizability and whether association-dissociation occur in the condensed-vapour phases, seem to enter the phenomenon and determine the sign and the magnitude of the ratio of the vapour pressures of isotopic molecules.

7.1.4 Fractionating Columns

The efficiency of refluxing and hence the separation yield is determined by the efficiency factor of the fractionating column. This latter depends on the dimensions of the column, and the nature of the packing used in filling the column. The essential desirable characteristics of an ideal column are :

(*a*) High throughput (flow rate),
(*b*) Large equivalent plate number *i.e.* low HETP,
(*c*) Low holdup, and
(*d*) Small pressure drop.

No single column may present all the above at their most favourable values. Usually a compromise is made to arrive at overall optimum conditions for a given separation problem. The more perfect the equilibrium on each plate, higher is the overall efficiency. The time needed for arriving at the equilibrium is inversely proportional to the holdup factor (rather to the ratio of holdup to throughput). We shall briefly discuss the more commonly used types of columns and packings.

(i) *The wetted-wall column*

This consists simply of a narrow long empty tube. The condensed liquid runs down in the form of a thin film along the inside wall of the tube while the vapour travels countercurrently up along the axis of the tube. At the optimum throughput, the HETP equals the radius of the tube, so that a column of 2 mm diameter and 1 m high would be equivalent to 1000 plates. The chief merit of the wetted-wall column is that the hold-up is small and it is self-wetting *i.e.* it needs no preflooding. However, the throughput being very low, a very large number of columns have to be operated in parallel to arrive at a significant yield. Using one hundred wetted-wall columns, Kuhn *et al*[3] separated the isotopic molecules of carbon tetrachloride, chloroform, methyl alcohol and benzene.

(ii) *The bubble-cap plates*

The fractionating column with bubble-cap plates has been in common use in the large scale distillation of water, alcohol, petroleum products and other solvents. Figure 7.1 shows a typical column with bubble-cap plates; the liquid flows down from plate to plate as the vapour ascends countercurrently after bubbling through the liquid layers which ensures intimate mixing of the two phases, a prerequisite for the establishment of equilibrium.

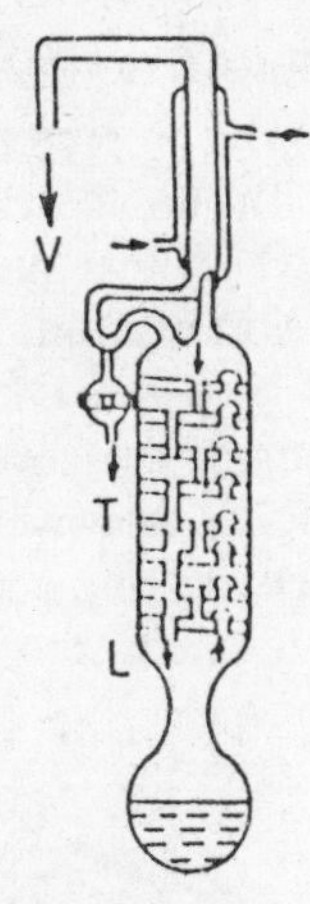

Fig. 7.1: **A bubble-cap fractionating column**—L:liquid; V:vapour, T:tap for test sample

Its main advantage is that its performance conforms to a simple theory so that yields can be predicted precisely. Variations in throughput has but a small effect on the efficiency. The major disadvantages are a large HETP, some 30-40 times that of a packed column (*vide infra*), and a large pressure drop at each plate. The bubble-cap fractionator is not used for isotope separations, except probably in the final finishing stage of a sample already heavily pre-enriched.

(iii) *Packed Columns*

Columns packed with a suitable material are more efficient than wetted-wall or bubble-cap plate fractionators. Since the HETP is some 30-40 times smaller and the pressure drop at each plate is lower, the hold-up is within tolerable limits. The liquid flowing down continuously wets the filling material surface and thus provides a large area of liquid film in contact with the vapour ascending countercurrently. The packing material continuously splits the vapour stream and generates turbulence which enhances lateral diffusion. This makes a larger throughput possible. The path of the vapour molecules is always irregular and randomly zig-zag. A ratio of 8:1 for the diameters of the column and of the individual packing units is found to be optimum. This ratio is found to provide an optimum balance between the free space available for mixing and the surface area of the packing material. A pre-flooding of the filled column enhances its efficiency. The actual performance depends greatly on the nature, size and shape of the packing. This last aspect has been much researched upon and some relevant findings are reported below.

7.1.5 Types of Packing

The column packing may be either of irregular random *dump type*, or of regular *preformed pattern type*. The former may consist of glass, metal, or ceramic beads or helices dumped into the column. This is the type of packing found in the columns commonly used in laboratory distillations but seldom used for isotope separation. The preformed pattern may be of an S-shaped Dixon wire gauze ring, or a helipak consisting of coils of inert material wire piled one above the other leaving enough space between adjacent turns to allow the formation of a thin liquid film by capillary action when wetted (Fig. 7.2). While with the Dixon ring the efficiency factor decreases with increasing throughput, the reverse is the case with helipak[1] which is an advantage. The HETP, however, increases with the column diameter in both cases. Both the Dixon ring and the helipak columns need to be preflooded. Several helipak columns were used in parallel by Kuhn *et al*[4] to obtain enriched heavy water.

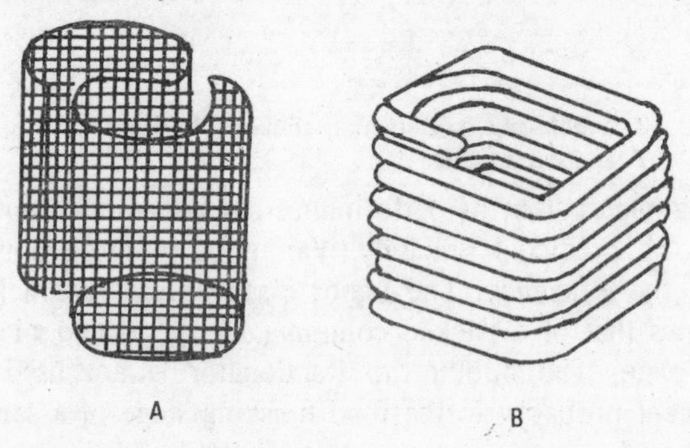

Fig. 7.2: **Types of packing**—A:Dixon wire gauze; B:Helipak

A variety of improved designs for the shape of packing are being constantly experimented upon : of these the *heligrid* or the spiral staircase like packing and the *spraypak* have been found very efficient. The latter is described below.

7.1.6 The Spraypak

The *spraypak* is a British patent invented by McWilliams, Prat, Dell and Jones[5] specially for use in the fractional distillation of heavy water. Here, the distillation column, rectangular in shape, is fitted with regular diamond-shaped cellular structure made up of fine expanded metal sheets (Fig. 7.3). It permits the liquid to run down as a smooth film closing the appertures. When the vapour breaks through the appertures, a spray forms which moves sideways perpendicularly to the liquid flow direction till it reaches another part of the structure. The spraypak functions as the equivalent of 33 per cent additional theoretical plates compared to a bubble-cap tower of the same height. This in its turn is equivalent to corresponding saving in construction materials and in energy consumption.

The special feature of the spraypak is its inherent self-distributing action on the liquid and vapour flows. When the spray density tends to fall, droplets are sucked in from adjoining regions till a uniform density is restored. The ratio of the vapour velocity to HETP for a spraypak being 3.6 times that for a tower with bubble-caps, the yield of the separated product

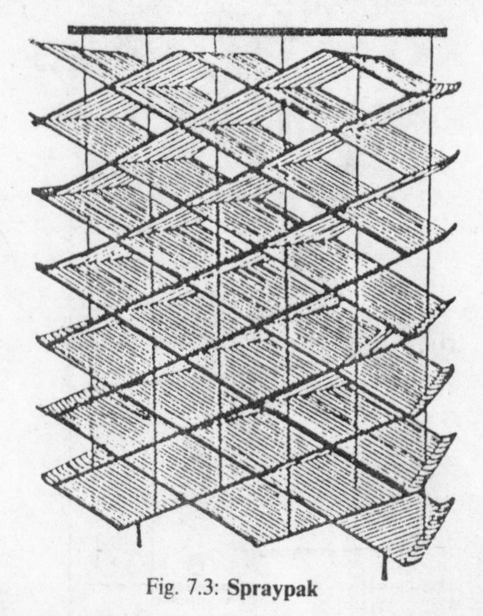

Fig. 7.3: **Spraypak**

From *Separation of Isotopes* by H. London, p.88, Plate I © George Newnes Ltd., 1961 : reporduced by perission of William Heinemann Ltd).

will also be greater in the same ratio.

London[1] has made several calculations in respect of the optimum sizes for the spraypak cells for the fractional distillation of different liquids in terms of their surface tension to density ratio $(\gamma/\rho)^{1/2}$.

7.1.7 Distillation Layout with Reflux

The larger the scale of the distillation plant, the more serious the problem of energy consumption. To minimize the latter, it is necessary to make the process as much reversible as possible by resorting to recompressing the vapour and refluxing it. A general layout of the plant with provisions for reflux is sketched in Fig. 7.4. The feed enters above the middle of the column. The vapour leaving the top of the column is passed through heat exchangers *HI* and *HII*, compressed and cooled to room temperature after passing through *HII* and returned to the boiler at the bottom of the column. A part of the liquid from the boiler is raised to the top of the column through *HI* and refluxed down the column as shown in the figure. The higher boiling isotopic species is withdrawn from the bottom of the column.

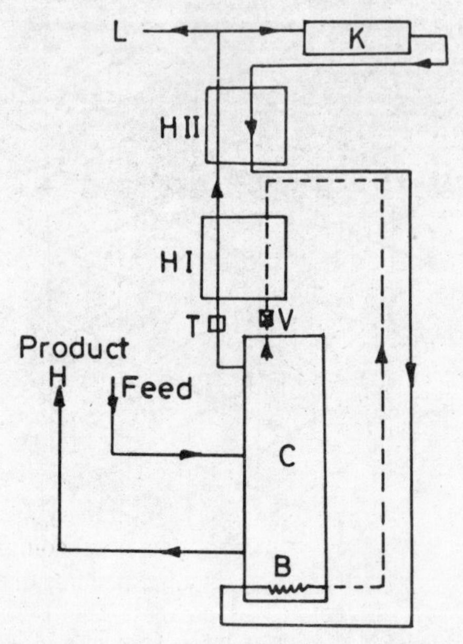

Fig. 7.4: **Distillation with reflux**—B:boiler; C:packed column; HI, HII:heat exchanges; T:trap; V:valve; K:condenser: H:heavy isotope; L:light isotope + waste

7.1.8 Applications

Some of the isotopes whose enrichment has been realized by the fractional distillation of the respective liquids, are shown in Table 7.2.

Table 7.2 : Some Isotope Separations by Fractional Distillation

Liquid distilled	Initial composition per cent	p_1/p_2 (Temp. /K)	Isotope enriched per cent	Reference
Hydrogen	HD 0.028	1.86 (292)	HD 5-10	6
Water	D 0.014	1.026 (373)	D 87-91	7
Water	^{17}O 0.039 ^{18}O 0.205	1.0065 (350)	^{17}O 1.5 and ^{18}O 0.65	7
Boron trifluoride (+ dimethyl ether)	^{10}B 20		^{10}B 83-95	8
Carbon monoxide	^{13}C 1.11	1.01 (70)	^{13}C 60	9
Neon	^{22}Ne 8.9	1.044	^{20}Ne 60; ^{22}Ne 2	10
Argon	^{36}Ar 0.34	1.006 (84)	^{36}Ar 0.6	11

Of the above, the separations of deuterium, heavy water and of boron-10 will be reverted to in a later chapter.

7.1.9 Evaluation

Fractional distillation is a simple technique to use but the yield is low in comparison with the energy consumed. Though adopted in the Manhatten Project, US, in the early period of 1942-45 for obtaining heavy water on a large scale, the same was thereafter abandoned in favour of other more efficient processes, besides the demand for heavy water fell in America due to their switching over to ^{235}U-based reactors*.

7.2 SOLVENT EXTRACTION

Solvent extraction, also known as *absorptive fractionation*, is another reversible physical statistical process suggested for the separation of gaseous isotopic molecules exhibiting a difference in their solubilities in a solvent which may be a non-reacting gas in the liquid state. The gaseous mixture to be separated is compressed to a high pressure to increase solubility and contacted with the liquid solvent countercurrently along a suitable column, as in fractional distillation. The reflux reactions consist in expanding the solution collecting at the bottom so that the solute gas escapes. This is then compressed and returned to the bottom of the column as reflux. The solvent depleted of the gas is pumped to the top of the column and let to flow down again (Fig. 7.5).

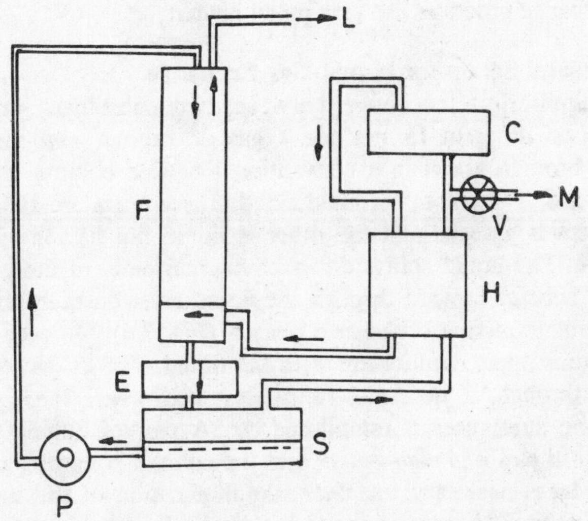

Fig. 7.5: **The solvent extraction cycle**—F:fractionating column; E: expansion chamber; H:heat exchanger; S:solvent, V:valve; C:compressor; L:less soluble output; M:more soluble output; P:solvent pump

The only application of solvent extraction to isotope separation seems to be the work of Augood[12] who effected a partial separation of HD from

* India and some other countries, however, continue to develop their heavy water production programmes.

hydrogen using liquid nitrogen as the solvent. HD was found to be more soluble than H_2 in liquid nitrogen, giving a separation factor of 1.02 at low pressures, the value falling with rise of pressure. The energy consumption was stated to be much higher than in the fractional distillation of liquid hydrogen. With liquid neon as the solvent, a much higher separation factor of the order of 3.24 accompanied by a great saving in energy was predicted by Augood. However, difficulties of gas purification at low temperatures did not allow the development of the process.

The use of hydrocarbon solvent mixtures of *n*-heptane and *n*-octane had also been examined without any positive result of importance[13].

7.3 CHEMICAL EQUILIBRIUM EXCHANGE PROCESSES

We shall now consider chemical equilibrium exchange between two substances in the same or in two different phases as liquid-gas or liquid-liquid*. There exist certain similarities between the chemical and physical equilibrium exchange as the fractional distillation, wherein exchange of the same substance between two phases occurs (§7.1). As pointed out earlier, all equilibrium processes are reversible and need much less energy for maintaining the equilibrium once attained, compared to the energy needed to carry out an irreversible process as diffusion (Ch. 6). Also, the separation factors attainable in chemical exchange processes are very much higher.

7.3.1 The General Set up for Liquid-Gas Exchange

A chemical equilibrium exchange between two substances or isotopic molecules in two different forms, one a gas or vapour, and the other a liquid, can be brought about in a conventional packed column used in the distillation process. The column is maintained at a temperature at which one of the substances is gaseous and the other is liquid but having appreciable vapour pressure. The liquid sprays down countercurrently to the gas-vapour mixture which bubbles upward through the liquid layer on each plate which permits the intimate mixing of the two phases (Fig. 7.6). On each plate the vapour gets into isotopic equilibrium with the liquid. The gas-vapour thus is made to pass through, if necessary, a catalyst bed where the equilibrium between the two substances is established**. A pair of bubble plate and catalyst bed constitutes *one plate stage* and the column comprises of a series of such plate stages necessary to effect a multiplication of the single stage process. Countercurrent flow is arranged by chemically refluxing the products at the two ends of the column.

7.3.2 Theory

At this point it is useful to recall the chemical isotope effects described under § 3.1.1. For the development of the theory of isotope exchange under

---

* The exchange between a stationary solid, and a moving fluid, constitutes a form of *chromatography* whose applications to isotope separation are considered in § 7.4.

** Where the two substances reach equilibrium quickly no catalyst would be needed.

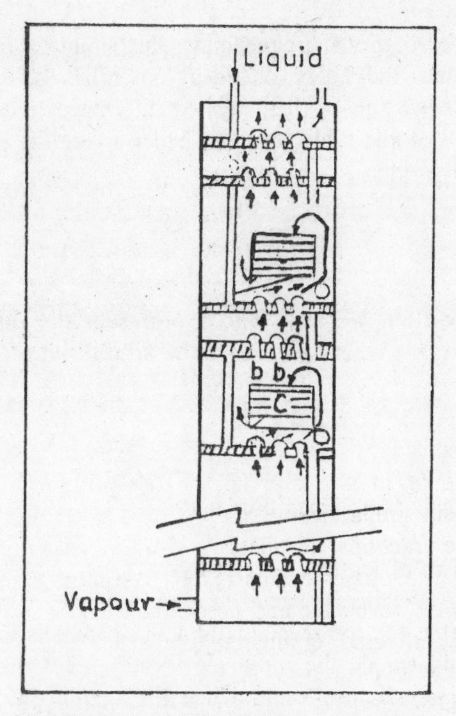

Fig. 7.6: **Liquid-vapour counter-current exchange**—b:bubble caps; c:catalyst bed; O:heating elements
(Reprinted with permission from *La Séparation des Isotopes by* M. Chemla and J. Périé, p. 222 © Presses Universitaires de France, 1974)

equilibrium, it is convenient to represent the reaction involving isotopic molecules simply as

$$A_1X + A_2Y \rightleftharpoons A_1Y + A_2X^*$$

for which the equilibrium constant, at a given temperature, is given by

$$\frac{[A_1Y]\,[A_2X]}{[A_2Y]\,[A_1X]} = K > 1 \qquad (7.5)$$

The slight excess of K over unity permits the enrichment of the isotope, in this case A_1, in one of the substances, AY, while the same is depleted in the substance AX. This deviation of K from unity is due to *the isotopic molecules* A_1X and A_2X, as also A_1Y, and A_2Y *not being totally identical.* There exists a small difference in the bond strengths of X in the A_1X and in A_2X molecules, as also in the bonds of Y in the A_1Y and in A_2Y molecules. These differences in their turn lead to differences in corresponding vibrational

* Since by convention, the concentrations of substances shown on the *right* side of the equilibrium form the *numerator* of the equilibrium constant expression, the direction of writing the reaction is so chosen as to make $K(=\alpha) > 1$. This would also make the enrichment factor ϵ positive, as $\epsilon = \alpha - 1$.

and possibly also rotational frequencies. This situation implies non-identical zero-point energies and hence chemical potentials of the pairs of isotopic molecules. The net result of these related differences is to alter the partition functions of the isotopic molecules and hence make the equilibrium constant deviate from unity, since

$$K = \frac{Z_{A2X}}{Z_{A1X}} \bigg/ \frac{Z_{A2Y}}{Z_{A1Y}} \qquad (7.6)$$

If, on our notation, we let x and y represent the mole fractions of the desired isotopic form before and after the equilibrium, respectively, we can write

$$K = \frac{y}{1-y} \bigg/ \frac{x}{1-x} = \alpha \qquad (7.7)$$

where α is the single stage separation factor.

7.3.3 Mechanism of Isotopic Exchange

For an isotopic exchange reaction to be possible, a suitable mechanism should be available for the reactants and the products to pass through. The mechanism should enable the rupture of bonds involving the isotopic atoms in the original reactant molecules when they come into close contact and also effect the formation of new bonds with the partner species exchanged. This process is naturally facilitated if the reactants can form an intermediate complex with loosened bonds. For instance, for the general reaction under study, the intermediate complex could be of the type

$$A_1X + A_2Y \rightleftharpoons A_1..X..A_2..Y \rightleftharpoons A_2X + A_1Y$$

This can end up by the rupture and reformation of new bonds to yield the exchange products (forward reaction) or reform the old bonds to yield initial reactants (back reaction), the probability of the forward reaction being slightly higher under the conditions of $K > 1$, as assumed.

In the reaction

$$^{14}NO + {}^{15}NO_2 \rightleftharpoons {}^{14}NO_2 + {}^{15}NO$$

studied by Leifer[14], the intermediate complex is

$$O^{14}N \ldots O \ldots {}^{15}NO$$

For an appreciable isotopic enrichment *i.e.* for a redistribution of the isotopic atoms significantly different from stoichiometric equipartition dictated by the relative proportions of the reactants*, it is necessary that at least one

---

* Consider the case where the initial molar ratio of ammonium ions to ammonia in the isotope exchange reaction

$$^{14}NH_4^+ + {}^{15}NH_3 \rightleftharpoons {}^{14}NH_3 + {}^{15}NH_4^+$$

is *n*. If on the attainment of equilibrium

Contd.

of the isotope atoms is loosely bound in its molecule. This condition sometimes may not be fulfilled and both the isotopic atoms may be so strongly bound in their molecules that they cannot participate in the exchange reaction. For instance, a mixture of heavy water vapour and hydrogen gas stays indefinitely without reaction. It is in such cases that a suitable catalyst would be helpful. In the example cited. the presence of finely divided nickel, or platinum, helps in breaking the bonds and their reformation, leading to isotopic exchange

$$\mathrm{DOD} + \mathrm{H_2} \xrightleftharpoons{\mathrm{Ni/Pt}} \mathrm{HOD} + \mathrm{HD}$$

The rates of isotopic exchange reactions vary from zero to high values depending on the nature of the system and the conditions of temperature, presence of complexing agents, *etc.* The underlying mechanisms of association dissociation, atom or electron transfer, *etc.* are considered in detail by Wahl and Bonner[15] and by Haissinsky[16].

7.3.4 Countercurrent Chemical Refluxing

It was stated in § 7.3.1 that to achieve a multiplication effect for the overall separation factor, the light and the heavy isotopic fractions collecting at the two ends of the column are reconverted to the initial chemical forms and returned to the column to recycle countercurrently. In illustration we present two examples of chemical refluxing in equilibrium isotope exchange reactions.

$$\text{(1)}\quad {}^{15}\mathrm{NH_3}\,(g) + {}^{14}\mathrm{NH_4^+}\,(aq) \rightleftharpoons {}^{14}\mathrm{NH_3}\,(g) + {}^{15}\mathrm{NH_4^+}\,(aq)$$

The ammonia gas enriched in ^{14}N reaching the top is reacted with hydrochloric acid solution and the ammonium chloride solution is made to drip down the column. At the same time, the ammonium chloride solution enriched in ^{15}N collecting at the bottom of the column is decomposed into ammonia gas by the action of hot caustic soda and the gas is made to rise again countercurrently to the solution.

$$\text{(2)}\quad \mathrm{H_2O} + \mathrm{HD} \rightleftharpoons \mathrm{HDO} + \mathrm{H_2}$$

The hydrogen gas, depleted in deuterium, escaping at the top is burnt with oxygen and the water formed flows back down the column again while the water, enriched in deuterium, collecting at the bottom is electrolysed into hydrogen and the gas made to re-ascend the column.

Similarly, appropriate refluxing reactions are adopted in all equilibrium exchange reactions. Some of these will be described where necessary. It should be noted that while all equilibrium reactions are reversible and

---

$$^{14}\mathrm{NH_4^+} / {}^{15}\mathrm{NH_4^+} = {}^{14}\mathrm{NH_3} / {}^{15}\mathrm{NH_4} = n$$

it is simply a stoichiometric equipartition effect, and $K = 1$. For a true isotope effect, the ratio $\lesseqgtr n$ and $K \neq 1$.

consume a minimum of energy, all chemical refluxing processes are irreversible and consume much energy and this makes the overall process expensive.

7.3.5 Dual Temperature Exchange

The purpose of each chemical reflux process is merely to push the equilibrium concentrations in the reverse sense to participate in recycling. This, as mentioned, is energy consuming. Also, in some cases no simple chemical process may be available for refluxing.

To avoid this, a brilliant alternative method was proposed, independently by : Rideal, by Harteck and Suess and by Spevack during the period 1939-45*. This is referred to as the *dual temperature exchange*** involving much less energy consumption. Since the value of the equilibrium constant K fixing the proportion of concentrations of reactants and products is dependent on temperature, an alteration of the temperature necessarily results in shifting the equilibrium in one direction or the other. The effect of temperature on the equilibrium constant is given by the van Hoff reaction isochore

$$\frac{d \ln K}{dT} = \frac{\Delta H}{RT^2} \tag{7.8}$$

The values of K for the reaction

$$H_2O + HD \rightleftharpoons HDO + H_2$$

for the temperatures 25, 80 and 600° C are 3.78, 2.79 and 1.30 respectively.

Thus, by carrying out the equilibrium exchange process in one column maintained at one temperature T with equilibrium constant K_T and transferring the enriched products to a second column, maintained at another temperature T' characterized by a different equilibrium constant ($K_{T'}$) the proportion of reactants to products change. We have thus achieved the equivalent of a reflux action without resorting to an irreversible chemical reaction. In other words, by carrying out the exchange reaction alternately in two columns at different temperatures, the necessary multiplication effect on the separation factor is realized. This is the principle of the dual temperature exchange. In choosing the two temperatures, it should be remembered that one of the isotopic forms has to be always in the vapour state and the other in the liquid state but with appreciable vapour pressure. This consideration limits the T-T' interval consistent also with a large change in the equilibrium constant. As a full use of this technique is made in the separation of heavy water, we shall revert to a detailed account of this in Ch. 9.

7.3.6 Applications

The chemical exchange method for isotope separation was first suggested by Urey[17] in 1935 and initially used by him for obtaining samples enriched in ^{13}C and ^{15}N. In the present time the method is used with success for the

---

* For detailed references, see *Separation of Isotopes* by H. London (George Newnes Ltd., London, 1961).

** Also referred to as bithermal exchange.

large scale production of heavy water of over 99.9 per cent deuterium content. Table 7.3 lists some of the well studied isotope exchange reactions. As examples of the application of chemical equlibrium exchange method, we describe here the separation of ^{15}N and ^{13}C isotopes; the separations of heavy water, ^{6}Li and ^{10}B being deferred to Ch 9.

Table 7.3 Equilibrium Constants for exchange reactions at 25° C

Element	Exchange reaction			K (theory)	Ref.
H	$H_2 + D_2$	$\rightleftharpoons$	$2\ HD$	3.33	16
	$H_2 + T_2$	$\rightleftharpoons$	$2\ HT$	2.56	
	$H_2O + D_2O$	$\rightleftharpoons$	$2\ HDO$	3.27	
	$H_2O + T_2O$	$\rightleftharpoons$	$2\ HTO$	3.42	
	$H_2O + HD$	$\rightleftharpoons$	$HDO + H_2$	3.78	
	$H_2O + HT$	$\rightleftharpoons$	$HTO + H_2$	6.19	
	$H_2O + DCl$	$\rightleftharpoons$	$HDO + DCl$	2.36	
				1.53 (17° C)	
	$NH_3 + HD$	$\rightleftharpoons$	$NH_2D + H_2$	3.60	
	$H_2O + HDS$	$\rightleftharpoons$	$HDO + DClH_2S$	2.22	
Li	$^{6}Li\ (Hg) + {}^{7}Li^{+}$	$\rightleftharpoons$	$^{6}Li^{+} + {}^{7}Li\ (Hg)$	1.05	
B	$^{10}BF_3\ (g) + {}^{11}BF_3.O\ (CH_3)_2(1)$	$\rightleftharpoons$	$^{11}BF_3\ (g) + {}^{10}BF_3.O\ (CH_3)_2(l)$	1.03 (100° C)	22 *a*
C	$^{12}CO_2 + {}^{13}CO$	$\rightleftharpoons$	$^{12}CO + {}^{13}CO_2$	1.02	
	$H^{12}CN + {}^{13}CN^-$	$\rightleftharpoons$	$^{12}CN^- + H^{13}CN$	1.03	
	$^{12}CO_3^{2-}{}_{(aq)} + {}^{13}CO_2$	$\rightleftharpoons$	$^{12}CO_2 + {}^{13}CO_3^{2-}{}_{(aq)}$	1.012	
	$^{12}CO_3^{2-} + {}^{14}CO_2$	$\rightleftharpoons$	$^{12}CO_2 + {}^{14}CO_3^{2-}{}_{(aq)}$	1.02	
N	$^{15}NH_3(g) + {}^{14}NH_4^+(aq)$	$\rightleftharpoons$	$^{15}NH_4^+(aq)\ {}^{14}NH_3(g)$	1.035	17*a*
	$^{14}NO\ (g) + H^{14}NO_3\ (l)$	$\rightleftharpoons$	$H^{15}NO_3\ (l) + {}^{14}NO(g)$	1.06	
	$HC^{15}N\ (g) + C^{14}N^-(aq)$	$\rightleftharpoons$	$C^{15}N^-(aq) + HC^{14}N\ (g)$	1.002	23
O	$C^{16}O^{16}\ O + H_2^{18}O$	$\rightleftharpoons$	$C^{16}O^{18}\ O + H_2^{16}O$	1.044	7
S	$^{34}SO_2 + H^{32}SO_3^-$	$\rightleftharpoons$	$H^{34}SO_3^- + {}^{32}SO_2$	1.019	17*a*
Cl	$^{35}Cl^{37}Cl + {}^{35}ClO_4^-$	$\rightleftharpoons$	$^{35}Cl^{35}Cl + {}^{37}ClO_4^-$	1.035	

(i) *The separation of* ^{15}N

(a) *Exchange between* NH_3 and NH_4^+ *ions*

Earlier, the isotope ^{15}N was mainly needed to meet the requirements of seientific research. However, of late it is being increasingly used for labelling nitrogen fertilizers used in agriculture for a better understanding of their role in promoting plant growth.

We describe first the process originally developed by Urey and Thode[18], consisting of six towers in series, wherein the ammonia gas rises countercurrently to a descending stream of 60 per cent ammonium nitrate solution, six such tapered columns being combined in a series cascade, the exchange reaction being

$$^{15}NH_3\,(g) + {}^{14}NH_4^+\,(aq) \rightleftharpoons {}^{15}NH_4^+\,(aq) + {}^{14}NH_3^+\,(g)$$

$$K = 1.035 \text{ at } 25°C$$

At each stage ammonia gas enriched in ^{14}N moves to the upper plate, while NH_4^+ ions enriched in ^{15}N flows down to the lower plate. To multiply the overall separation effect, the ammonia gas leaving at the top of columm II is admitted to column I at its bottom and the ammonium nitrate solution collecting at the bottom of column I is pumped to the top of column II and so on (Fig. 7.7 (*a*)) across a tapering cascade of six columns. At the end of

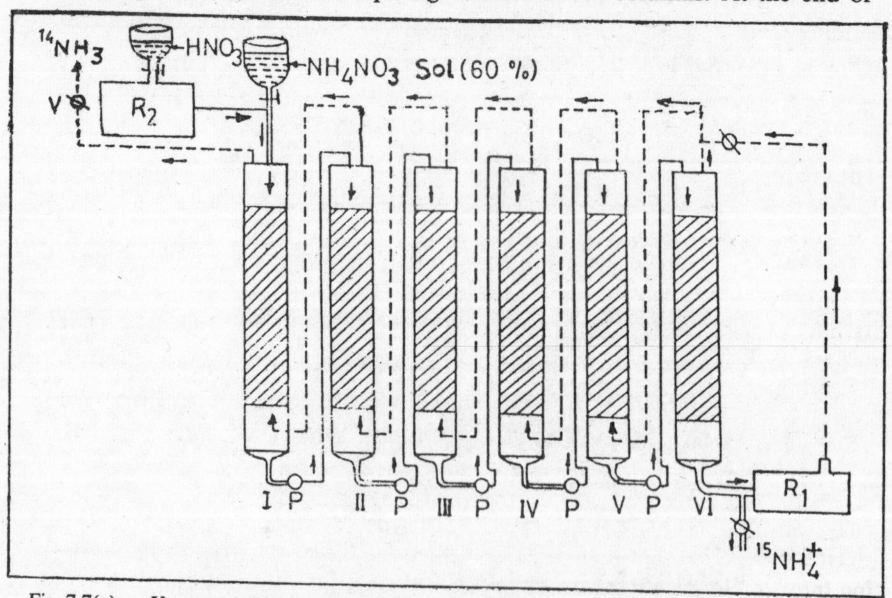

Fig. 7.7(a): **Urey and Thode's cascade for separating ^{15}N by chemical exchange between NH_4NO_3**

I, II... VI: packed columns in series.—R_1: Refluxer to convert NH_4NO_3 solution to NH_3 gas; R_2: Refluxer to convert NH_3 gas to NH_4NO_3 solutions—liquid flow; --- gas flow; P: pump; V: valve.

(from H. C. Urey and H.G. Thode, *J. Chem. Phys.*, 1937, 5, 856) the sixth column the ammonium nitrate solution collecting is decomposed to NH_3 gas by treatment with hot caustic soda solution. The ammonia gas generated in the refluxer is fed back to column V and thence to IV and finally to column I. In a second refluxer at the top of the first column the ammonia gas is reacted with hydrochloric acid and the ammonium chloride solution formed continues the next cycle from I... VI column meeting countercurrently ammonia gas from VI... I column. The end products $^{14}NH_3$ and $^{15}NH_4^+$ are tapped off from the top of the I and the bottom of the VI columns respectively.

The portion described constitutes one section and Thode and Urey set up three such sections, in what is termed a *tapering cascade*, the number of columns and their lengths and diameters falling off. While the first section had six columns, the second had three and the last only two columns. The packing materials were 6 or 5 mm "Berl" saddles in the first section and "Fenske" spirals in the other two sections. Each section corresponded to 150 theoretical plates. Operating at a pressure of 8 cm Hg, they achieved a separation factor of 1.023 against the theoretical value of 1.035. The yield was 1.3 g of ^{15}N of an isotopic purity of 75 atom per cent.

The replacement of ammonium nitrate by carbonate solution is stated to result in a higher efficiency[19] as this permits the operation as a *closed reflux,* in which the chemicals produced at one end of the plant are used for forming the substances needed for effecting reflux at the other end, thus the reflux reactions at the bottom of the last column are

$$NH_4HCO_3 + Ca(OH)_2 \rightarrow NH_3 + CaCO_3 + 2H_2O \qquad (a)$$

The chalk formed is calcined to yield lime and CO_2

$$CaCO_3 \rightarrow CaO + CO_2 \qquad (b)$$

$$CaO + H_2O \rightarrow Ca(OH)_2 \qquad (c)$$

The ammonia and water vapour of (a) and CO_2 of (b) are reacted to reform ammonium carbonate solution.

$$NH_3 + CO_2 + H_2O \rightarrow NH_4HCO_3$$

Thus, the only material continuously used up is ammonia gas and heat energy.

Holmberg[8,20] is reported to have further simplified the process by converting into simple *exchange distillation.* Here the ammonium carbonate solution runs down a packed column where it is decomposed by heat into NH_3 and CO_2 and these gases ascend up the column countercurrently. At the top these are made to recombine to form ammonium carbonate. With a single column of 1 m height, Holmberg reported a separation factor of 1.15 at 60 °C and atmospheric pressure.

(ii) *Exchange between* HNO_3 *and* NO

Separation of ^{15}N has been accomplished with a greater efficiency by Spindel and Taylor[21] employing the exchange reaction between nitric acid and nitric oxide.

$$H^{14}NO_3\,(l) + {}^{15}NO\,(g) \underset{K=1.06}{\rightleftharpoons} H^{15}NO_3\,(l) + {}^{14}NO\,(g)$$

The reaction is catalysed by the nitrous acid which is a side product. Using a stainless steel packed column of 5 m height, which is much shorter than the sum of the heights of Urey and Thode's columns, they obtained 99.9 per cent pure ^{15}N. Further improvements were successfully adopted by Gowland and Johns[22].

(iii) *The Indian plant*

Noting that the uptake of nitrogen by plants from the added fertilizers is generally below 30%, the need for a better understanding of the mechanism of N-uptake becomes evident. There can be no better way of this than by using fertilizers enriched in ^{15}N. With this imperative need in view, the Bhabha Atomic Research Centre, Bombay, streamlined the original Spindel-Taylor process just described. Two columns are used in cascade with provision for reflux at both ends, as shown in Fig. 7.7(*b*). The product at the bottom end where ^{15}N enriched nitric acid collects is decomposed by SO_2 into NO which is made to ascend, the reflux reaction being

$$2HNO_3(l) + 3SO_2 + 2H_2O \rightarrow 2NO(g) + 3H_2SO_4(l).$$

At the top end where the depleted NO escapes it is reoxidized by O_2 by the reaction

$$4NO(g) + 3O_2 + 2H_2O \rightarrow 4HNO_3(l)$$

and the HNO_3 is refluxed down.

To build a steady state gradient in ^{15}N in the columns the plant has to work continuously for a few days. The pilot plant is reported to have yielded 200 g of nitric acid bearing 50 atom per cent of ^{15}N per year. The process know-how is now transferred to the Rashtriya Chemicals and Fertilizers Ltd.

(iv) *The separation of* ^{13}C

The interest in ^{13}C, of natural abundance 1.11 per cent, arises from the fact that it is a stable isotope which in the enriched form is a valuable tracer in the study of certain biological-chemical problems where the system may be sensitive to the 200 keV radiation from ^{14}C, the otherwise more common and convenient tracer. In addition, being paramagnetic ^{13}C is often used in NMR studies where a greater sensitivity or a more spreadout of the spectrum

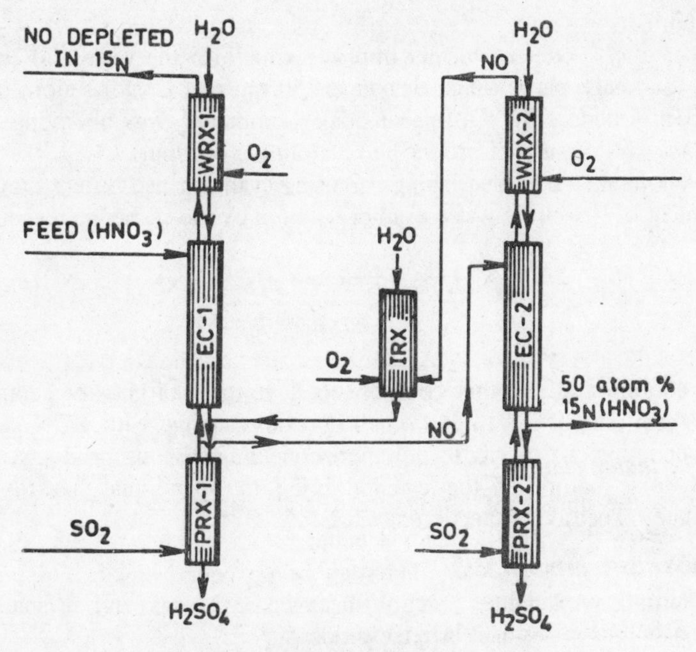

Fig. 7.7(b) EC:Exchange columns; PRX:Product refluxers; WRX:Waster refluxers; IRX:intermediate refluxer (from *Nuclear India,* 1985, *23*(10), p. 4, with permission)

is considered more desirable than the one provided by the straight proton resonance.

(a) *Exchange between* HCN *and* CN^- *ions*

The separation of ^{13}C is based on the isotope effect in the chemical equilibrium between hydrogen cyanide gas and sodium cyanide solution.

$$H^{12}CN\,(g) + {}^{13}CN^-\,(aq) \rightleftharpoons H^{13}CN\,(g) + {}^{12}CN^-\,(aq)$$

$$K = 1.012 \text{ at } 25°C$$

The hydrogen cyanide gas enriched in ^{13}C ascends the column countercurrently to the descending solution of sodium cyanide enriched in ^{12}C. The process is similar to the one described for the separation of ^{15}N.

In their first attempt Urey and coworkers[23] used a 3 m high packed column in which the HCN gas flowed upward countercurrently to the 20 per cent descending solution of NaCN . The HCN gas enriched in ^{13}C reaching the top was refluxed by caustic soda and the NaCN solution formed made to descend the column, while the NaCN solution enriched in ^{12}C collecting at the bottom was decomposed by sulphuric acid and the HCN generated was made to ascend again. An enrichment factor of 1.84

was reported.

Becker *et al*[24] effected further improvements over the years and came out with a four-stage plant which delivered 250 mg of ^{13}C of 12 atom per cent purity. In America, the Eastman-Kodak Company[25] was operating a plant for many years capable of producing 65 atom per cent pure ^{13}C.

The above method of enriching ^{13}C using cyanides also brings about some enrichment of ^{15}N in the waste cyanide solution by the exchange reaction.

$$HC^{15}N\ (g) + C^{14}N^-\ (aq) \rightleftharpoons HC^{14}N\ (g) + C^{15}N^-\ (aq)$$

$$K = 1.002 \text{ at } 25°C$$

The column and all other components have to be of glass or ceramics, as metals form complex cyanides which do not exchange with HCN gas. Also the tendency of HCN gas to polymerize yielding viscous products and the lethally toxic nature of the cyanide bring in operational hazards which necessitate expensive protection devices.

(b) *Exchange between* CO_2 *and* HCO_3^- *ions*

The alternative exchange reaction between CO_2 gas and a solution of bicarbonate ions is safe as it is free cyanides.

$$H^{12}CO_3^-\ (aq) + {}^{13}CO_2\ (g) \rightleftharpoons H^{13}CO_3^-\ (aq) + {}^{12}CO_2\ (g)$$

$$K = 1.012 \text{ at } 25°C$$

As in the other exchange process, the ^{13}C concentrates in the solution phase. Refluxing is effected by acidifying the bicarbonate solution collecting at the bottom and by absorbing the CO_2 gas at the top by caustic soda. The reaction is, however, not favoured for scaling up because of the very long period of time needed for the equilibrium and consequently the high HETP involved. Even a catalyst packed column of 6 m height is found to be equivalent to less than 10 plates.

7.4 CHROMATOGRAPHY

Chromatographic* separation is another reversible statistical countercurrent chemical exchange process between two phases as in distillation, but with the difference that one of the phases is stationary and the other is a fluid. When a solution containing two or more species of solute passes through a section of the column constituting the stationary phase, an absorption or exchange takes place which tends to establish an equilibrium between the mobile and the fixed phases. If the equilibrium values of the concentrations of the

* The term *chromatography* is used universally in the sense of an analytical technique with no relation to the colour properties of the substances analysed. Hence the use of the term is etymologically unfortunate.

different solutes, say two isotopic molecules, are different, an enrichment of one of the species occurs in the flowing solution and of the other in the fixed solid phase. At every stage the constituent species of the mixture distribute between the mobile and the stationary phases.

Chromatography in its different forms constitutes a powerful analytical tool in the separation of the constituents of all types of mixtures in the laboratory as well as in industry. Its use in the separation of isotopic molecules in however limited.

7.4.1 The Separation Factor

Expressing the chromatographic separation as an equilibrium reaction between the two isotopic species in the solution and stationary phases

$$A_1^+{}_{(sol)} + A_2^+{}_{(st)} \rightleftharpoons A_1^+{}_{(st)} + A_2^+{}_{(sol)}$$

$$\frac{[A_1^+{}_{(st)}]}{[A_1^+{}_{(sol)}]} \Bigg/ \frac{[A_2^+{}_{(st)}]}{[A_2^+{}_{(sol)}]} = K = \alpha$$

If we express the concentrations in terms of mole fractions, x and y, the former for the isotopic species A_1 in solution initially and the latter for the same species in the solid phase at the end of the separation, we arrive at the usual expression

$$\frac{y}{1-y} = \alpha \frac{x}{1-x}$$

This is illustrated in the next section in the case of the separation of the lithium isotopes.

Chromatographic separations yield data which permit the calculation of the enthalpy and entropy changes accompanying the exchange reaction. The extremely low values of -9.5 J mole^{-1}, and -6.7×10^{-3} J mole^{-1} K^{-1} for the Δ H and Δ S for the exchange reaction ^{6}Li $-$ ^{7}Li show the highly reversible nature of the equilibrium exchange[1].

7.4.2 Applications

(i) *The separation of* ^{6}Li

Before the theory had been developed, Taylor and Urey[26] applied the chromatographic technique to separate the isotopes of lithium and potassium using a 30 m high column loaded with a zeolite ion exchanger. The exchange equilibrium involved is

$$^6Li^+{}_{(aq)} + {}^7Li^+{}_{(zeol)} \rightleftharpoons {}^6Li^+{}_{(zeol)} + {}^7Li^+{}_{(aq)}$$

where the subscripts *aq* and *zeol* stand for the ion concentrations in the solution (aqueous medium) and fixed to the solid stationary phase (zeolite). The ^{6}Li was found to enrich the zeolite phase at the front. If x and y be the mole fractions of the $^6Li^+$ in solution and in the zeolite phase after equlibrium is established

$$\frac{y}{1-y} = K\frac{x}{1-x}$$

hence, the single stage separation factor $\alpha = K$. The actual enrichment observed by Taylor and Urey was much less than the expected value.

Later, Glueckauf, Barker and Kitt[27] made a near breakthrough using a column of zeo-karb HI sulphonated coal eluted by acetic acid. The exchange reaction.

$$[Li^+ + Ac^-]_{aq} + \underset{\text{(H form)}}{HR} \rightleftharpoons \underset{\text{(Li form)}}{LiR} + HAc^*$$

$$K = 60{,}000$$

proceeds virtually irreversibly with a K value of the magnitude shown. A meter high column functioned as the equivalent of 10^4 theoretical plates, and the $^6Li^+$ was heavily depleted in the front boundary, the $^6Li/^7Li$ ratio in the first 0.5 cm^3 of the eluant being of the order of one per cent compared to its natural abundance of 7.5 per cent. Such a nearly irreversible reaction mentioned earlier does not appear to be consistent with the low separation factor of 1.002 5 suggested by the ratio of the concentrations.

$$\left[\frac{^7Li^+}{^6Li^+}\right]_{(aq)} = 1.002\,5\left[\frac{^7Li^+}{^6Li^+}\right]_{st}$$

It has been reported by Merz[28] that the sense of the above equilibrium gets reversed, that is, it is the ^7Li which gets preferentially adsorbed, if zirconium phosphate is the ion exchanger. The reversal of the equlibrium is indicated by the K value falling below unity. The actual value reported by Merz with a zirconium phosphate ion exchanger was 0.999 975.

(ii) *The separation of* ^{10}B

Yoneda, Uchijima and Makishina[29] obtained partially enriched ^{10}B by passing a solution of boric acid on the anion exchanger-Amberlite CG-400-I in the OH^- form. The ^{10}B was preferentially adsorbed by the ion exchanger in the presence of glycerol following the reaction

$$H_2{}^{10}BO_3^-{}_{(aq)} + H_2{}^{11}BO_3^-{}_{(resin)} \rightleftharpoons H_2{}^{11}BO_3^-{}_{(aq)} + H_2{}^{10}BO_3^-{}_{(resin)}$$

The ^{10}B content in the solution phase had fallen from 19.7 to 14.8 per cent.

---

* The symbol Ac^- is used for the acetate ion, CH_3COO^-.

(iii) *The separation of* ^{15}N

The equilibrium between NH_4^+ ions and aqueous solution of ammonia is really a two-stage equilibrium

$$NH_4^+{}_{(res)} \rightleftharpoons NH_4^+{}_{(aq)} \rightleftharpoons NH_4OH_{(aq)}$$

The first stage is an ion exchange reaction for which the separation factor, as in other similar cases, is known to be small. The second stage being between two chemically different species may be expected to yield a larger separation factor for the ^{14}N-^{15}N exchange. This was shown to be the case by Spedding, Powell and Svec[30] who, employing the second stage equilibrium, were able to obtain substantial amounts of highly enriched ^{15}N. A 1.5 m high column loaded with 100-200 mesh Dower-*X*-12 was used. The maximum enriched sample had a $^{15}N/^{14}N$ ratio of 0.81, beginning with natural nitrogen with a ratio of 3.6×10^{-3}

$$^{14}NH_{4(res)}^+ + {}^{15}NH_4OH_{(aq)} \rightleftharpoons {}^{15}NH_4^+{}_{(res)} + {}^{14}NH_4OH_{(aq)}$$

$$\alpha = K = 1.0257$$

The novel feature of this equilibrium is the self-regeneration of the reactant species at the end of each cycle and this advantageous situation enabled the use of a single 1.5 m column over and over again as the equivalent of some 200 m length. After some 40 cycles [$= 1/(\alpha - 1)$], a steady state sets in and no further enrichment would be possible.

The use of a cation exchanger *Nalcite HCR* in the cupric form[31] gave an enrichment factor $\epsilon = 0.022$, which is nearly of the same order as with Dowex-*X*-12, the equilibrium in this case being:

$$4NH_{3(aq)} + Cu^{2+}R \rightleftharpoons Cu(NH_3)_4^{2+}R$$

(iv) *The separation of* ^{235}U

Employing a 90 cm high column loaded with Dowex 50 in the Cu^{2+} form, and eluted by 0.005 *M* diammonium magnesium EDTA solution*, Spedding and Powell[31] are reported to have obtained 4 to 5 per cent enriched ^{235}U, corresponding to a separation factor of 1.000 8.

7.4.3 Gas Chromatographic Separations

A limited amount of work on the separation of isotopes using gas chromatographic methods has been reported, mainly of the isotopes of hydrogen and neon.

(i) *The separation of deuterium*

Using a 44 cm long column of 8 mm diameter, loaded with 20 g of palladium black supported on 6 g of asbestos, Glueckauf and Kitt[32] effected

* EDTA for ethylene diamine tetraacetate.

the separation of hydrogen isotopes

$$2HD \rightleftharpoons H_2 + D_2$$

$$K = 1.8 \text{ at } 25°C$$

The palladium catalyses the above reaction besides preferentially adsorbing the hydrogen isotope H_2. Both helium and hydrogen were used as the carrier gas.

The same authors[33] used an alternative technique of driving the mixed gas (40 per cent D_2 as HD) over the column. Instead of using a carrier gas, they heated gradually the rear end of the column containing the hydrogen-helium mixture. In this process, 16 *l* (STP) of hydrogen mixture was displaced which came out in three bands as shown in Table 7.4. Also the column used was a tapering one with the diameter decreasing from 3.45 to 1.1 cm which lowers the HETP.

Table 7.4 Gas Chromatographic separation of deuterium by Pd black; initial mixture 40% D_2 as HD[33]

	Front band	Middle band	End band
Elution period*/min	75-105	105-120	beyond 120
Volume of gas eluted/*l*	5.8	2.5	6.5
Composition/% D_2	97.5	12.1	1.6

(ii) *The separation of* ^{22}Ne

Gluekauf, Barker and Kitt[27] adopted a similar gas chromatographic process for the separation of the neon isotopes on a small scale of about one millilitre (STP), but by a reverse process of *cooling*. A tube filled with activated charcoal and nitrogen, and connected to a supply of neon, was progressively cooled from the rear end by liquid nitrogen (Fig. 7.8). As soon as the nitrogen was completely adsorbed, neon was admitted at the rear end. Successive samples collected at the front end were analysed by a mass spectrometer. The first half millilitre sample had 80-90 per cent ^{22}Ne showing that the lighter isotope ^{20}Ne is preferentially adsorbed on the charcoal The adsorption exchange corresponds to a separation factor of 1.002.

$$^{20}Ne_{(g)} + {}^{22}Ne_{(ads)} \rightleftharpoons {}^{20}Ne_{(ads)} + {}^{22}Ne_{(g)}$$

$$K = 1.022, (-196°C)$$

7.4.4 Evaluation

The chief merit of chromatography as a technique for isotope separation is the high separation factors for light elements specially when used as an ion exchanger. With a properly chosen column material and solvent it is possible to realize the equivalent of 10^4—10^5 theoretical plates per metre length of the

*Since in this techinique the elution was effected by the progressive heating of the rear end of the column, the period shown here refers to the time since the beginning of heating. In the first 74 minutes only helium was coming out.

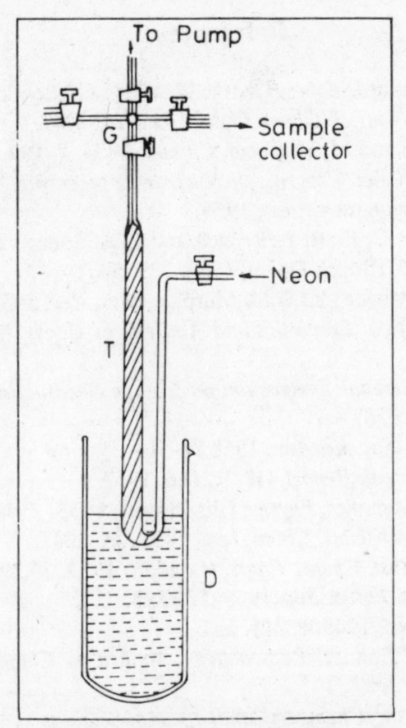

Fig. 7.8: **Separation of ^{22}Ne by gas chromatography**—T:tube filled with activated charcoal and nitrogen; G:gas pipette; D: Dewar flask with liquid notrogen (from E. Gluekauf, K.H. Barker and G.P. Kitt, *Faraday Soc, Disc.*, 1949, *7*, 199).

column. However, chromatography suffers from the following drawbacks:

(i) Since the equilibrium is between the stationary and the mobile phase, the operation has to be discontinuous.

(ii) Not all the theoretically possible plates become effective, as the enrichment occurs mainly in the front and the rear bands. The effective plate number is hardly 10 per cent of the theoretical. With a provision for refluxing and recycling, the same column can be used over and over again until a steady state sets in, which happens after about $1/\epsilon$ cycles.

(iii) The very slow nature of equilibrium attainment is a big hurdle. This does not apply for gas chromatographic separation.

(iv) A difficulty arises from the fact that the difference in the free energies of isotopic ions is nearly the same in the stationary and the mobile phases, which means the enrichment factor would be close to zero. This difficulty would not exist if the isotopes to be separated are in different chemical forms in the two phases. For instance, if they are adsorbed in the ionic form on the stationary phase and moved in the molecular or complexed form in the mobile phase, reasonably high separation factors would be possible, as evidenced in the NH_4^+ (resin)—ammonium hydroxide (aqueous) system.

References

1. H. London, *Separation of Isotopes* (George Newnes Ltd., London, 1961).
2. J. Bigeleisen and E.C. Kerr, *J. Chem. Phys.*, 1955, *23,* 2442.
3. W. Kuhn, P. Baertschi and M. Thurkauf, *Chimia*, 1954, *8*, 109, 145.
4. P. Baertschi and W. Kuhn, *Progress in Nuclear Engineering Series IV, Technology and Engineering,*Vol. 1 (Pergamon Press, 1956).
5. J.A. McWilliams, H.R.C. Pratt, F. R. Dell and D.A. Jones , *Trans. Inst. Chem. Engrs.*, (London), 1956, *34*, 17 (British Patent, 1952, 729, 587).
6. H.C. Urey, F.G. Brickwedde and G.M. Murphy, *Phys. Rev.* 1932, *40*, 1.
7. I. Kirschenbaum, *Physical Properties and Analysis of Heavy Water* (McGraw-Hill Book Co., N. Y., 1951).
8. *Proceedings of International Symposium on Isotope Separation* (North Holland Publishing Co., Amsterdam, 1958).

9a. T.F. Johns, *Proc Phys. Soc.* London, 1958,*71*, 701.

9. T.F. Johns, *AERE Harwell Report*, GP/R2166, 1957.
10. W.H. Keesom and J. Haantjes, *Physica* (Eindhoven) 1935, *2*, 986
11. K. Clusius and H. Mayer, *Helv. Chim. Acta*, 1953, *36*, 2045.
12. D.R. Augood, *Trans. Inst. Chem . Engrs.* (London) 1957, *35*,394.
13. S.K. Lachowiez, *Trans. Farad. Soc.*,1955, *51*,1198.
14. E. Leifer, *J. Chem. Phys.*, 1940, *8*, 301.
15. A.C. Wahl and N.A. Bonner, *Radioactivity Applied to Chemistry* (Wiley, New York, 1951).
16. M. Haissinsky, *Nuclear Chemistry and Its Applications* (Addison-Wesley, Reading (Mass), 1964).

16.a G.T. Armstrong, *USAEC, Report NBS,* 2626 (1953)

16.b M. Green, G.R. Martin, *Trans. Farad. Soc.*, 1952 ,*48*, 416.

16.c G.T. Armstrong, F.G. Brickwedde and R.B. Scott, *J. Chem. Phys.* 1953, *21*, 1297.

17.a H.G. Thode, R.L. Graham, and J.A. Ziegler, *Canad. J. Res.* 1945, *23B*, 40.

17. H.C. Urey and L.J. Greiff, *J. Amer. Chem. Soc.*, 1935, *57*, 321.
18. H.G. Thode and H.C. Urey, *J. Chem. Phys.*, 1939, *7.*, 34

18.a C.R. Bailey and B. Topley., *J. Chem. Soc.*, 1936. 921

19. A.A. Palko, L.L. Brown, G.M. Begun and E.F. Joseph, *US AEC Report*, ORNL-2138 (1956).
20. K.E. Holmberg (British Patent 736, 459)
21. V. Spindel and T.J. Taylor, *J. Chem. Phys.*, 1955, *23*, 981; 1956, *24*, 626.
22. L. Gowland, T.F. Johns, *AERE*, Harwell Report Z/R, 2629 (1959).

22.a M. Kilpatrick, C.A. Hutchinsom Jr., E.V. Taylor, and C.M. Judsom, *US AEC Tech. Inform*, NNES III-5, Oak Ridge (1952).

23. C.A. Hutchinson, D.W. Stewart and H.C. Urey, *J. Chem. Phys.* 1940, *8*, 532.
24. E.W. Becker, K. Bier, S. Scholz and W. Vogel, *Z. Naturf.*, 1952, *7 a*, 664.
25. D.W. Stewart, *Nucleonics*, 1947, *1*, 18.
26. T.I. Taylor and H.C. Urey, *J. Chem. Phys.*, 1938, *6*, 429.
27. E. Glueckauf, K.K. Barker and G.P. Kitt, *Faraday Soc. Trans.*, 1949, *7*, 199.
28. E. Merz, *Z. Elektrochem.*, 1959, *63*, 288.
29. Y. Yoneda, T. Uchijima and S. Makishima, *J. Phys. Chem.*, 1959, *63*, *12*, 2057.
30. F.H. Spedding, J.E. Powell and H.J. Svec, *J. Amer. Chem. Soc.* 1955, *77*, 6125.
31. F.H. Spedding, J.E. Powell, *US AEC Report,* ISC-475, 1954, 1958.
32. E. Glueckauf and G.P. Kitt, *Vapour Phase Chromatography* (Ed. Desty, Butterworth Sci. Publ. 1957).
33. E. Glueckauf amd G.P. Kitt, *Proc. Symp. Isotope Separations* (North Holland, Publ. Co., Amsterdam, 1957)

Chapter 8

SPECIAL IRREVERSIBLE METHODS

In this chapter we present some other methods of isotope separation which differ in their nature, underlying principles and importance as a practical tool. These methods are all irreversible statistical processes though some are specific in action. Leaving out the methods which did not progress beyond the laboratory stage, we limit ourselves only to

(1) Molecular distillation,
(2) Electrolysis,
(3) Ionic migration,
(4) Photochemical including laser methods, and
(5) Biological processes.

8.1 MOLECULAR DISTILLATION

Molecular distillation, or sublimation, is an irreversible statistical process, taking place at such low temperatures where the vapour pressure of the substances is very low of the order of 0.1 pascal ($\sim 10^{-3}$ mm Hg)*. Under these conditions, the mean free path is so long that once the molecule evaporates from the solid (or liquid) it has virtually zero probability of colliding with molecules of its own kind and returning to the condensed phase, hence it is an irreversible process. The technique of molecular distillation is widely used in the purification of high grade oils needed for diffusion pumps, as well as in the isolation of certain essential substances present in food materials in a high degree of purity.

8.1.1 Theory of Molecular Distillation

The mass μ of a substance of molecular weight M evaporating from a unit surface of the solid or liquid in unit time is given by the Langmuir equation[1].

$$\mu = p\,C\left(\frac{M}{2\pi\,RT}\right)^{1/2} \tag{8.1}$$

where p is the vapour pressure at temperature T and C is the molar

---

* The SI unit of pressure is the pascal, 1 Pa = 1 newton/m^2 = 7.5×10^{-3} mm Hg (or torr).

concentration of the substance in the condensed phase. Hence, the number of moles evaporating is

$$n = \mu / M = pC\,(2\mu\; M\; RT)^{-1/2} \tag{8.2}$$

If the substance consisted of a mixture of two isotopes of masses, M_1 and M_2, the ratio of their moles evaporating from unit area in unit time is given by

$$\frac{n_1}{n_2} = \frac{C_1}{C_2}\sqrt{\frac{M_2}{M_1}} \tag{8.3}$$

The vapour pressures of the isotopic molecules at low temperatures being nearly the same, the p terms in Eq. 8.2 cancel out. In the usual notation of representing the concentration of the desired isotope in the condensed phase before sublimation by its mole fraction x and by y its mole fraction in the distillate, we have

$$\frac{y}{1-y} = \frac{x}{1-x}\sqrt{\frac{M_2}{M_1}} = \alpha\,\frac{x}{1-x} \tag{8.4}$$

The elementary separation factor once again comes out to be in accord with the $M^{1/2}$ law, as in all irreversible statistical processes.

8.1.2. The Process of Molecular Distillation

The liquid or the solid sample is taken in a vial A and frozen hard to liquid nitrogen temperature and the space above is evacuated through a vacuum manifold V, to which are also connected a series of receiving tubes B, C, D,... each through a three-way stopcock S. All these tubes can be either chilled or warmed by changing the bath surrounding them (Fig 8.1). Holding B at liquid nitrogen temperature, the contents of A are carefully thawed or warmed up to a temperature enough to generate a small vapour pressure,

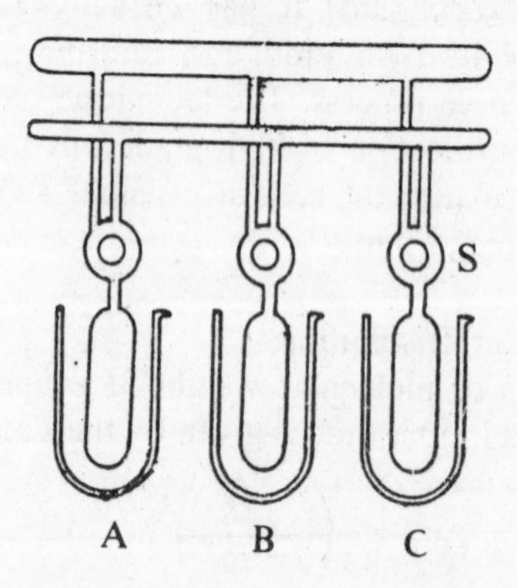

Fig. 8.1: **Molecular distillation**—A, B, C: liquid nitrogen cooled collector tubes joined to vaccum manifold

such that the material distils over and condenses in B, the fraction collecting being enriched in respect of the isotope of higher vapour pressure. When enough sample collects in B, its contact with A is closed. The tube C is chilled and the contents of B warmed, so that the second stage molecular distillation occurs between B and C. The process is consecutively repeated over several stages till the sample collecting in the end tube is sufficiently enriched in the isotope of higher vapour pressure and the residue in A is depleted in respect of this isotope.

Though commonly used for isolating a component of high purity, the process of molecular distillation is rarely used for isotope enrichment, the yields being extremely low.

8.1.3 The Brewer-Madorsky Gravity-Controlled Countercurrent Distillation Still

Brewer and Madorsky[2] designed a still for molecular distillation wherein the equivalent of refluxing is accomplished automatically by gravity feeding. The arrangement consists of a series of cells located at increasing heights as the steps of a staircase (say positive slope). The floor of each cell is at a high enough temperature being heated by a controlled current. This serves as a distiller while the top of each cell is a sloping roof (negative slope), overlapping adjacent cells. The roofs are all glass-plated and maintained at a low temperature by circulating cold alcohol vapour at about −20° C. This enables the condensation of the vapour underneath (see Fig. 8.2). The vapour rising from the surface of the *i*th cell condenses on the roof and the drops formed slide down and drop into the $(i + 1)$th cell. The level of liquid in no cell is allowed to rise above a fixed level. The levels are so fixed that the excess liquid when formed must overflow from the $(i + 1)$th to the previous (*i*th) cell. This is simply accomplished by the correct positioning of the opening communicating the adjacent cells. The entire unit is inside an evacuated glass vessel ($p < 10^{-5}$ mm Hg). Under these gravity controlled flows, the forward flow rate of vapour equals the reverse flow rate of the

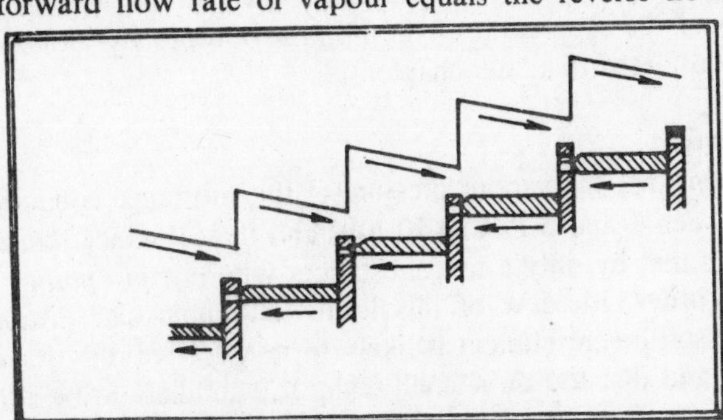

Fig. 8.2: **The Brewer-Madorsky still for gravity-controlled molecular distillation.**
(after A.K. Brewer and S.L. Madorsky. *J. Res. Nat. Bu. Stand,* 1947, *38,* 129, with permission from the National Bureau of Standards, Washington, D.C.)

liquid. The fraction enriched in the isotope with the higher vapour pressure moves forward and upward (from cell $i \rightarrow (i+1) \rightarrow (i+2) \rightarrow \ldots.$) while the fraction enriched in the isotope of lower vapour pressure moves in the reverse sense (from cell $\ldots. \rightarrow (i+2) \rightarrow (i+1) \rightarrow i$). This is equivalent to countercurrent mixing.

8.1.4 Applications

(i) *Separation of* ^{204}Hg

Brönsted and Hevesy[3] are credited to have been the first to have effected a partial enrichment of the mercury isotope ^{204}Hg (natural abundance 6.8 per cent) by molecular distillation of the metal at 40-60° C in an evacuated double walled vessel, the collector wall being cooled by liquid air. The process had to be continued over a prolonged period till 2.7 litres of mercury were reduced to 0.2 ml, when the ^{204}Hg was enriched by a factor of 1.11. This was long before the Brewer-Madorsky still had been developed.

Later, using in a cascade of ten cells of the gravity fed countercurrent type designed by themselves, Brewer and Madorsky[2] achieved a separation factor of 1.008 5 in the enrichment of ^{204}Hg by distilling mercury at 55-100° C and condensing at −18 °C under a pressure of $\sim 10^{-5}$ mm Hg. The theoretical value for α by Eq. 8.4 is, however, 1.014 9 [$= \sqrt{(204/198)}$]. The separation was even poorer when Madorsky *et al*[4] attempted to use a cascade of 150 cells.

(ii) *Separation of* ^{37}Cl and ^{40}K

Subsequently, Hevesy and coworkers studied the enrichment of isotope of chlorine (^{37}Cl) by the distillation of a mixture of water and hydrochloric acid[5] and of potassium ^{40}K by the evaporation of the metal[6].

(iii) *Separation of* ^{6}Li

The Brewer-Madorsky still has been continuously improved by different groups of workers, notably by Lang *et al*[7] and by Johns *et al*[8] to adopt the same for the separation of the high priced ^{6}Li (natural abundance 7.5 per cent). Details of the separation of the lithium isotopes by molecular distillation will be considered in a later chapter.

8.1.5 Evaluation

The requirement that the vapour pressure of the substance condensing on the ceiling be between 1 and 5 Pa (10-50 μm) and the substance remain liquid at this pressure is met by only a few substances whose *triple point** pressure is of the above order. In view of this limitation, molecular distillation as a technique for isotope enrichment is likely to be effective only for isotopes of light elements and that too on a small scale. It is unlikely to be adopted on a large scale.

---

* The *triple point* is a specific point in the pressure-temperature phase equilibrium diagram where the three phases solid-liquid-vapour are simultaneously in equilibrium with one another. In the case of water, for instance, it is 273.16 K and 4.579 mm Hg pressure.

8.2 ELECTROLYSIS

Having established the existence of deuterium[9], a heavy isotope of hydrogen (D or ^{2}H) in 1932, Urey argued that electrolysis of water which contains 0.015 per cent deuterium, would be the simplest way of isolating the isotope. The gas discharged at the cathode of an electrolytic cell is distinctly poorer in deuterium content than the water electrolysed. A high value of 8-10 had been reported for the separation factor, defined as[10]

$$\alpha = \frac{(H/D)\ \text{electrolytic gas}}{(H/D)\ \text{water}} \tag{8.5}$$

The actual value of the separation factor depends on several factors as, the temperature, nature of the cathode metal, its hydrogen overvoltage*, its surface roughness, current density, *etc.* The lowest separation factor reported is about 3.5, which is high enough indeed and this minimum value is close to the equlibrium constant for the exchange reaction

$$HD + H_2O \underset{K = 3.7\ (25°C)}{\rightleftharpoons} H_2 + HOD \qquad (a)$$

8.2.1 Theory of Isotope Separation by Electrolysis

The precise mechanism leading to isotope separation by electrolysis cannot be said to be well understood. The earliest idea centered round the marked difference in the ionic mobilities of the H^+_{aq} and D^+_{aq} ions, the values being 36.2 and 22.1 in units of $10^{-4} cm^2\ V^{-1} s^{-1}$ respectively. Large as this difference is, the view is now abandoned since our ideas of the basic process of electrolysis of water have undergone a change. The H^+ ion is no longer considered a participant in the electrolysis of water, specially in the presence of small amounts of alkalies or salts. The primary product of electrolysis envisaged is the *hydrated electron* (e^-_{aq})**

$$\text{Water} + e^- \rightarrow e^-_{aq} \qquad (b)$$

Being short-lived, the hydrated electron decays to yield hydrogen and OH^-_{aq} ions

---

* The hydrogen overvoltage of a metal is the potential needed to be applied in *excess* of the theoretical value for the *visible* evolution of hydrogen gas, when that metal is used as the cathode in electrolysis. While theoretically only 1.23 V are needed to electrolyse water, some 0.7-1.0 volt in *excess* is needed with most metals. This varies with current density and the pH of the medium, besides the nature of the metal.

** The *hydrated electron* is now recognized as a distinct species of matter consisting of an electron surrounded by 3 or 4 water molecules, the entire cluster moving together. Symbolized as e^-_{aq}, it has characteristic properties as mean radius, mean life, absorption spectrum, standard electrode potential, mobility, free energy of formation, *etc.*[11]

$$e^-_{aq} \rightarrow 0.5\ H_2 + OH^-_{aq} \qquad (c)$$

One may say, the overall reaction of electrolysis is the direct reduction of water by an electron.

$$H_2O + e^- \rightarrow 0.5\ H_2 + OH^-_{aq} \qquad (d)$$

Urey considered thermodynamic and kinetic isotope effects (§ 3.1 and 3.2) to be at the origin in leading to isotopic enrichment in the electrolysis of water. The former relates to a shift in the equilibrium of the exchange reaction (*a*) to make $K > 1$. This is wholly accountable by the difference in the zero-point energies of H_2 and HD (Table 4.4). The separation factor should equal the equilibrium constant as in all exchange reactions. The observed factor being two to three times larger, it may be attributed to the role of the kinetic isotope effect. A series of consecutive and concurrent reactions are involved between the application of the potential which initiates electrolysis and the liberation of hydrogen gas from the cathode surface, reactions (*a*), (*b*) and (*c*) being followed by

at cathode

$$M + H \rightarrow M \ldots H \qquad (e)$$

$$M + D \rightarrow M \ldots D \qquad (f)$$

$$M \ldots H + H \ldots M \rightarrow 2M + H_2 \qquad (g)$$

$$M \ldots H + D \ldots M \rightarrow 2M + HD \qquad (h)$$

It has not been possible to identify the specific reaction(s) displaying the kinetic isotope effect. However, on the general consideration that the *deuterium overvoltage* for a given metal is slightly in excess of the *hydrogen overvoltage* for the same metal, the reaction (*e*) may be expected to proceed relatively faster than reaction (*f*), and in any case the reaction (*g*) will dominate over (*h*) on account of the far superior concentration of the species M...H over that of M...D. These considerations are adequate to explain the main observation that the hydrogen gas liberated from the cathode surface of any metal is impoverished in deuterium content compared to the residual water after electrolysis. This should be particularly true if the potential applied is not very much in excess of the overpotential involved.

8.2.2 Cathode Hydrogen Overvoltage and Separation Factor

We shall now consider the large differences in the separation factor observed with different metals used as the cathode in the electrolysis of water. The important property, which varies from metal to metal in its functioning as a cathode in electrolysis, is its hydrogen overvoltage. This last varies from nearly zero for *platinized platinum**, palladium and gold to around 0.8 to 1.4 volt for Cd, Sn, Pb, Zn and Hg. The general observation has been that the separation factor, as defined by Eq. 8.5, runs counter to the hydrogen overpotential of the cathode metal (see Table 8.1), the case of platinized platinum being abnormal.'

Table 8.1 Cathode Hydrogen Overpotential and Separation Factor in the Electrolysis of Water [16]

Metal	Overvoltage/V		α
Current density	0.1 A cm^{-2} (14)	1.0 A cm^{-2} (15)	
Hg	1.18	1.41	
Zn	1.06	-	
Pb	1.20	1.40	3-4
Sn	-	1.06	
Al	-	1.00	
Cu	0.82	0.80	
Ni	0.89	-	9.51
Ag	0.90	0.92	
Pt (smooth)	0.39	0.55	6-14
C (graphite)	-	-	6.7
Fe	0.82	0.75	9.9
Au	0.77	-	
Pd	-	0.11	
Pt (platinized)	0.055	-	3.6**

This has been elegantly explained by Conway[12] on the basis of the general theory of absolute reaction rates. It is assumed that with metals of high hydrogen overpotential as Hg, Zn and Pb, the adsorption complex M....H formed (reaction (*e*)) is characterized by a weak bond such that the H atom of the bond readily exchanges with a D atom of a heavy water molecule as shown below

$$M \ldots H + D-OH \rightarrow M \ldots D + H_2O \qquad (i)$$

This will be followed by loss of D as HD by reaction (*h*) and this represents a depletion of the residual water in respect of deuterium, and hence a poor separation factor close to the thermodynamic value (~ 3.7). On the contrary, where the metal needs a smaller hydrogen overpotential, as Fe, Ni, Cu, Ag, the M .... H bond is considered to be much stronger resisting exchange reaction of the type (*i*). In other words, less M...D is formed, and hence less HD is lost, and this results in a higher separation factor of the order of 6-14.

* Platinum, or any other metal, can be "*platinized*" by taking a very clean form of the metal as the cathode in a 2 per cent solution of chloroplatinic acid (H_2PtCl_6) and electrolyzing for 5 min, the anode being pure platinum and the potential applied 3 V. The electrodes are then taken out, and electrolysis continued in very dilute sulphuric acid. It was the suggestion of Kohlrausch (in 1875) that platinized platinum is more reversible and less polarizable.

** Plantinized platinum behaves in an abnormal manner.

8.2.3 Applications

(a) *Concentration of heavy water*

In view of its relative simplicity, electrolysis of water was the only method in use for the production of heavy water till abot 1940. The plant at Rjukan, Norway, held at that time some 165 litres of heavy water which constituted the world's largest stock at that time.*

Since the single stage separation factor is 3.5 or even higher, it would appear that a cascade of a small number of stages, about 15, would be all that is needed to arrive at 99.9 per cent D_2O. Practical difficulties, however, stand in the way of realizing such a cascade. The major factor is the loss of a vast amount of hydrogen and the associated deuterium in the early few stages of the cascade, where refluxing by burning all that hydrogen to water (for recycling) is impracticable on account of the cost involved. At best, the project can be worked for hydrogen as the main product and heavy water as a by-product**.

The Canadian plant built during the war time was shut down subsequently. Similar plants were also set up in U.S. and in England, but only to produce enriched heavy water for use in other finishing processes, as distillation or chemical exchange. Details of the cell used for electrolysis and the composition of the water used are available in literature[18]. We shall revert to the separation of heavy water in a later chapter.

Electrolysis is not used for the separation of any other isotope.

8.3 IONIC MIGRATION

Methods of isotope separation based on slight differences in velocities of ionic migration of isotopic ions under an electric field as in the processes of electromigration, electrophoresis, *etc.* will now be considered. For a long time there was some uncertainty as to whether it is the size or the mass of ion which determines its mobility in an electric field. If it depended on the size, there would be no isotope effect in electromigration, as all isotopic ions have practically the same radii. Lindemann[19] was convinced that it is the ionic mass which is relevant. However, the experiments of Kendall and Crittenden[20] on the electromigration of a solution of potassium chloride in agar-agar gel showed no isotope effect for the isotopes of chlorine. This failure was in all probability due to the insensitivity of the mass spectrometer used in revealing small variations in the $^{37}Cl/^{35}Cl$ ratio, for the same method was successfully used by Kendall and White[21] later to separate ion pairs of like mobilities as Ba^{++} and Ca^{++} and SCN^- and I^-, as well as the ions of rare earths[22].

---

* Realizing the role of heavy water in nuclear research, Prof. Frédérie Joliot-Curie obtained this entire stock for France. However, when the fall of France became imminent in June 1940, the stock was secretly transported to Cambridge, England, through his colleagues von Halbon and Kowarski[17].

** The amount of hydrogen to be released is some five to six thousand times the deuterium to be recovered.

Interest in the technique was revived during the Second World War when Brewer, Strauss and Madorsky[23] achieved partial enrichment of the isotopes of potassium, chlorine, copper and uranium by electromigration. Recognizing that in aqueous media the hydration of ions greatly offsets the difference in ionic velocities in the dry state, Klemm[24] in Germany and Chemla[25] in France used molten electrolyte media for isotope separations by electromigration and obtained significantly higher separation factors, notably in the case of isotopes of lithium. Lunden[23] was able to effect marked enrichment of isotopes of more than fifteen elements by the electromigration of their ions in a medium of their fused salts coupled with countercurrent refluxing (*vide infra*).

8.3.1 **Theory of Ionic Migration**

Though the difference in the mobilities of isotopic ions is the major factor leading to their partial separation in electromigration, contributions due to differences in the degree of dissociation and in complex formation also have to be taken into account. No simple model appears to be adequate in explaining electromigration in aqueous media where the ions are invariably hydrated, nor in fused salt media where the ions are relatively more free from solvation effects. A model envisaging electrophoretic movement of ions as vibrating particles jumping from site to site on acquiring necessary energy for overcoming the effective potential barrier, appears to be on the whole acceptable, since even in the so-called "dry" media (solid or fused state), the migrating ions may not be wholly free from effects of the surrounding ions. It is necessary to take into account effects due to the relative difference in ionic velocities $\Delta v/v$ as well as the difference in the degrees of dissociation of the two isotopic ions. Following treatment is based on these twin ideas.

Let the electrolyte consist of two isotopic ions of mole fractions of y (light) and $(1 - y)$ (heavy) and a common opposite ion. Let γ be the valency of the isotopic ions. If v_l and v_h be the velocities of the light and heavy isotopic ions and u that of the common opposite ion, the molar transport (T) of the two ionic species are given by

$$\left.\begin{aligned} T_l &= \frac{y\, v_l}{u + v} \times \frac{I}{\gamma F} \\ T_h &= \frac{(1 - y)v_h}{u + v} \times \frac{I}{\gamma F} \end{aligned}\right\} \qquad (8.6)$$

where I is the total current and F the Faraday constant and v is the mean velocity of the isotopic ions given by

$$v = yv_l + (1 - y)\, v_h$$

The total transport of the two isotopic species is

$$T = T_l + T_h$$

$$= \frac{v}{u+v} \times \frac{I}{\gamma F}$$

$$= \frac{t.I}{\gamma F} \tag{8.7}$$

where t is the transport number of the isotopic ions, considered same for the two species. This total ionic transport is balanced by an equal reverse transport (T') of the electrolyte counterflow in countercurrent electromigration. The result is the same in simple ionic migration with electrolyte at rest *i.e.* there is no countercurrent flow. In the latter case the coordinate system has to be looked upon as moving with a velocity such that the net total transport is zero.

$$T' = -T \tag{8.8}$$

If x and $(1 - x)$ be the mole fractions of the *total* light and heavy isotopic species *i.e.* inclusive of dissociated and undissociated fractions, we have for the electrolyte transport for the light and heavy species

$$\left.\begin{aligned} T'_l &= -xT \\ T'_h &= -(1-x)\,T \end{aligned}\right\} \tag{8.9}$$

The mole fractions of the *total* (x) and of the ionic species (y) are related through the equilibrium constant for the isotope exchange

$$^{6}LiNO_3 + {}^{7}Li^{+} \rightleftharpoons {}^{7}LiNO_3 + {}^{6}Li^{+}$$

$$K = \frac{y}{1-y} \Big/ \frac{x}{1-x} \tag{8.10}$$

8.3.2 Simple Process Factor

Since K is only slightly in excess of unity, it can be shown that the separation factor, referred to in this context as the *simple process factor*, equals

$$\alpha = 1 + \left[\frac{\Delta v}{v} + K - 1\right]$$

$$= \frac{\Delta v}{v} + K \tag{8.11}$$

and the enrichment factor

$$\epsilon \equiv (\alpha - 1) = \left[\frac{\Delta v}{v} + K - 1\right] \simeq \frac{\Delta v}{v} \tag{8.12}$$

8.3.3 Klemm's Relative Mass Effect Factor

Klemm[24] emphasized that the isotopic effect in electromigration in dry (solid or fused salt) media depends not only on the above simple process, or enrichment factor (Eq. 8.11; 8.12) but on the *relative mass effect factor* μ defined as

$$\mu = -\epsilon(\Delta M / M) = -\frac{\Delta v}{v} \Big/ \frac{\Delta M}{M} \qquad (8.13)$$

This indicates that the isotope effect depends not only on the mass difference ΔM but on the absolute mass as well. For electromigration in aqueous medium, μ is in the range -0.01 to -0.07, except for hydrogen isotopes for which the value is some 100 times greater. The observed relative mass effect values for some isotopic pairs are given in Table 8.2 for electromigration in aqueous and in Tables 8.3 and 8.4 for fused electrolyte media.

Later, Klemm derived empirical expressions for the relative mass effect in terms of purely the masses of the migrating ions and of those of the surrounding medium, without involving the relative velocity terms $\Delta v/v$. For the electromigration of cations of mass m^+

$$\mu^+ = -0.5 \left[1 + \frac{m^+}{2.1\, m^-}\right]^{-1} \qquad (8.14)$$

where m^- is the mass of the anion. Results on the cation migration in fused LiCl, $CdCL_2$, TlCl and $ZnCl_2$, as well as in solid AgI, seem to be in accord with the above empirical expression.

In interpreting the mass effect of the migrating ion, Klemm envisages two types of jumps : spontaneous and those induced by the surrounding lattice. The mass effect μ is considered a function of the fraction $Z_s/(Z_s + Z_i)$, where Z_s and Z_i are the spontaneous and induced jumps in unit time

$$\mu = -0.5\, t \frac{Z_s}{Z_s + Z_i} \left[1 + \frac{m}{m'}\right]^{-1} \qquad (8.14\,(a))$$

where m is the mass of the migrating ion of transport number t, and m' is the mass of the other ion surrounding it.

8.3.4 HETP in Electromigration

While electromigration builds up a concentration gradient along the field direction, the reverse flow of the electrolyte due to diffusion tends to neutralize the effect partially. This gives rise to the concept of the *height equivalent of a theoretical plate* (HETP) in the process of electromigration. When equilibrium is attained, the equilibrium enrichment factor is given by

$$\epsilon_0 = \frac{y}{1-y} \Big/ \frac{y_0}{1-y_0} = \exp(L/H) \qquad (8.15)$$

where y_o and y are initial and equilibrium mole fractions of the ion, L the length of the tube and H the HETP. When diffusion is the only cause of the back flow (ideal case), H is simply related to the self-diffusion coefficient D by

$$H = D/\bar{v} \qquad (8.16)$$

where $\bar{v}$ is the mean transport velocity between v_l and v_h. Usually the minimum value of H is such that it provides around 2000 theoretical plates per metre length of the plain capillary tube. However, if a packed tube were to be used, the ideal packing would correspond to $H \approx$ grain diameter.

8.3.5 Optimum Duration of Electromigration

Klemm[27] had further shown that where the concentration of the desired isotope changes from initial C_o to C_t (or its mole fraction from y_o to y_t) after a time t of electromigration, the enrichment factor is given by

$$\epsilon = \alpha - 1 = \left[\frac{C_t}{C_0} - 1\right] = 2\,\frac{\Delta v}{v}\,\frac{vt}{\sqrt{\pi D t}} \qquad (8.17)$$

where D is the diffusion coefficient. This relation shows that for maximum separation, the electromigration time has an optimum value. As the enrichment factor increases with migration distance vt, it decreases with the diffusion length $\sqrt{Dt}$. While the former is a displacement in the field direction increasing the separation, the latter is a spreading in and against the field direction, leading to a mixing of the components.

8.3.6 Applications: Aqueous Medium with Countercurrent Electromigration

We present here some applications of electromigration in aqueous media, including in agar gel, to the separation of some selected isotopes as typical examples

(a) *Separation of potassium isotopes*

Brewer and Madorsky[23] succeeded in obtaining highly enriched samples of ^{39}K and ^{41}K by the electromigration of potassium chloride solution in a U-shaped tube using electrodes of platinum gauze, with an arrangement for controlled counterflow of hydrochloric acid from a reservoir above the cathode compartment. The tube was packed with glass beads or sand held under pressure (see Fig. 8.3). The counterflows were so adjusted that the speed of H^+ ions was intermediate between those of $^{39}K^+$ and $^{41}K^+$ ions

$$v\,^{39}K^+ > vH^+ > v\,^{41}K^+$$

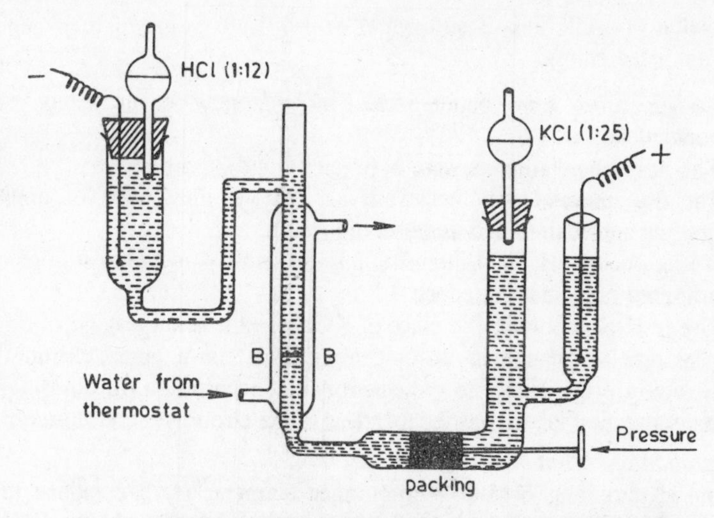

Fig 8.3: The Brewer-Madorsky cell for the separation of isotopes of potassium by countercurrent electromigration. (after A.K. Brewer and S.L. Madorsky, *J, Res. Nar. Bur Stand.*, 1947, *38*, 137, with permission from the National Bureau of Standards, Washington, D.C.)

Under these conditions, the slower moving $^{41}K^+$ ions were washed back while the faster moving $^{39}K^+$ ions go ahead building thereby an isotopic concentration. Under ideally controlled conditions, a sharp visible stationary "moving" boundary gets established which ensures non-mixing of the isotopic ions as they separate out. This involves a delicate control of the concentrations of HCl and of the KCl solutions, such that

$$C_{HCl} / C_{KCl} = t_H^+ / t_K^+$$

where the *t* s are the transport numbers of the ions indicated. Since in the absence of counterflow the boundary moves through a volume $V = t_{K^+} / C_2$ towards the cathode per faraday of electricity, to hold the boundary stationary a volume V of HCl should be made to flow counter to KCl flow per faraday of electricity. Thus by controlling C_{HCl} and/or C_{KCl}, the boundary can be held stationary.

A proper control, as monitored by the boundary, reduces the hold back of the enriched material and also ensures continuous flushing of the cathode. The counterflow has to be reliable and automatic as the experiment may last for several hundred hours. Automatic devices for controlling and stabilizing the boundary are described in the original papers[23,30]. The HCl and KCl solutions have to be initially deaerated to avoid release of air bubbles during electrolysis which give rise to local extra heating effects which disturb the control.

Brewer and Madorsky carried out many experiments with tubes of varying dimensions and lasting over varying periods from 70 to 800 h. The best enrichment obtained corresponded to $^{39}K / ^{41}K$ ratio of 24, as against the natural value of 14.2. This is equivalent to $\alpha = 1.69$. Some of their general findings are listed below:

(1) Longer tubes gave better yields, a horizontal set up being more convenient,
(2) The best tube diameters were between 12 and 25 mm.
(3) Packing material was irrelevant as long as they were of uniform grading and maintained under compression,
(4) Variations of pH, field intensity, current density and temperature over a limited range had no effect.
(5) Use of K_2SO_4 solution in place of KCl gave the same results.
(6) The rate of change in isotopic abundance for a given current was inversely proportional to the electrolyte concentration. In other words, the separation factor was independent of the electrolyte concentrations.

(b) *Separation of Uranium Isotopes*
Using an 18 cm long 0.43-0.78 mm inner diameter plain capillary tubes, London studied the enrichment of uranium isotopes by electromigrating solutions of uranyl nitrate and uranyl chlorate at 68°C with countercurrent flow of sodium nitrate acidified with nitric acid. At the end of 546 hours electromigration the cathodic fraction showed a 38 per cent enrichment of ^{235}U corresponding to a simple process factor of 1.8×10^{-4}. Details are considered in a later chapter.

(c) *Other isotopes*
Other isotopic separations effected by electromigration in aqueous media include those of H-D, ^{24}Mg-^{26}Mg, ^{35}Cl-^{37}Cl and ^{63}Cu-^{65}Cu. The results are summarized in Table 8.2.

Table 8.2 Enrichment factors and relative mass effects in isotope separation by electromigration in aqueous media

Isotopes	Electrolyte	$\epsilon/10^{-3}$	$-\mu/10^{-3}$	Ref.
H–D	H_2SO_2	1200	1800	23
^6Li–^7Li	$LiNo_3$	3.8	25	25
^{24}Mg–^{26}Mg	$MgSO_4/MgBr_2$	2.4	30	23.25
^{35}Cl–^{37}Cl	NaCl	0.9	11	23
^{39}K–^{41}K	KCl/K_2SO_4	3.85	77	23
^{63}Cu–^{65}Cu	$CuSO_4$	0.7	22.4	23

8.3.7 Applications : Fused Salt Media with Countercurrent Electromigration

The first electromigration in the fused salt medium was by Klemm, Hintenburger and Hoernes[33] in 1947 for enriching the isotopes of potassium and lithium. Working in Klemm's laboratory, Lunden[26] developed the technique of electromigration in molten salt media, with countercurrent flow, to effect the partial separation of isotopes of over fifteen elements. A plant for the commercial production of ^7Li was set up by Klemm[34] by countercurrent electromigration in fused lithium salts. These studies provided, in addition, the values of relative ionic velocities of many isotopic ions. It was shown, however, by Chemla, Bonin and Süe[35-37] that relative ionic velocities can also be obtained by the technique of simple ionic migration of a narrow zone in solid crystals, aqueous or fused salt media (§ 8.3.6).

We shall describe the process of countercurrent electromigration in fused salts with reference to the separation of chlorine, bromine and lithium isotopes as typical examples.

(a) *Separation of Isotopes of Chlorine and Bromine*

The cell used by Lunden[26,33,38,39] for the separation of ^{35}Cl and ^{37}Cl by countercurrent electromigration in fused zinc chloride is shown in Fig. 8.4. It

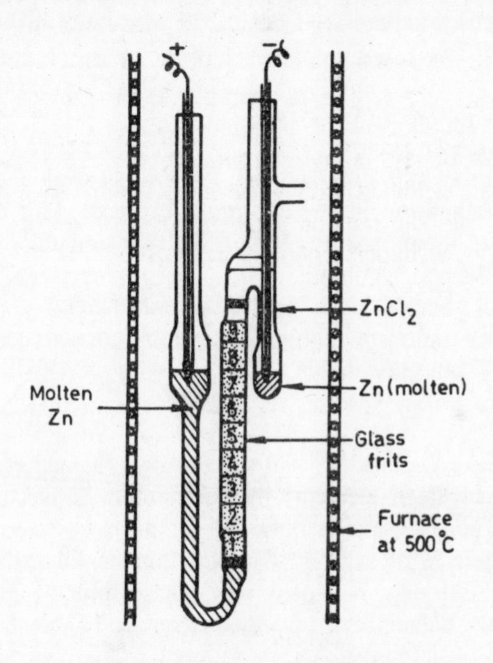

Fig. 8.4: **Lunden's supramax cell for the separation of isotopes of chlorine by countercurrent electromigration in fused zinc chloride.**[26,34]

(after A.Lunden, *Thesis,* Gotberg/1956)

consisted of a U-shaped supremax glass packed with 0.1 mm glass frits to minimize convection currents, the entire assembly being mounted in a furnace maintained at 500° C. The cell was filled with molten $ZnCl_2$ under vacuum. Both tne anode and the cathode were molten zinc. The chlorine liberated at the anode reacted with zinc to reform zinc chloride, the chlorine being thus refluxed. The net reaction was the transportation of metal zinc from the anode to the cathode compartment, the amount of zinc chloride being conserved. When all the chlorine is refluxed, $ZnCl_2$ as a whole flows towards the cathode with a rate intermediate between the velocities of the isotopic chloride ions:

$$v^{35}Cl^- > vZnCl_2 > v^{37}Cl^-$$

The slower moving heavier $^{37}Cl^-$ ions are washed back while the faster moving lighter $^{35}Cl^-$ ions go ahead of the counterflow. An isotopic concentration is thus built up, in the same way as in the electromigration of aqueous solution of KCl described earlier. At the end of eight days, the experiment was arrested and the anode limb cut up into short segments and the ratio of $^{35}Cl/^{37}Cl$ determined in each segment. Results showed a relative mass effect of $-\mu^- = 43 \times 10^{-3}$, which is much higher than the value of 37×10^{-3} observed in the electromigration of aqueous KCl solution (Table 8.2).

Results on the separation of isotopes of chlorine and bromine by electromigration in other fused halides are similar. The maximum mass isotope effect $-\mu^-$ was 86×10^{-3} is fused thallous chloride in a column packed with alumina.

(b) *Separation of Lithium Isotopes*

Lunden[26,39] used a slightly different cell for the separation of isotopes of lithium by electromigration in fused lithium bromide with countercurrent refluxing of Li^+ ions, monitored by the position of a boundary between fused LiBr and $PbBr_2$ in the anode column.

The details of Lunden's work and of a commercial plant set up by Klemm[34] for the separation of lithium isotopes by countercurrent electromigration in fused LiCl are presented in Ch. 9.

(c) *Other Isotopes*

As already mentioned, Lunden[33,39] had successfully obtained enriched samples of isotopes of over fifteen elements by countercurrent electromigration in their fused salts. The mass effects observed in the electromigration of some isotopic ions in their fused salts by Klemm, Lunden and collaborators are listed in Table 8.3. It may be noted that the μ values for the fused salt media are distinctly higher than in aqueous media (Table 8.2) confirming

Table 8.3 Relative mass effects in isotope separation by electromigration in fused salt media[28]

Cations			Anions		
Isotopes	Fused medium	$-\mu^+/10^{-3}$	Isotopes	Fused medium	$-\mu^-/10^{-3}$
$^{6}Li-^{7}Li$	LiCi	141	$^{35}Cl-^{37}Cl$	$ZnCl_2$	43
	LiBr	148		TlCl	86
	$LiNO_3$	50		$PbCl_2$	52
$^{39}K-^{41}K$	KNO_3	37			
$^{63}Cu-^{65}Cu$	CuCl	80	$^{79}Br-^{81}Br$	$PbBr_2$	44
$^{64}Zn-Zn$	$ZnCl_2$	78			
	$ZnBr_2$	110			
$^{204}Tl-^{205}Tl$	TlCl	40			
$^{206}Pb-^{208}Pb$	$PbCl_2$	24			

Klemm's theory, pointed out earlier, that the isotope effect in the migration of ions is higher when the same are free from solvation.

8.3.8 Zone Electromigration

The details of the columns used for the separation of the isotopes of the alkali metals by zone electromigration in aqueous agar gel and in fused alkali nitrate media are given in Table 8.4.[30, 31, 40a]

Table 8.4 Simple zone electromigration: Experimental conditions

	Aqueous medium (Fig. 8.5a)	Fused salt medium (Fig. 8.5b)
Column	100 cm of 1.5-2% agar gel + 1% NH_4HO_3	50 cm of fused $(Na+K)NO_3$ spread on asbestos paper 30 mg/cm^2
Initial zone width	5 mm	5 mm
Temperature	25° C	250–350° C
Field	1.5 V/cm	10 V/cm
Current	≃35 mA	≃350 mA
Duration	24 h	3–4 h
Ionic migration	60–70 cm	25–40 cm

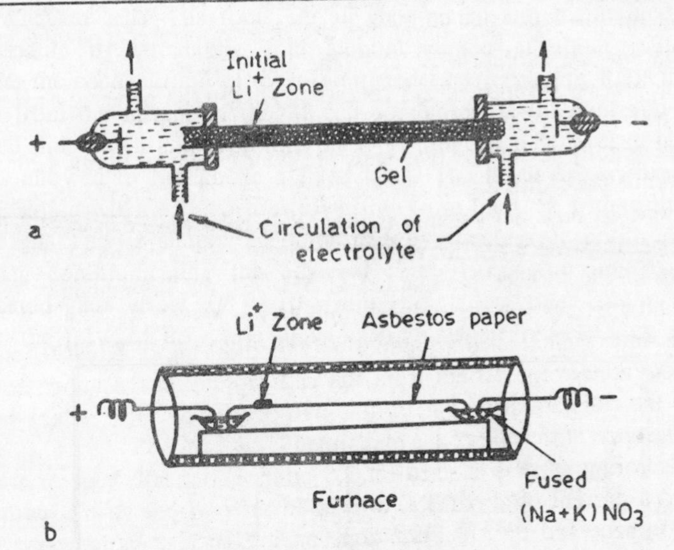

Fig. 8.5: **The Arnikar-Chemla set-up for simple ionic migration (a) in agar gel (b) in fused (Na + K) NO_3**

At the end of the experiment, the column is withdrawn and the relevant active region* is cut into segments of 5 mm length and analysed by a mass spectrometer.

---

* Where the isotope mixture is non-radioactive, the active region is located by the position of radioactive tracer as ^{22}Na incorporated in the initial zone.

The data of the simple migration experiments, with no countercurrent reflux, also permit a calculation of relative ionic mobilities. The resulting enrichments are of the same order as in more elaborate methods using larger amounts with countercurrent electromigration.

8.3.9 Application to the Separation of Isotopes in Agar Gel

The method was used by Chemla and Arnikar[25,31] for effecting an enrichment of ^{6}Li and ^{7}Li by 24 hour electromigrations in agar gel medium. A value of 0.0036 was obtained for $\Delta\, v/v$ for the two ions[30], in excellent agreement with later results of Fuoss[32] obtained with pure ^{6}Li and ^{7}Li isotopes. Details are presented in § 9.2.

8.3.10 Applications to the Separation of Isotopes in Fused Salt Medium

The method was adopted for the separation of the isotopes of lithium, potassium, rubidium and cesium (artificial mixture of ^{131}Cs and ^{137}Cs) by electromigration in a medium of fused sodium nitrate at 350° C. The separation of ^{85}Rb and ^{87}Rb is described here in detail as a typical example of electromigration in a fused salt medium.[40,40a]

1. *The Preparation of the Asbestos Strip*

Commercial asbestos paper of 0.3 mm thickness is sprayed all over with 1:1 nitric acid till the carbonate impurities are decomposed, care being taken to ensure that in the evolution of CO_2 the paper does not tear off or form holes. The process is repeated 2-3 times and finally the paper is washed with distilled water. The sheet is cut into 60 cm long and 12 mm wide strips. While still slightly moist the two ends are gently bent at about 120° which would permit their dipping in the electrode cells. The strip is then calcined over a flame to burn off organic matter. It is then placed horizontally in a tubular electric furnace and covered over with molten $NaNO_3$. After about two hours in the furnace maintained at 350 °C the impregnation becomes uniform and the coverage is usually around 30 mg $NaNO_3$ per cm^2.

On the impregnated strip are deposited some 20 mg of $RbNO_3$, in a 5-8 mm narrow zone some 10 cm from the end dipping in the anode compartment. To the zone are also added a few microcuries of ^{86}Rb* which helps in locating the zone at the end of the experiment.

The electromigration is effected for 2.5 hours under 600 V dc or about 10 V/cm and a current of about 350 mA**. In the end, the strip is withdrawn from the furnace and the rubidium zone on it is located as indicated by the activity of ^{86}Rb tracer. This zone is cut into a series of 5 mm wide segments and the total rubidium content in each determined by a GM counter, the specific activity of the tracer used being known. Finally, the ratio of $^{85}Rb/^{87}Rb$ in each segment is determined by an Atlas Werke mass spectrometer provided with a solid ion source. The distribution of total rubidium and the isotope ratio are shown in Fig. 8.6*a* as a function of distance from the origin.

---

* ^{86}Rb is a radioactive isotope of half-life 18.8 days decaying to ^{86}Kr.

** This corresponds to a dissipation of about 3.5 W/cm^2 whereas in electrophoresis in aqueous medium even when well cooled it is not possible to put in more than 1 W/cm^2.

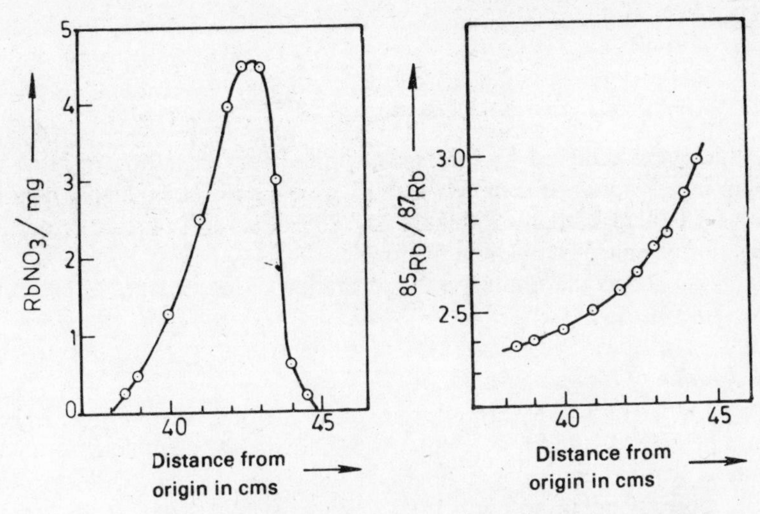

Fig 8.6(*a*): **Separation of isotopes of rubidium by simple ionic migration in fused rubidium nitrate**[40]

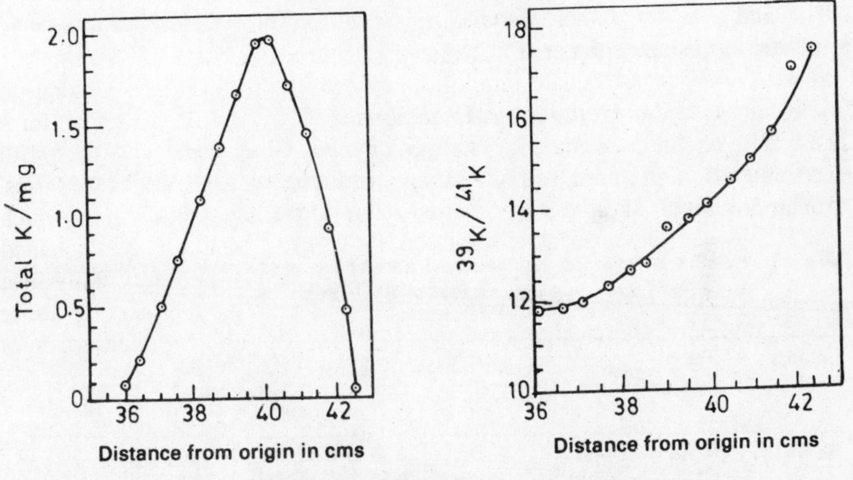

Fig. 8.6(*b*): **Separation of isotopes of potassium by simple ionic migration in fused potassium nitrate**[40]

The ratio $^{85}Rb/^{87}Rb$ in the cathode and anode ends of the zone are 3.00 and 2.40 respectively, the value for the initial sample being 2.61. This shows that the $^{85}Rb^{+}$ ion has a higher mobility compared to the heavier isotope ion. These results provide data in regard to the displacement of the peak (l=42.86 cm) from the origin under the conditions of the experiment, as well as the distribution of the two isotopic ions after migration. A replotting of these results show a displacement Δl of 0.6 mm in the centres of gravity of the two curves. Hence, we have for the relative difference of ionic mobilities

$$\frac{\Delta v}{v} = \frac{\Delta l}{l} = \frac{0.06}{42.86} = 0.0014$$

and for the relative mass effect

$$-\mu^+ = \frac{\Delta v/v}{\Delta m/m} = 0.062$$

The enrichment achieved by the above simple ionic migration in fused salt medium in 2.5 hours corresponds to 1.15 which is distinctly higher than the value of 1.014 realized by Ramirez[41] by 200 hours of countercurrent electromigraion in aqueous rubidium hydroxide at 80° C.

Fig. 8.6 *b* shows the results on the separation of the isotopes of potassium by the same method.

(b) *Separation of Isotopes of Lithium*

Following four hours of electromigration of 30 mg of $LiNO_3$ in fused eutectic mixture of $NaNO_3$ (45.5) + KNO_3 (54.5) under same conditions as for the separation of rubidium isotopes described above, we obtained enrichments of isotopes corresponding to the $^7Li/^6Li$ ratios of 7.20 and 16.20 for the cathodic and anodic ends of the zone respectively against the ratio 11.5 for the initial sample[40]. This corresponds to separation factor of 1.41, $\Delta v/v$ = 0.014 and $-\mu^+$ = 0.089. Details of separation of lithium isotopes are presented in the next chapter.

(c) *Separation of Isotopes of other Alkali Metals*

The results on the separation of isotopes of other alkali metal ions by zone electromigration in fused sodium nitrate medium studied by Chemla and Arnikar are similar and these are summarized in Table 8.5.

Table 8.5 Relative isotopic ion mobilities and mass effects in the zone electromigration in fused $NaNO_4$ medium under a field of 10 V/cm

Isotopes	Temp /°C	Time /h	Mean migration /cm	Δl /cm	$\Delta v/v = \Delta l/l$ /10^{-3}	Mass effect $-\mu^+/10^{-3}$	Ref
$^6Li-^7Li$	250	4.0	26.0	0.36	14.0	89	30
	300	3.5	23.4	0.24	10.2	67	30
	350	3.0	21.1	0.19	9.0	59	30
$^{22}Na-^{24}Na$	350				10.0	100	37
$^{39}K-^{41}K$	350	3.0	40.0	0.108	2.74	54.8	30
$^{85}Rb-^{87}Rb$	350	2.5	42.9	0.06	1.4	62.0	30
$^{131}Cs-^{137}Cs$	350	3.0	41.0	0.11	2.7	60	30

8.3.11 Electromigration in Solids

Some preliminary work on electromigration in solids was reported by Klemm[42] who electrolysed α-AgI between a silver cathode and graphite anode at 230°C with a current density of 2 A/cm^2. At the end of 24 hours, a four per cent decrease in the ratio of $^{107}Ag/^{109}Ag$ at the anode end was observed, showing thereby that the lighter isotopic ion had migrated faster His results indicated a mass effect of $\mu^+ = -0.11$.

Another experiment of importance was of Chemla and Süe[35] who effected the electromigration of an artificially prepared radioactive mixture of ^{22}Na and ^{24}Na, deposited initially on the anode surface of a 3 mm thick single crystal of NaCl.

8.4 PHOTOCHEMICAL METHODS

To understand the principle of photochemical separation of isotopes, let us consider a general case of a photochemical reaction in which one of the reactants is made up of isotopic atoms or molecules whose peak absorptions differ slightly. In such a case, it should be possible in principle to effect a separation of the two isotopic species photochemically. All that is necessary is to irradiate the reactant mixture by light of the precise frequency absorbed by one of the species, the rest of the radiation being filtered out. Only this particular isotopic species will be excited and hence susceptible to be transformed into a product species, while the other isotopic reactant species remains unaffected. The product and reactant should be separable by chemical means. This would be an irreversible specific, or a non-statistical, process.

8.4.1 Isotopic Spectral Shifts

In the case of *isotopic atoms* as the reactants, the difference in their atomic spectra lies in the hyperfine structure of *uv* or the visible region due to a difference in the *reduced mass of the electron* and the nucleus concerned, (vide Eqs. 1.2 and 1.3). This relative frequency change would be of the order of 10^{-6}. In the case of neon isotopes ^{20}Ne and ^{22}Ne with a mass difference of 10 per cent, the frequency shift is of the order of 2.5×10^{-6}. This relative shift decreases further for heavier elements. However, where *isotopic molecules* are involved, the shift in the molecular oscillations is larger as it arises from a difference in the *reduced masses of the atom* and the rest of the molecule. The spectra of interest are the vibrational and rotational bands in the *ir* region. A mass variation of three per cent leads to a one per cent shift in the absorption frequency of the isotopic molecules. Thus, the photochemical method of isotope enrichment becomes practicable with isotopic molecules as reactants from this point; but larger bandwidths offset this advantage (see Table 8.5).

8.4.2 Photosynthesis and Photodecomposition

Broadly speaking most photochemical reactions involve either photosynthesis or photodecomposition. The principles of isotopic enrichment in the two cases are briefly described below.

(i) *Photosynthesis*

Suppose an element consists of two isotopic atoms A_1 and A_2 which absorb at γ_1 and γ_2 and further suppose that the excited species A^* can react with some other reactant B to form the product AB. Now if the reaction mixture is irradiated by radiation of frequency γ_1 from which all other lines are filtered out, following reactions will occur.

$$A_1 + A_2 \xrightarrow{h\gamma_1} A_1^* + A_2$$

followed by

$$A_1^* + B \rightarrow A_1B$$

The product A_1B can be separated from the unaltered A_2. Examples are cited below.

(ii) *Photodecomposition*

If a molecule AB consists of two isotopic species A_1B and A_2B which absorb slightly different frequencies γ_1 and γ_2, as before, the mixture is irradiated by radiation of frequency γ_1 from which γ_2 is filtered out, We have then the following photodecomposition

$$A_1B + A_2B \xrightarrow{h\gamma_1} A_1 + B + A_2B$$

As before A_1 can be separated from A_2B chemically. Examples are cited below:

However in reality, the occurrence of several side reactions with the products of both decomposition and synthesis reactions, a clear separation of the isotopic species becomes very difficult and imperfect. This is all the more so as secondary reactions in general do not distinguish between isotopic atoms or molecules.

8.4.3 Essential Requirements

For a successful isotopical separation by photon absorption, following essential conditions have to be met.

(1) There must exist a significant isotope shift in the absorption spectrum of the reactants.
(2) A suitable photochemical reaction should be available in which only one of the isotopic species, the one which is excited, participates to yield the product.
(3) A complete and quick separation of the product from the unaltered reactant should be possible.
(4) There should be no side reactions between the product and reactants.
(5) An intense light source should be available from which it is possible to filter out the rest of the spectrum outside the narrow band needed to excite one of the isotopic species only. Also, the light source should operate at such a low temperature that the Doppler broadening is narrower than the width of the spectral band to be utilized.

The requirements in respect of the light source are best met only with tunable lasers, as will be shown below. However, we shall briefly point out some of the early attempts at isotope enrichments in the pre-laser era, before considering laser light induced separations.

8.4.4 **Applications : Early Attempts**

(a) *Separation of Chlorine Isotopes*

Merton, Bowen, Hartley and Ponder[43] are credited to have been the first to attempt a separation of isotopes of chlorine by irradiating HCl gas by light specifically absorbed by ^{37}Cl. No positive result was reported.

(b) *Separation of Mercury Isotopes*

Zuber[44] irradiated a mixture of mercury vapour and oxygen by the well-known filtered line 253.7 nm of mercury, which excites selectively ^{200}Hg and ^{202}Hg. He succeeded in increasing the proportion of these two isotopes by a factor of four*.

A more successful experiment was that of Gunning[45], who irradiated a mixture of water and mercury vapour by the light emitted by previously isolated ^{202}Hg* under electrodeless discharge. This led to the formation of HgO containing 90 per cent ^{202}Hg.

(c) *Separation of Carbon Isotopes*

Photolysis of carbon monoxide by the *uv* line 206.24 nm (from the iodine spectrum) results in the enrichment of ^{13}C in the product C_3O_2 and ^{12}C in the CO_2. The mechanism envisages the reactions[46]

$$^{13}CO + {}^{12}CO \xrightarrow{h\gamma} {}^{13}C + {}^{12}CO_2$$

$$^{13}C + {}^{12}CO \xrightarrow{M} {}^{13}C = {}^{12}C = O$$

$$^{13}C = {}^{12}C = O + {}^{12}CO \xrightarrow{M} O = {}^{12}C = {}^{13}C = {}^{12}C = O$$

(d) *Separation of Uranium Isotopes*

Several attempts were made by Urey[28] to obtain enriched uranium by irradiating vapours of UF_6 and UCl_6 by the hyperfine structure lines of uranium. The results were all negative due once again to secondary reactions which resulted in a remixing of the products.

8.4.5 **Laser-Induced Separations**

With the advent of tunable lasers, their use in the separation of isotopes is becoming more common. As pointed out under § 8.4.3, it is only the tuned laser beam which fulfils the requirements of a radiation source for use in isotope separation.

8.4.6 **The Laser Action**

Normally, the number of atoms N_0 in the lowest energy level (the ground state) is much larger than the number in any higher energy level N_E as per the Boltzmann distribution ($N_E = N_0 e^{-E/kT}$, Eqs. 2.4 and 2.7). Suppose it

---

* Naturally occurring mercury consists of seven stable isotopes from ^{196}Hg to ^{204}Hg, ^{197}Hg and ^{203}Hg being radioactive.

becomes possible by some means to invert the situation *i.e.* make $N_E > N_o$. The resulting system would be metastable. This would be possible in a system which permits three conditions, *viz.* (i) it should be possible to excite an appreciable fraction of the atoms (or molecules) in the ground state, energy E_o, to a higher energy state, E', by "pumping" (exposing to) high intensity radiation of frequency corresponding to $E' - E_o$; (ii) the excited atoms (or molecules) deexcite immediately (within about 10^{-8} s) to an intermediate state of energy E, such that $E_0 < E < E'$, and (iii) the state E is metastable with a long life, *i.e.* the transition from there to the ground state, $E - E_o$, is forbidden. Under these conditions, the population of the intermediate state, E, builds up and may exceed that of the ground state. Subsequently, if the inverted system is stimulated by even a single photon of light of the exact frequency corresponding to the difference between the two levels $(E - E_o)$, all the atoms will in one act fall down to the ground state each emitting a photon of *same* monochromatic *frequency* and in the same phase. This is *L*ight *A*mplification by *S*timulated *E*mission of *R*adiation (the LASER), whose intensity can be extremely high. The laser action was first postulated by Einstein in 1917.

For photoseparation of isotopes, only the gas phase is practicable, the lines for solids and liquids being too broad. Again, the gas has to be at a low enough temperature where the Doppler broadening, which is proportional to $\sqrt{T}$, does not exceed the isotope shift. At the same time, the temperature cannot be lowered beyond a point, as a minimum vapour pressure (about 1 torr), is needed to provide the necessary concentration of the absorbing species. The absorption characteristics of isotopic atoms of some elements and of some isotopic molecules are presented in Table 8.6.

Table 8.6: Absorption characteristics of isotopic atoms of some elements and of some isotopic molecules[47]

(a) Isotopic atoms

Atoms	Absorption/ nm	Isotope shift/cm^{-1}	Doppler line width /cm^{-1}
^{6}Li - ^{7}Li	323.26	0.35	0.026
^{10}B - ^{11}B	249.77	0.175	0.44
^{200}Hg - ^{202}Hg	253.65	0.179	0.04
^{235}U - ^{238}U	424.63	0.280	0.055

(b) Isotopic molecules

Molecules	Absorption /μm	Isotope shift/cm^{-1}	Rotational band width /cm^{-1}
$^{10}BCl_3$ - $^{11}BCl_3$	10.15	9	25
$^{14}NH_3$-$^{15}NH_3$	10.53	24	71
$^{48}TiCl_4$-$^{50}TiCl_4$	20.06	7.6	18
$^{235}UF_6$-$^{238}UF_6$	16.05	0.55	16

The isotope shifts included in Table 8.6 show that the values for isotopic atoms of Li, Hg and U are well above their Doppler line widths. In other cases flow cooling of the species and/or simultaneous excitation of more than one line or band is resorted to, to overcome the inadequate isotope shift.

8.4.7 Some Photoinduced Reactions

The more important photoinduced reactions used in the separation of isotopes are:

(*a*) Two-photon ionization,
(*b*) Two-photon dissociation,
(*c*) Single-photon predissociation,
(*d*) Single-photon chemical reaction, and
(*e*) Multi-photon dissociation

These processes are described below in relation to the separation of isotopic atoms of interest (A) from the other isotope (A′).

(a)Two-photon ionization

Here the atoms of the isotope we are interested (A) present in a beam emerging from an oven mixed with other isotopic species (A′) are irradiated simultaneously by light of two wavelengths λ_1 and λ_2 so selected that the former photon (λ_1, usually in visible) selectively excites A to A* but not A′. The second photon (λ_2, usually in *uv*) ionizes A* to A, the species A′ remaining unaffected (Fig. 8.7 *a*)

$$A + A' \xrightarrow{\lambda_1} A^* + A'$$

$$A^* \xrightarrow{\lambda_2} A^+ + e^-$$

The ionic species A^+ is easily extracted out by an electric or a magnetic field and thus separated from the species A′.

Though not considered economical energywise, the method had been used in the separation of isotope ^{235}U (Ch. 10).

(b) Two-Photon Dissociation

A mixture of isotopic molecules BA and BA′ is, as in above, simultaneously irradiated by light of two chosen wavelengths λ_1 and λ_2 such that the former selectively excites BA to BA* without affecting BA′. The second photon acting on BA* dissociates it into B and A (Fig.8.7 *b*)

$$BA + BA' \xrightarrow{\lambda_1} BA^* + BA'$$

followed by

$$BA^* \xrightarrow{\lambda_2} B + A$$

* Normally, the excited state has a life-time of the order of 10^{-8} s before deexciting to the ground state. In some cases, where the transition is forbidden, the life-time may be much longer and the state is then referred to as a *metastable* one.

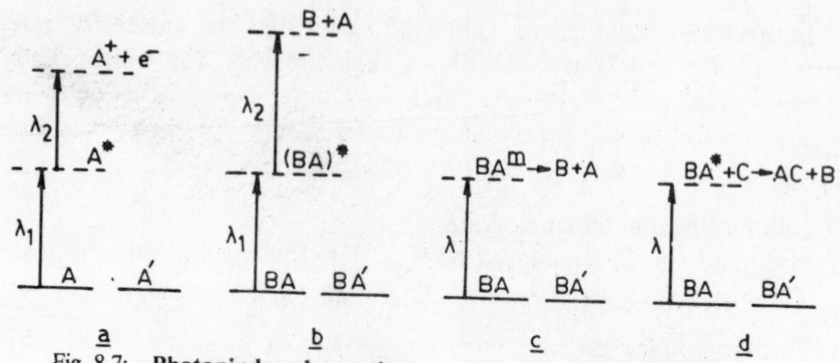

Fig. 8.7: **Photoninduced reactions**—(a) Two-photon ionization; (b) Two-photon dissociation; (c) Pre-dissociation; (d) Photochemical reaction.

By chemical analysis, the isotopic species A is separated from the unaltered BA′. Since here too, two photons are needed the process is energywise uneconomical. This method has also been used in the separation of uranium (Ch. 10).

(c) One-Photon Predissociation

In some cases, the photoexcitation of the molecule BA may result in a metastable state $(BA^m)^*$. If this happens to cross over to an energy level corresponding to an unbound state of the molecule, the same will dissociate into the constituent atoms, in this case B + A. This is referred to as *Photopredissociation* (Fig. 8.7 *c*)

$$BA + [BA'] \xrightarrow{\lambda} BA^m \longrightarrow B + A + [BA']$$

BA′ remains unaltered, so that A can be separated from BA′. The process is energy conserving as only one photon is used. The application of this process for the separation of the isotopes of bromine is described in the next section.

(d) Single-Photon Chemical Reaction

If the selectively photoexcited isotopic molecule BA* were to meet an atom or molecule C with which it can react chemically, before deactivating to the ground state, a new photochemical product would be formed (AC) and this can be separated from the unreacted BA′ (Fig. 8.7*d*); thus

$$BA \xrightarrow{\lambda} BA^* \xrightarrow{+C} AC + B$$

Since only a single photon is needed, the process is energy conserving, though not many examples are available where it is used for separating isotopes.

(e) Multi-Photon Dissociation

Generally, the vibrational energy levels of a polyatomic molecule are equally spaced apart*. On irradiation by an infrared laser precisely tuned to the

* The equal spacing holds good only for the lower few levels (*vide* § 1.4).

resonance energy level $h\nu$, absorption of photons can occur sequentially one step at a time pushing the molecule up the ladder of evenly spaced vibrationally excited states of quantum numbers $0 \rightarrow 1 \rightarrow 2 \rightarrow 3 \rightarrow$, until the highest bound state is crossed which must result in the dissociation of the molecule. Thus,

$$(BA)_0 \xrightarrow{\lambda} (BA)_1^* \xrightarrow{\lambda} (BA)_2^{**} \xrightarrow{\lambda} (BA)_3^{***} \xrightarrow{\lambda} \ldots B + A$$

where the subscripts 0, 1, 2, 3,...represent vibrational quantum numbers, (Fig. 8.8). The other isotopic molecule BA′, not being in resonance with the radiation employed, stays unaltered and can be chemically separated.

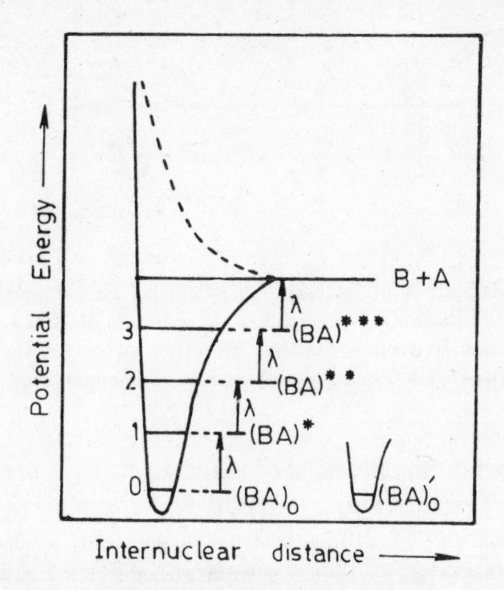

Fig. 8.8: **Multiphoton dissociation**

$$BA + BA \xrightarrow{nh\nu} B + A + BA'$$

The separation of ^{34}S from the more abundant ^{32}S is realised by this multiphoton absorption process (next section). The multi-photon dissociation method has been also used in the separation of uranium isotopes and compared with all other methods, photochemical or otherwise, is found to present the best performance (Ch. 10).

8.4.8 Photodeflection of Isotopes

This is purely a physical process arising from resonant absorption and reemission of laser light by the desired isotopic species in an atomic or molecular beam consisting of two or more isotopic species A, A′,... by a laser beam of frequency ν matching the resonance absorption of one of the isotopic species, say A, and in a direction perpendicular to the atomic beam. Only the atoms (or molecules) A in resonance with the radiation capture the photons and receive thereby an additional momentum $h\nu/c$* in the direction of the laser beam *i.e.* perpendicular to the particle beam. Soon the excited particles deexcite by resonance emission of photons. This emission is, however, isotropic, no direction being preferred. Thus, a net deflection in the path of the isotopic species involved in the resonance absorption-reemission process occurs. For a net spatial separation of the isotopic species $Nh\,\nu/c$ should be greater than the transverse momentum spread of the initial beam (Fig. 8.9). This method had been used to photodeflect a selected isotope of barium from the vapour of natural barium metal (see next section).

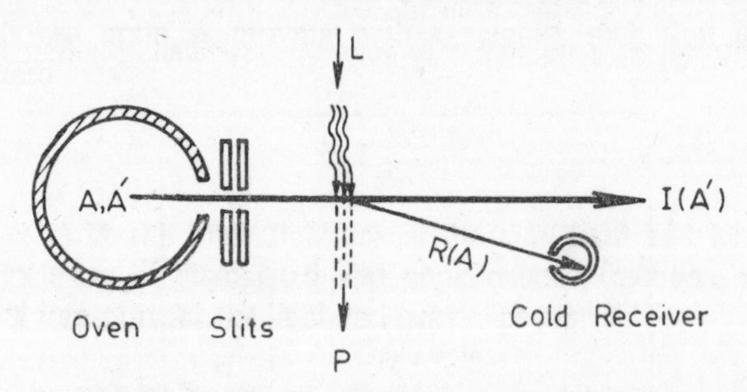

Fig. 8.9: **Photodeflection of isotopes**—A, A′: initial isotopic mixture; I: initial direction of momentum of beam A A′; L: Laser of resonance frequency for A; P: direction of photoninduced momentum of A·R: direction of resultant momentum of A.

8.4.9 Laser Applications

One of the most important applications of tuned lasers is in the separation of isotopes. Laser induced separations are found to be much cheaper than by the use of other methods. A computation shows that where 5 MeV of electrical energy is needed per ^{235}U atom separation by diffusion, the energy required for laser induced separation is of the order of a few keV per ^{235}U atom[40] As more special lasers, of rare gas halide type and continuous wave tuning devices are being perfected, their use in isotope separations also will expand.

Reserving the use of lasers in the separation of heavy water or deuterium, boron and uranium to later chapters, a brief account will be given here of their use in the separation of a few other isotopes, though none of the separations achieved can be said to be ready for being scaled up to a commercial production level.

(a) *Separation of Isotopes of Sulphur*

The multiphoton infrared photodissociation of sulphur hexafluoride to separate ^{34}S (natural abundance 4.2 per cent) was studied by Robinson[49]. The vapour of sulphur hexafluoride with hydrogen was irradiated by 200 ns pulses of CO_2 laser of energy 1-2 J, tuned to match the resonance vibration frequency

* The momentum of a photon of frequency ν is given by $h\nu/c$ where h is Planck's constant and c the velocity of light.

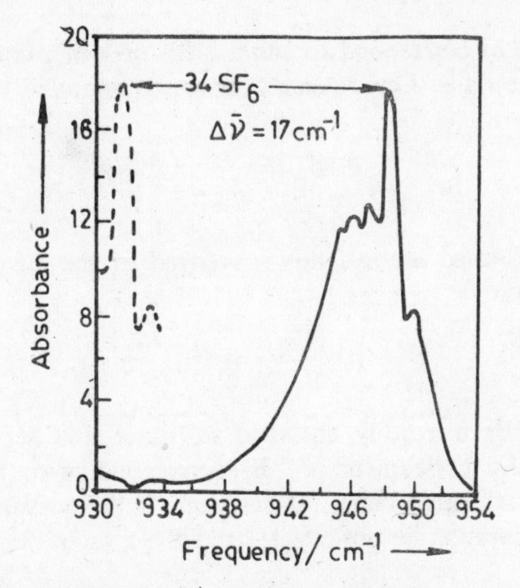

Fig. 8.10 : Absorption spectra of $^{32}SF_6$ in the region 10.6 μm.
(The figure occurs in *Industrial Applications of Lasers*, by J.F. Ready: The Academic Press Inc., 1978. It is reproduced here with the permission of Dr. B.B. Snavely, the original author of the figure and Academic Press, Inc.)

of $^{32}SF_6$, which is around 946 cm^{-1} (in the 10.6 μm *ir* region) (Fig. 8.10)[49a] A successive capture of the laser photons of this frequency pushed the $^{32}SF_6$ molecule up the vibrational manifold till it dissociated (§ 8.4.6(*e*)). The fluorine set free was scavenged by the hydrogen present. The vibrational frequency of $^{34}SF_6$ being much lower (around 930 cm^{-1}), remained unaffected. Mass spectrometric analysis of the residual SF_6 showed an increase in the $^{34}S/^{32}S$ ratio by a factor of 33 over the initial value.

(b) *Separation of Isotopes of Titanium*

Natural titanium is a mixture of five isotopes, ^{46}Ti to ^{50}Ti, of which the heaviest isotope ^{50}Ti is only 5.3 atom per cent in natural abundance. However, this particular isotope has the lowest neutron capture cross section of 0.179 barn compared to the 7.8 barns of the most abundant isotope ^{48}Ti. As a structural material ^{50}Ti would be more advantageous than natural titanium in the construction of nuclear reactors. Since there is an isotope shift of 7.6 cm^{-1} between the *ir* absorption lines of $^{48}TiCl_4$ and $^{50}TiCl_4$ (Table 8.5*b*), their separation is feasible as in the case of sulphur isotopes described above.

(c) *Separation of Isotopes of Bromine*

Natural bromine is a mixture of two isotopes in nearly equal amounts, ^{79}Br (50.69) and ^{81}Br (49.31 atom per cent). The method of single step photopredissociation (§ 8.4.6*c*) was adopted by Leone and Moore[51] to separate the bromine isotopes. Bromine vapour was irradiated by a frequency-doubled Nd-YAG laser emitting the green line 558 nm which could be tuned to within ~ 1 cm^{-1} to correspond to either $^{79}Br_2$ or $^{81}Br_2$, both of which have excited metastable states. On crossing over to the unbound level the molecule dissociates

$$^{81}Br_2 \xrightarrow{\lambda} {}^{81}Br_2^m \rightarrow {}^{81}Br + {}^{81}Br$$

The ^{81}Br atoms formed are instantly scavenged by the HI initially added to the bromine vapour.

$$^{81}Br + HI \rightarrow H^{81}Br + 0.5\, I_2$$

The product $H^{81}Br$ is readily absorbed in water and separated from the unaltered $^{79}Br_2$. An enrichment of ^{81}Br corresponding to 80 per cent was obtained, the initial value being 50 per cent. If the scavenging by HI and final chemical separation be not effected rapidly, a scrambling reaction of the type

$$^{81}Br_2^m + {}^{79}Br_2 \rightarrow {}^{81}Br_2 + {}^{79}Br_2^m$$

occurs which results in the formation of some $H^{79}Br$ and subsequent fall in the proportion of $^{81}Br_2$ in the final product.

(d) Separation of Isotopes of Barium by Photodeflection

Natural barium consists of ^{134}Ba (2.4), ^{135}Ba (6.5), ^{136}Ba(7.8), ^{137}Ba (11.2) and ^{138}Ba (71.9 atom per cent)*. Bernhardt[52] separated these isotopes, one at a time, by irradiating a beam of vapour of barium metal atoms issuing from an oven, by a transverse beam of 1.5 mW laser of wavelength 553.57 nm. The laser could be tuned to the precise wavelength of resonant absorption of any of the desired isotopes of barium. Bernhardt *et al.* reported in the photodeflected beam (§ 8.47), ejecting out 10^{14} atoms per second a threefold enrichment over the natural abundance.

Fig. 8.11 illustrates the increasing use of lasers in the separation of isotopes.

8.5 BIOLOGICAL PROCESSES

Plants, animals and microbes are known to display isotope effects in the uptake of materials around them in their normal metabolic processes. Usually, the molecules containing the lighter isotope is utilized more rapidly than those containing the heavier species. Thus, the preference is for H_2O to HDO, $^{12}CO_2$ to $^{13}CO_2$, ^{14}N to ^{15}N and ^{16}O to ^{18}O. The isotope effect in the case of water refers only to the water used up in building the cell materials, and not to the free water externally absorbed. In the last there is no isotopic discrimination[53]. No doubt, this effect may be of physical origin in part due to relatively faster diffusion through the cell material of the lighter isotope

* The element also contains 0.1 per cent each of ^{130}Ba and ^{132}Ba.

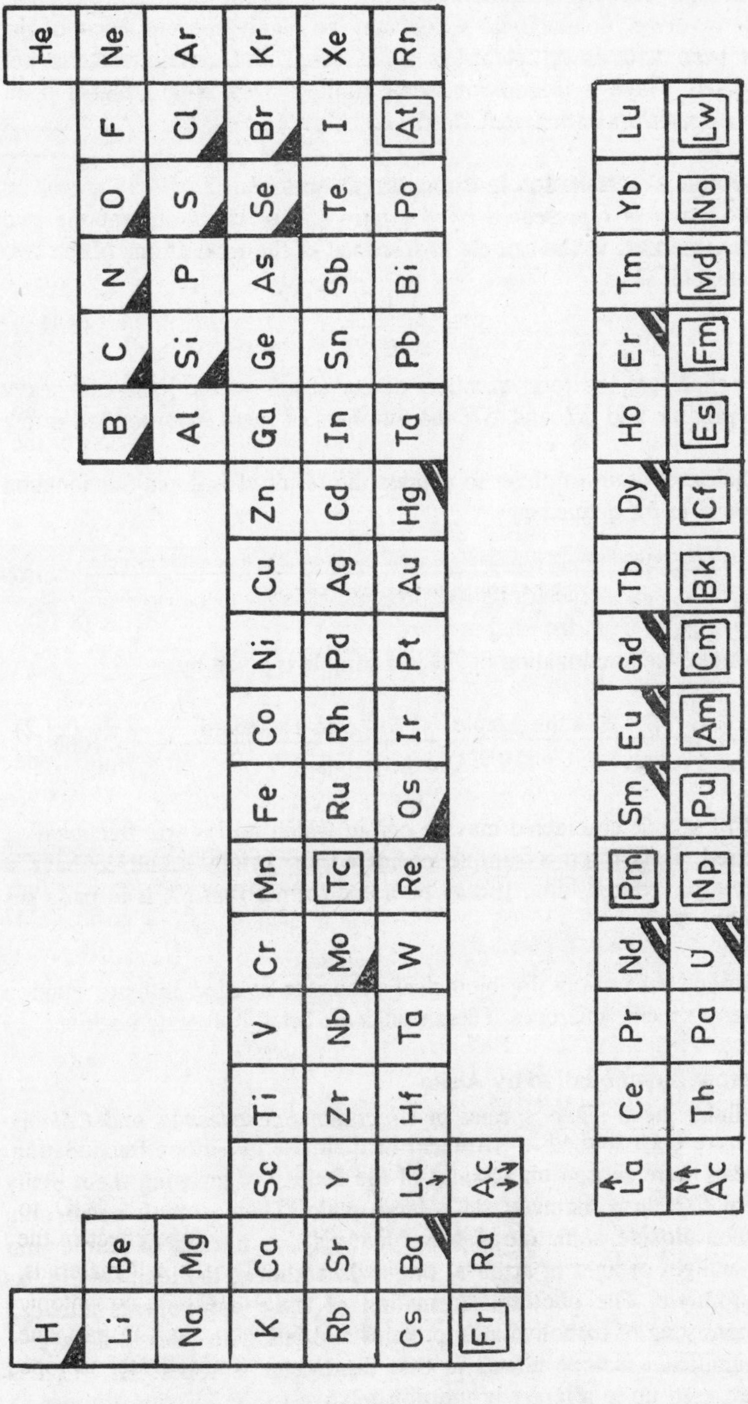

H																	He
Li	Be											B	C	N	O	F	Ne
Na	Mg											Al	Si	P	S	Cl	Ar
K	Ca	Sc	Ti	V	Cr	Mn	Fe	Co	Ni	Cu	Zn	Ga	Ge	As	Se	Br	Kr
Rb	Sr	Y	Zr	Nb	Mo	Tc	Ru	Rh	Pd	Ag	Cd	In	Sn	Sb	Te	I	Xe
Cs	Ba	La	Hf	Ta	W	Re	Os	Ir	Pt	Au	Hg	Ta	Pb	Bi	Po	At	Rn
Fr	Ra	Ac															

La	Ce	Pr	Nd	Pm	Sm	Eu	Gd	Tb	Dy	Ho	Er	Tm	Yb	Lu
Ac	Th	Pa	U	Np	Pu	Am	Cm	Bk	Cf	Es	Fm	Md	No	Lw

Fig. 8.11: **Isotopes separated by the use of lasers, *via* (*a*) atomic route ◩ (*b*) molecular route ◣ Man-made elements ☐**
(Courtesy : Dr. J.P. Mittal, BARC).

than the heavier one. Also where an enzyme or microbe catalyzed exchange reaction is involved, the catalytic action may be highly specific. Most of the work has been with lower plants as algae, fungi and certain bacteria not always related. There is in addition, some limited work with a higher plant as barley in deuterium enrichment, described below.

8.5.1 Isotopic Fractionation in Biological Processes

The fractionation is represented by the ratio (S) of the atoms of the two isotopes incorporated in the sample as fractions of the total atoms of the two species available. Thus

$$S = \frac{\Delta L/L}{\Delta H/H} \tag{8.18}$$

where L and H are the total numbers of the atoms of the light and heavy isotopes available and ΔL and ΔH the numbers of these incorporated in the sample.

Microbiologists seem to prefer to express the result of isotopic fractionation of a nuclide X as δX defined as

$$\delta X = \frac{\text{Ratio of } X \text{ to other isotopes in sample} - \text{same ratio found in a standard sample}}{\text{Same ratio in the standard sample}} \times 1000 \tag{8.19}$$

As an example, the fractionation of ^{34}S in a sample is given by

$$\delta\, ^{34}S = \frac{^{34}S/^{32}S \text{ in the sample} - ^{34}S/^{32}S \text{ in a standard}}{^{34}S/^{32}S \text{ in the standard}} \times 1000$$

The standard sample considered may be one in which no isotopic fractionation had occurred. It is often a sample of meteoritic origin found to have a constant isotopic composition. It may be noted further that δX is in parts per 1000 (symbol ‰)

Most isotope separations by biological processes studied hitherto employ either algae or specific microbes. These results are briefly presented below :

8.5.2 Isotope Fractionation by Algae

The unicellular green algae, species of *Chlorella, Scenedesmus* and *Chlamydomonas*, have been studied in particular in their use in isotope fractionation. These species were chosen on account of the facility of growing them easily to a size of 5-20 μm diameter on a large scale. Their growth needs only water, carbon dioxide, some nitrogeneous materials, as nitrates or ammonium salts, and sunlight or light of artificial origin. In addition, the temperature has to be appropriate. The phototransformation of these components into cell materials consisting of carbohydrates, proteins and fats with release of oxygen, involves significant isotope effects in that the lighter isotopes ^{1}H, ^{12}C, ^{14}N and ^{16}O are used up in a larger proportion relative to the heavier isotopes D, ^{13}C, ^{15}N and ^{18}O.

Further, some of the techniques normally used in chemical industry are available for scaling up isotope separations effected by these algae and the process appears therefore a promising one, specially for the enrichment of deuterium and ^{13}C isotopes.

Applications: (a) Separation of Deuterium

Reitz and Bonhoeffer[53] found that when different species of algae were allowed to grow in water containing from 12-70 per cent D_2O, isotope exchange occurred corresponding to a separation factor $(\Delta H/H)/(\Delta D/D)$ (Eq. 8.18) varying from 1.58 to 2.57. This isotope exchange was only in respect of the water combined in the cell materials, and not in respect of the free cell water imbibed which always had the same composition as the surrounding medium.

A result of interest and partly puzzling is that of Weinberger and Porter[54] who obtained the same separation factor of 2.5 for the species *Chlorella Pyrenoidosa* growing in water containing five per cent D_2O, as in water containing 0.01 to 4.0 mCi of tritium. On general consideration, the factor $(\Delta H/H)/(\Delta T/T)$ should have been greater than $(\Delta H/H)/(\Delta D/D)$. The same authors[55] further showed that the ether soluble components of the cell material, as lipids showed a greater isotope fractionation compared to the methanol soluble components as sugars and amino acids. The precise mechanisms of these fractionations are not known.

The efficiency of isotopic separation by algae, expressed in terms of the number of stages needed to double the initial deuterium content of water is 100 if the single stage separation factor is 2.5. The corresponding number of stages is 200 if the latter factor is 1.4.

(b) *Separation of* ^{13}C

Algae, as mentioned before, discriminate between the isotopes of carbon as well. The uptake of ^{12}C is known to be three times faster than that of ^{13}C, and some 12-14 times faster than that of ^{14}C [56]. The latter result is abnormal as, in terms of relative mass differences, only a six fold increase is expected.

Based on this high separation factor, Calvin and Weigl[57] set up a pilot plant for obtaining large amounts of carbon dioxide enriched in ^{13}C.

8.5.3 **Microbial Processes of Isotope Fractionation**

It is well-known during the past 30 years, that certain microbes are able to act selectively on a mixed inorganic substrate and promote a chemical transformation of one or a few components of the mixture leading to their concentration, dispersion or fractionation.[58,59] As illustration may be cited the selective oxidation of arsenopyrite by *Thiobacillus ferroxidans*[60], of microbiological leaching of copper in chalcopyrite by the same bacillus[61]*, and the

---

* Bacterial leaching processes contribute about five per cent into yearly world copper production, the main reaction being[61]

$$2CuFeS_2 + 8.5\,O_2 + H_2SO_4 \xrightarrow{\text{bacteria}} 2CuSO_4 + Fe_2(SO_4)_3 + H_2O$$

preferential reduction of MnIV over FeIII in the ferromanganese nodules on sea beds[62].

In addition to the above important role of microbes in hydrometallurgical operations, certain bacteria are also known to bring about isotopic fractionation. As an example may be cited the preferential assimilation of $^{12}CO_2$ over $^{13}CO_2$ and in the reduction of $^{32}SO_4^{2-}$ over $^{34}SO_4^{2-}$ by *Desulphovibrio Desulphuricans*[63]. Our present knowledge seems to indicate that the faculty of isotope fractionation is confined to a small number of specific microbes which do not seem to be related to one another[59].

8.5.4 Bacterial Growth and Functioning

The selected bacteria can be grown easily in a liquid culture containing sugars, aminoacids and vitamins and some other organic substances. No exposure to light is needed in most cases.

Several bacteria containing the enzyme *Hydregenase* reduce various substances to hydrogen, while some reduce catalytically oxygen rich substances and make available the energy liberated to transform carbon dioxide to other cell building materials. In all these reactions the light isotope of hydrogen reacts two to three times faster than the deuterium present.

In general, in all the isotope fractionation reactions, the microbes may react, either directly if they carry particular enzymes, or only indirectly and passively if they are of a nonenzymatic nature. In general, where microbes do function as isotope fractionators, they metabolize substrates containing the lighter isotopes as H, ^{12}C, ^{14}N, ^{16}O and ^{32}S in preference to those containing the heavier isotopes as D, ^{13}C, ^{15}N, ^{18}O and ^{34}S.[59,64–66]

8.5.5 Applications

(a) *Separation of Deuterium*

Farkas, Farkas and Yudkin[67] were the first to show that the exchange reaction

$$H_2O\,(liq) + HD\,(gas) \rightleftharpoons HDO\,(liq) + H_2$$

$$K = 3.87\ (25°C)$$

can be bacteria catalysed.

This was observed in the H_2 evolved during the decomposition of sodium formate by *Escherichia Coli*. The equilibrium constant is as high as 3.87 at 25 °C, the liquid phase being enriched in deuterium. However, the reaction is very slow and the time for the attainment of equilibrium is very long, being of the order of hours, as reported by Linday[68]. She studied the possibility of separating the depleted gaseous product from the D_2O enriched water by the use of different bacteria containing the enzyme *Hydrogenase*. Some of the results were apparently confusing. In the reduction of methylene blue, the bacterium *Hydrogenomonas Facilis* did not distinguish between H and D isotopes. A similar result of no discrimination in the rates of uptakes of hydrogen isotopes was observed by Franke and Monch[69] in the hydrogenation of fumaric acid to succinic acid by *Escherichia Coli*[69], though the net amount of H added was more than that of D.

(b) *Separation of Sulphur Isotope* ^{34}S

Natural sulphur consists mainly of ^{32}S and ^{34}S in the proportion of 95.1 and 4.2 atom per cent, giving for the $^{32}S/^{34}S$ a value ranging from 21.3 to 23.2. The ratio for sulphur of meteoritic origin is virtually constant at 22.22. Hence meteoritic sulphur is taken as the standard for reference in Eq. 8.19[59].

Thode *et al*[70] appear to have been the first to report the reduction of $^{32}SO_4^{2-}$ to H_2S by *Desulphovibrio desulphuricans* at a rate about one per cent faster than the reduction of $^{34}SO_4^{2-}$. The maximum value for $\delta^{34}S$ (Eq. 8.19) observed by Jones and Starkey[64] was −12.9‰ for incubation for 68 ha at 28 °C*. The exchange equilibrium postulated by Thode *et al.*[71] is

$$^{32}SO_4^{2-} + H_2\,^{34}S \rightleftharpoons {}^{34}SO_4^{2-} + H_2\,^{32}S$$

$$K = 1.085\ (0\,^\circ C)$$
$$= 1.074\ (25\,^\circ C)$$

A much larger value of −46 ‰ for $\delta\,^{34}S$ was reported by Kaplan and Rittenberg[72] for the reduction of SO_4^{2-} to S^{2-} induced by the same species of bacterium but in the presence of ethyl alcohol. Curiously with a different bacterium *Thiobacillus Concretivoros*,the $\delta^{34}S$ is + 19 ‰ indicating a reverse enrichment, namely that of ^{34}S. No isotopic fractionation occurs if the substrate is simple elementary sulphur.

The microbial reduction of sulphate to sulphide is envisaged by Thode and Harrison[73] as proceeding in two stages, each induced by a different enzyme :

(i) The entrance of the SO_4^{2-} into the microbe cell to form APS (adenosine 5′-phosphosulphate), after the SO_4^{2-} is activated by ATP

$$SO_4^{2-} \xrightarrow{\text{ATP sulphurylase}} \text{APS} + \text{pyrophosphate}$$

This reaction goes to completion as the pyrophosphate gets hydrolysed.

(ii) The reduction of APS by the breaking of an S-O bond to form the sulphite and AMP (adenylic acid)

$$\text{APS} + 2e^- \xrightarrow{\text{APS reductase}} \text{AMP} + SO_3^{2-}$$

Of these, the first reaction has no isotope effect, while the second displays a

* From the definition of δx as the isotope fractionation factor (Eq. 8.19), a negative value for δx means enrichment of the lighter isotope relative to the standard reference and a positive value for δx means an enrichment of the heavier isotope. Hence the sample is enriched in ^{32}S here.

kinetic isotope effect. The final reduction of SO_3^2 to S^2 proceeds in three stages each aided by a different enzyme. A mechanism of this type is not considered adequate to explain the value of $-46‰$ for $\delta\ ^{34}S$ observed by Kaplan[71].

Besides the reduction of SO_4^{2-} to S^{2-}, there are several other microbial reductions of inorganic sulphur compounds which are accompanied by an isotopic fractionation[59].

References

1. I. Langmuir, *Phys. Z.,* 1913, 14, 1273.
2. A.K. Brewer and S.L. Madorsky, *J. Res, Nat. Bur. Stand.* 1947, *38,* 129.
3. J.N. Brönsted and G.H. Heversy, *Nature,* 1920, *106*, 144
4. S.L. Madorsky, P. Bradt and S. Strauss, *USAEC Report,* AECD, 1913 (1948).
5. J.N. Brönsted and G. Hevesy, *Nature,* 1921, *107,* 619.
6. G. Hevesy, *Acta Chem. Scand.* 1949, *3*, 1263.
7. D.M. Lang, *et al. Proc. Internat. Symp. on Isotope Separations* (N. Holland Publ. 1958).
8. T.F. Johns *et al.,* ibid.
9. H.C. Urey, F. Brickwedde and G.M. Murphy, *Phys. Rev.,* 1932, *40,* 1
10. M.P. Appleby and G. Ogden, *J. Chem. Soc.,* 1936, 163.
11. E.J. Hart and M.Anbor, *The Hydrated Electron,* Wiley Interscience, 1970.
12. B.E. Conway, *Proc. Roy. Soc.,* 1958, *A 247,* 400; 1960, *A 256,* 128.
13. S. Glasstone, K.J. Laidler, and H. Eyring, *The Theory of Rate Processes,* McGraw-Hill Book Co., New York, 1941.
14. W.J. Moore, *Physical Chemistry,* Longmans Green & Co., London, 1963.
15. N.J. Selley, *Experimental Approach to Electrochemistry,* Edward Arnold, London, 1977.
16. M. Chemla, and J. Périé. *La Séparation des Isotopes,* Presses Universitaire de France, Paris, 1974.
17. H.D. Smyth, *Rev. Mod. Phys.,* 1945, *17,* 351.
18. GM. Murphy, *Production of Heavy Water,* McGraw-Hill Book Co., New York, 1955.
19. F.A. Lindemann, *Proc. Roy. Soc.,* 1921, *99A,* 102.
20. J. Kendall and E.D. Crittendon, *Proc. Nat. Acad. Sci.,* U.S.A., 1923, *9,* 75.
21. J. Kendall and J.F. White, *ibid,* 1925, *11,* 393.
22. J. Kendall and B.L. Clarke, *ibid,* 393.
23. A.K. Brewer, S.L. Madorsky and S. Strauss, *J. Res. Nat. Bur. Stand.,* 1947, *38,* 137, 185 1948, *41,* 41.
24. A. Klemm, *Z. Naturforsch,* 1951, *69,* 487. *Z. Elektrochem.,* 1954, 58, 609.
25. M. Chemla, *Proc. Internat. Symp. on Isotope Separations,* N. Holland Publ. 1957-58, 288.
26. A. Lunden, *Thesis, Univ. Goteburg,* 1956
27. S. Klemm, *J. Chim. Phys.,* 1952, *49,* C 18
28. L.G. Longworth and D.A. MacInnes, *Prog. on Report on Separation of Uranium Isotope by CC Electrolysis A 164,* S-8, Rockefeller, Institute for Med. Res., 1942.
29. H. London, *Separation of Isotopes,* G. Newnes, London, 1961.
30. H.J. Arnikar, *Thesis,* Univ. Paris, 1958, *Ann. de Physique,* 1959, *13*(4), 1291.
31. H.J. Arnikar, *J. Inorg. Nucl. Chem.,* 1959, *11,* 249.
32. R. Fuoss (*Private Communication*)
33. A. Klemm, H. Hintenberger and P. Hoernes, *Z. Naturf.,* 1947, *2a,* 245.
34. A. Klemm, *Proc. Internat. Symp. Isotopes, N.* Holland Publ. 1957-58.
35. M. Chemla and P. Süe, *Compt. Rendus,* 1953, *236,* 2397.
36. A. Bonnin, M. Chemla and P. Süe, ibid, 1955, *241,* 40.
37. M. Chemla and A. Bonnin, *ibid,* 1288.

38. A. Lunden and M. Herzog, *Z. Naturf.*, 1956, *11a*, 520.
39. A. Lunden, *ibid*, 590.
40. C. Mangalo, H.J. Arnikar and M. Chemla, *Comp. Rendus.*, 1947, *244*, 2796
40a. H.J. Arnikar and M. Chemla, *Radioisotopes in Scientific Research,* Vol. II, Pergamon Press, London, 1958, p.421.
41. E.R. Ramirez, *J. Amer. Chem. Soc.*, 1954, *76*, 6237.
42. A. Klemm, *Naturwiss.* 1944, *32*, 69.
43. T.R. Merton, E.J. Bowen, H. Hartley and A.O. Ponder, *Phil. Mag.*, 1922, *43*, 430.
44. K. Zuber, *Helv. Phys. Acta*, 1936, *9*. 285.
45. H.E. Gunning, *Canad J. Chem*, 1958, *36*, 89
46. W. Spindel, *Isotope Effects in Chemical Processes,* Amer. Chem. Soc., Washington, D.C., 1969.
47. J.F. Ready, *Industrial Applications of Lasers,* Academic Press, New York, 1978
48. *Chemical and Biochemical Applications of Lasers* (Ed. C.B. Moore, Acad. Press, New York, 1977)
49. C.P. Robinson, *Ann. New York Acad. Sci.*, 1976, *267*. 81.
49a. W.T. Tsang and S. Wang, *Appl. Phys. Lett.*, 1976, *28*, 596.
50. B.B. Snavely, *IEEE/OSA Conference on Laser Engineering and Applications.* Washington, D.C., 1975.
51. S.R. Leone and S.R. Moore, *Phys. Rev. Lett.* 1974, *33*, 269.
52. A.F. Bernhardt *et al. App. Phys. Letters,* 1974, *25*, 617.
53. O. Reitz and K.F. Bonhoeffer, *Z. Phys. Chem.*, 1935, *A 172*, 369; *A 174*, 424.
54. D. Weinberger and J.W. Porter, *Science,* 1953, 117, 636.
55. D. Weinberger and J.W. Porter, *Arch. Biochem. Biophs.*, 1954, *50*, 160.
56. F.E. Wickman, *Nature,* 1952, *169*, 1051.
57. M. Calvin and J.W. Weigl, *Concentrating Isotopic Carbon,* U.S. Patent, 1952, (2,602,047)
58. L.C. Bryner, J.V. Beck and D.G. Wilson, *Ind. Eng. Chem.* 1954, *46*, 2587.
59. H.L. Ehrlich, *Geomicrobiology*, Marcell Dekker. Inc. New York, 1981.
60. H.L. Ehrilich, *Econ. Geol.*, 1964, *59*, 1306.
61. A.E. Torma, P.R. Ashman, T.M. Olsen, adn K. Bosecker, *Metals,* 1979, *5*(33), 479.
62. H.L. Ehrlich, *Interuniversity Programme of Research of Ferromanganese Deposits on Ocean Floor,* National Science Foundation, Washington, D.C., 1973.
63. R.N. Doetsch and T.M. Cook, *Introduction to Bacteria and their Ecobiology,* Univ. Park Press, Baltimore, Maryland.
64. G.E. Jones, and R.L. Starkey, *App. Microbiology,* 1957, *5*, 111.
65. C. Emeliani, J. Hudson, E.A. Shinn and R.Y. George, *Science,* 1978, 627.
66. R.P. Wellman, F.C. Cook, and H.R. Krawse, *ibid,* 1968, *161*, 269.
67. A. Farkas, L. Farkas and J. Yudkin, *Proc. Roy. Soc.*, 1934, *B 115,* 373.
68. E.M. Linday and P.J. Syrett, *J. Biochem. Microbiology,* 1960, *2*, 225.
69. W. Franck, and J. Monsch, *Biochem. Z.*, 1948, *319*, 807.
70. H.G. Thode, H. Kleerekoper and D. McElcheran, *Research,* 1951, *4*, 581.
71. A.P. Tudge and H.G. Thode, *Canad. J. Res.*, 1950, *29B*, 507.
72. I.R. Kaplan and S.C. Rittenberg, *J. Gen. Microbiol,* 1964, *34*, 195.
73. A.G. Harrison and H.G. Thode. *Trans. Frad. Soc.*, 1957, *53*, 1; 1958, *54*, 34.

Chapter 9

LARGE SCALE SEPARATION OF STABLE ISOTOPES

Having surveyed the more important methods of isotope separation developed so far, we propose to consider the application of these methods to the large scale separation of isotopes used in the present atomic age. These include (*a*) stable isotopes needed to maintain the atomic power plants and nuclear reactors developed so far, (*b*) isotopes which serve as nuclear fuels and (*c*) radioisotopes needed for research and for use in medicine, industry and agriculture. We shall be limiting to only the more widely used methods employed in the large scale production of these isotopes.

The stable isotopes needed in the present age in large amounts and in a state of high degree of isotopic purity are heavy hydrogen ^{2}H, light lithium ^{6}Li, light boron ^{10}B, and heavy ^{50}Ti. Ironically, all these happen to be the less abundant isotopes. In the case of lithium, however, the residue after ^{6}Li is separated is ^{7}Li which too finds important application in special reactors. While these are covered in the present chapter, the separation of nuclear fuels and radioisotopes will be treated in the subsequent chapters.

9.1 HEAVY HYDROGEN

Heavy hydrogen ^{2}H, or deuterium D, occurring to the extent of 0.015 atom per cent is mainly used in the form of heavy water to moderate or slow down fission neutrons from an initial energy of a few MeV to less than one eV. Though in an elastic collision with a D atom, the maximum fraction of the initial kinetic energy a neutron loses is only about 89 per cent compared to its collision with an H atom*, the latter, *i.e.* light water, cannot be used as a moderator if a heavy loss of neutrons by H(n,γ) D reaction is to be

---

* The classical expression for the fraction of the initial kinetic energy $\Delta E/E_o$ transferred by a neutron to the atom of mass number A it collides with, varies from zero to a maximum given by $\Delta E/E_o = 4\,A/(A+1)^2$ as the scattering angle varies from zero to 180°. This fraction is maximum for H(A=1) and 0.89 for D(A=2). Also the *slowing down* length *i.e.* the mean distance a fission neutron travels in the medium is 5.3 cm in light water and 11.2 cm in heavy water.

avoided. This loss is due to the high neutron capture cross section of 332 mb of the H atom compared to 0.53 mb of the D atom.

A calculation of the neutron reproduction, or the multiplication factor k_∞, (under condition of the reactor size being adequately large to prevent the escape of even a single neutron, fast or thermal), shows the factor passes through a maximum as the ratio of the uranium atom to that of the moderator N_U/N_M is varied. The maximum for k occurs for a ratio of 4-10 for H_2O and 150-500 for D_2O. Hence, all nuclear reactors employing natural uranium as the fuel generally employ heavy water as the moderator*.

Of the different methods developed for the separation of heavy water we shall describe only four methods; (1) fractional distillation of hydrogen, (2) fractional distillation of water, (3) electrolysis and (4) chemical equilibrium exchange.

9.1.1 Fractional Distillation of Hydrogen

The sample of enriched hydrogen which was used by Urey in his celebrated work which led to the discovery of deuterium in 1932, was obtained by the fractional distillation of liquid hydrogen.

With 157 ppm of heavy water ($^2H_2{}^{16}O$), and with other isotopic water molecules with ^{18}O in trace amounts, water can be certainly considered as the most "ideal" solution, the "solutes" obeying Raoult's law strictly. Representing the mole fraction of the species H_2O, D_2O and HDO by x, y and z respectively, we may write for the exchange equilibrium:

$$H_2O + D_2O \rightleftharpoons 2HDO$$

$$K = \frac{[HDO]^2}{[H_2O][D_2O]} = \frac{z^2}{xy} = 4$$

the classical value, whence $z = 2(x \cdot y)^{1/2}$.

Assuming the vapours behave as ideal gases, the total vapour pressure of water at a given temperature should be, by Dalton's law,

$$P = p_x^o x + p_y^o y + p_z^o z$$

where p_x^o, p_y^o and p_z^o are vapour pressures of the "pure" components H_2O, D_2O and HDO. Fig. 9.1 is the vapour pressure diagram for the ideal solution. The fractional distillation of water to separate D_2O would proceed with an elementary separation factor given by

$$\alpha = \frac{Y(1 - Y)}{Y'(1 - Y')}$$

where Y and Y' are the mole fractions of D_2O in the liquid and the vapour

* Reactors employing uranium enriched in ^{235}U to about 3-5 per cent, as the Apsara reactor of Bombay, or the power reactors of Tarapur, use light water as the moderator.

phase. From the simple stoichiometry involved ($x + y + z = 1$), we have

$$Y = (1/2\,z) + y$$

and

$$(1 - Y) = (1/2\,z) + x$$

The values for the single stage separation factor are

$$\begin{aligned} \alpha \simeq P^0\,H_2O\;/\;P^0\,D_2O &= 1.11\;(50°C) \\ &= 1.08\;(70°C) \\ &= 1.05\;(100°C) \end{aligned}$$

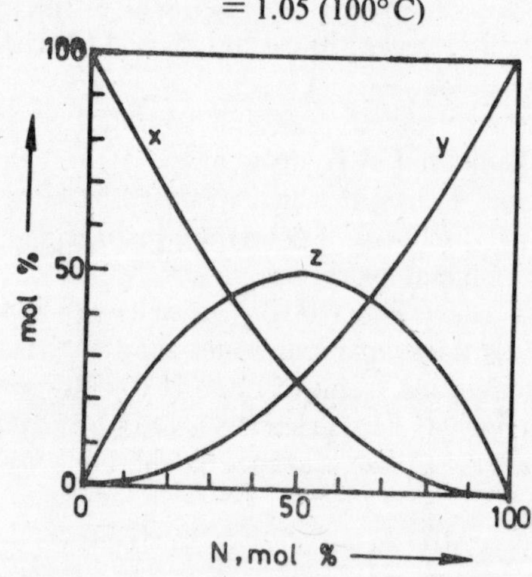

Fig. 9.1: Mole fractions of the species H_2O (x), D_2O (y) and HDO (z). as functions of the mole fraction [D]/([D] + [H]) the liquid phase. assumed to be an ideal solution.

The empirical equation of Bigeleisen[2,3] applicable to the *normal* isotope effect (see Sec. 7.1.3),

$$\ln\alpha = \ln\frac{P_1}{P_2} = \frac{A}{24}\cdot\left[\frac{h}{2\pi KT}\right]^2\left[\frac{1}{m_1} - \frac{1}{m_2}\right] \qquad (7.4)$$

reveals two features which makes the distillation of liquid hydrogen an attractive process. These are (i) the variation of the separation factor log α with $1/T^2$ and (ii) its dependence on the *atomic* masses m_1 and m_2 of the isotopes. The ratio of the vapour pressures p_{H2}/p_{HD} (equal to the separation factor) is 1.73 (Table 7.1) at the boiling point temperature of 20.4 K. This is an encouragingly high value needing just 34 plate equivalents to yield nearly pure D_2 beginning with natural hydrogen. Numerous difficult technological problems have, however, to be overcome in manipulating hydrogen at such a low temperature. The very low density of 0.07 g/ml of liquid hydrogen and the blocking of the unit by the deposition of solid impurities at temperatures below 65 K are the major hurdles to b. surmounted. Any rise of temperature results in a marked lowering of the vapour pressure ratio. The value falls from 1.73 at 20.4 K to 1.45 at 26 K, *i.e.* for a five degree rise of temperature.

Detailed studies by Clusius and Starke[4] in Germany showed that H_2, HD and D_2 fractions can be separated by distilling liquid hydrogen at atmospheric pressure. The power needed for the process was assessed to be no more than about three per cent of that consumed in the separation by electrolysis of water.

The plant designed by Clusius and Starke

Natural hydrogen containing 0.028 per cent of HD at a little over atmospheric pressure is cooled to below the dew point and fed to the first distillation tower (Fig. 9.1a). Here a first separation occurs and most of the HD-free hydrogen is removed from the top, while a fraction containing 5-10 per cent HD settles down. The latter is pumped to the top of a smaller secondary tower. Here nearly pure HD collects in the middle part of the tower. This fraction is moved to a heat exchanger where it is warmed upto room temperature. The gaseous HD then enters a catalytic exchange reactor where disproporanation occurs

$$2HD \rightleftharpoons H_2 + D_2$$

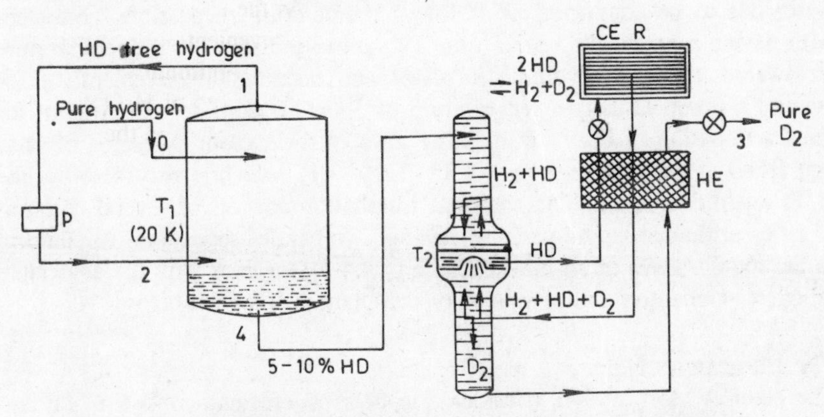

Fig. 9.1(a): **Clusius-Starke Plant for the separation of D_2 by the distillation of hydrogen** T_1: primary tower; T_2: secondary tower; HE: heat exchanger; CER: catalytic exchange reactor; P: compressor; O: natural H_2 (cold); 1: HD-free H_2 (high pres); 3: pure D_2; 4: 5-10% HD.
(after K. Clusius and K. Starke, *Z. Naturf., 4A, 549*), from *Separation of Isotopes* Ed. by H. London, George Newnes Ltd., 1961, p. 86; reprinted by permission from William Heinemann Ltd).

The equilibrium mixture of the three components is once again cooled to liquid hydrogen temperature in the heat exchanger and returned to the bottom half of the second tower. Here finally the fraction H_2-HD separates from the nearly pure liquid D_2. This last settles at the bottom of the tower. The H_2-HD fraction is recyled to the top of the same tower to recover HD by repeating the cycle.

Several innovations and improvements have since then been effected over the earlier design of Clusius and Starke. Plate type column is now considered more advantageous over a packed column. The hydrogen needed is generally taken from the synthetic ammonia plant, which contain troublesome impurities. The other alternative is electrolytic hydrogen which India earlier planned to utilise[5]. This latter source is expensive and only limited amounts of the gas would be available.

9.1.2 Fractional Distillation of Water

Fractional distillation of water was the first method developed under the *Manhattan Project** for the production of heavy water of high purity in the U.S. over the period 1942-45. Three factories were set up which even as they reached a stage of progress in producing enough heavy water to meet the needs of the time, were shut down in October 1945, or modified to work the more economical chemical exchange process for obtaining the same product. Because of the technical simplicity of the process and of the potential scope for its development, the original fractional distillation plant set up at Morgantown will be briefly described.

Since all isotope effects fall with rise of temperature, the distillation of water has to be conducted at as low a temperature as possible, consistent with having a practically usable value of vapour pressure. As the temperature is lowered, the molar volume increases and becomes inconveniently large beyond a point. Usually a temperature of 50 °C is considered optimum for the fractionation of heavy water. The ratio of the vapour pressures of H_2O and HDO at this temperature is 1:05 (Table 7.1), which is also therefore the H/D separation factor. The complete installation consists of several columns of different diameters and packings arranged as parallel-series units to conform to the requirements of an ideal cascade (§ 4.4.4), together with the associated reboilers, condensers and pumps. A typical plant is described below.

The Morgantown Plant

The plant at Morgantown, considered to be most efficient, consisted of a ten-stage cascade of distillation towers, with five towers in parallel in the first and in the second stages, making a total of 18 towers. The diameters of the towers of the first six stages decrease from 4.57 to 0.46 m, while the last stage towers were all of 0.25 m diameter. Also the former set of large diameter towers were of bubble-cap plate type and the rest were packed with 16 × 16 mm ceramic rings. The total towers in the ten series stages were equivalent to 757 theoretical plates. Feed steam at about 10.5 kg/cm^2 pressure enters at the top of the first tower and the slightly enriched water collecting at the bottom of this tower is pumped to the top of the next tower and so on. Also the vapour from the top of the second tower is returned to the bottom of the first tower (Fig. 9.2); similarly from other even numbered

* This was the cryptic code name used by the U.S. government for all isotope separations during the period of the War.

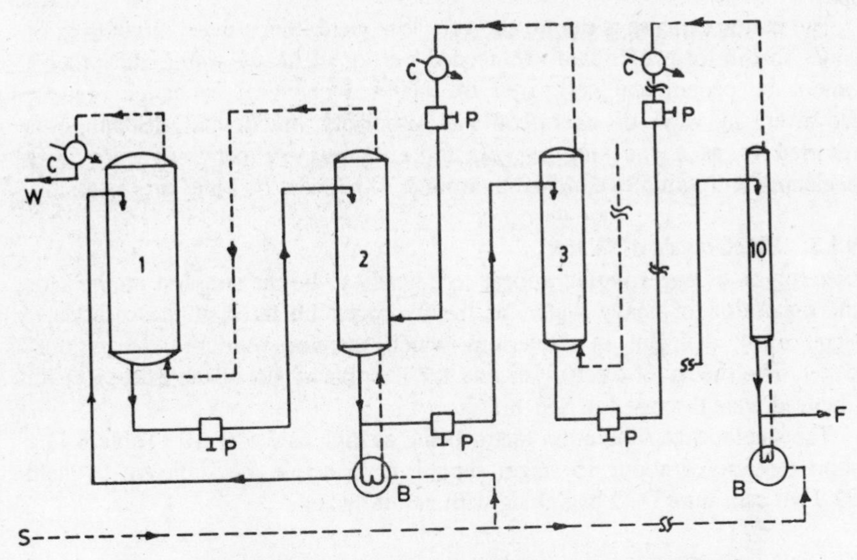

Fig. 9.2: **Morgantown heavy water plant** 1, 2,... 10: Series of towers; B: Boiler; C: Condenser; P: Pump; S: Steam; F: Final product (89 atom % D); W: Waste water (0.0139 % D); —: Liquid;---: Vapour
(from *Separation of Isotopes*, Ed. by H. London, George Newnes Ltd., 1961, p. 80; reprinted by permission from William Heinemann Ltd).

towers too. The vapour escaping from odd number towers is condensed and the liquid returned to the middle of the preceding tower. The enrichment in D_2O increases from initial 0.0143 atom per cent to 89 per cent at the end of the last tower; the stage-wise enrichment is shown in Table 9.1.

Table 9.1: Typical stage-wise enrichment of D_2O in a 10-stage distillation plant

Stage	0	3	4	5	6	7	8	9	10
D_2O/atom %	0.0143	0.117	1.4	3.8	10.0	11.5	21.2	56.4	89.0

The net input and output in a typical operation were:

Input 88 500 kg of water/h of D_2O content = 0.0143 per cent

Output 0.4 kg of water/h of D_2O content = 89.0 per cent

The purity, recovery and production rate of D_2O in the Morgantown and similar plants, as initially set up, were about 50 per cent of the design expectations. This defective performance was traced to leakage of water between the tower plates and the wall. When this was largely overcome, the performance improved to close on 75 per cent of design expectations. The use of Spraypacks (see Sec. 7.1.6) in place of bubble-caps was expected to increase the steam velocity by a factor of 5-7.

The major expense is due to the very low yield. Since over 220,000 kg of water in the form of steam are needed per kg of heavy water, the process cannot be economical compared to catalysed chemical exchange process. However, in view of its operational simplicity, fractionnal distillation is resorted to as a *finishing process* for obtaining 99 per cent pure D_2O beginning with samples enriched to around 10 per cent by other processes.

9.1.3 Electrolysis of Water

Electrolysis of water would appear technically to be the simplest method for the separation of heavy water, as the hydrogen liberated at the cathode is significantly deficient in deuterium which therefore enriches the residual water. The theory of electrolysis and the possible origin of the isotope effect involved were discussed in Sec. 8.2.

The single stage separation factor being as high as 3.5 to 10 (Table 8.1) a short cascade of about 15 stages should, in principle, be sufficient to yield 99.9 per cent pure D_2O beginning with natural water.

The cell design

Generally, a vertical cylindrical cell is preferred wherein the central anode is separated by a porous diaphragm of nickel wire-reinforced asbestos from the outer casing metal which serves as the cathode. This last is water cooled to keep the temperature down. The electrolyte often recommended is a solution of 25 per cent by weight of potassium carbonate, as potassium ion has relatively the highest conductance ($\lambda° = 74.5\ \Omega^{-1}\ cm^2\ eq^{-1}$). Steel has been found to be the best cathode material on account of its relatively low hydrogen overpotential and consequently associated with a high separation factor close on 10 (Table 8.1). Also steel can withstand the action of the alkaline solution provided the temperature rise is not much. A Canadian plant of this type was set up during the War, but later shut down[6]. Also in the pre-War years, electrolysis of water was the principal process adopted to meet the limited needs of the time.

Limitation

The attractive feature of the process arising out of the high separation factor limiting the cascade to around 15 stages is wholly offset by the huge wastage in the form of loss of 97 per cent of hydrogen and the associated amount of HD, separated from water by electrolysis at great cost. Recycling or refluxing by burning to water all this hydrogen, amounting to 20 000 kg per kg of D_2O, is inconceivable as all the heat generated during the combustion may not be harnessable. This apart, the risk of explosion in an operation with amounts of hydrogen of this magnitude is real. Refluxing by combustion may be practicable only from the fourth or the fifth stage of the cascade of electrolysis, *i.e.* after the enormous loss of all the gas of the earlier stages. Fig. 9.3 illustrates the stage to stage progress of electrolysis in terms of the total amount of water, the D_2O content and the flux of the latter, as well as the loss of hydrogen (+ HD), in the early stages *i.e.* before the recycling begins.

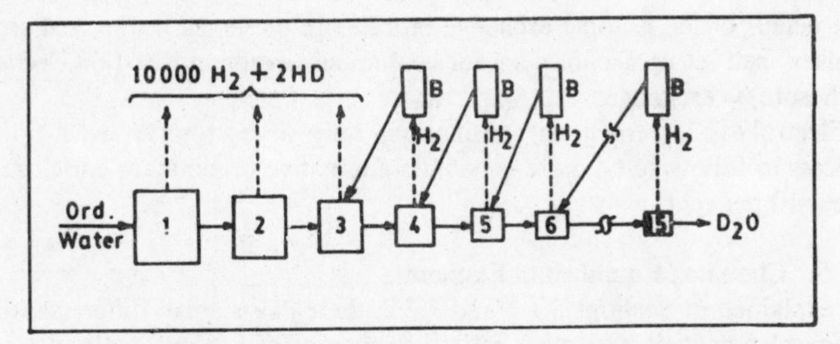

Fig. 9.3: **Production of heavy water by electrolysis in cascade:** 1,2,...15 electrolysis cells B: burner

Cell	1	2	3	4	5	6	...15
H_2O/kg	10000	2750	750	285	108	42	0.5
D_2O %	0.015	0.04	0.1	0.27	0.72	1.9	99.7
D_2O/kg	1.49	1.082	0.785	0.785	0.785	0.787	0.488

(Reprinted with permission from *La Se'paration des Isotopes* by M. Chemla and J. Périe, p.220 © Presses Universitaires de France, 1974)

The net input and output in a typical operation were:

Input: 10 000 kg of water of D_2O content 0.014 per cent; Energy ~ 62 MWh;

Output: 0.488 kg of water of D_2O content 99.7 per cent; 9715 kg of hydrogen gas + 285 kg waste water, together containing about 1 kg of D_2.

A process of this kind may become viable only if the major product hydrogen gas is all used up in some useful process, but this too is unlikely as hydrogen prepared by chemical process, as catalytic decomposition of water, would be many times more cheap than electrolytic gas.

9.1.4 Electrolysis with Isotopic Exchange

The serious drawback of the electrolytic process described in the previous section is the loss of close on 67 per cent of the deuterium initially present in the water, as HD with the hydrogen gas which cannot be refluxed. A considerable part of this deuterium can be recovered by coupling the electrolysis cell with an isotopic exchange unit[7]. In this set up, the escaping HD gas is made to interact with water vapour. The recovery of deuterium results in the isotopic exchange reaction

$$HD + H_2O\,(g) \rightleftharpoons H_2 + HDO\,(g)$$
$$K\,(=\alpha) = 3.78 \text{ at } 25\ ^\circ C$$

The details of the isotopic exchange process are presented in the next section. Walker[8] had set up a pilot plant for continuous operation based on electrolysis with isotopic exchange.

Electrolysis, as fractional distillation, may at best serve as a finishing process to follow some more economic alternative preliminary enrichment to about 10 per cent.

9.1.5 Chemical Equilibrium Exchange

As explained in Sections 3.1.1 and 7.3.2, there exists small differences in the chemical potentials and other related properties of isotopic molecules which make the equilibrium constant deviate slightly from unity. This circumstance permits a separation of isotopes by contacting countercurrently two different chemical substances containing the isotopic species. Numerous examples of this were cited in Sec. 7.3. The application of this method to the separation of heavy water employing the following isotope exchange reactions which are in commercial use is described below:

$$HD\,(g) + H_2O\,(g) \rightleftharpoons H_2\,(g) + HDO\,(g)\text{: } K = 3.78,\ 25\ ^\circ C$$

$$HDS\,(g) + H_2O\,(l) \rightleftharpoons H_2S\,(g) + HDO\,(l)\text{: } K = 3.22,\ 30\ ^\circ C$$

$$HD\,(g) + NH_3\,(l) \rightleftharpoons H_2\,(g) + NH_2D\,(l)\text{: } K = 3.60,\ 25\ ^\circ C$$

$$NH_2D \xrightarrow{+O} HDO$$

(a) The HD-H_2O gas phase exchange

The isotopic exchange between HD and water vapour does not take place except in the presence of a catalyst as finely divided platinum or nickel. Secondly, the temperature has to be high enough to ensure that all the water is in the vapour state, as liquid water inhibits the catalytic action. At the same time, the temperature cannot be so high as to be characterized by too low a separation factor, which falls with rise of temperature (Table 9.2). To meet these requirements, Barr[6] in 1945, designed a catalytic exchange tower described below.

Table 9.2: Variation with temperature of the separation factor (= equilibrium constant) for the exchange reaction HD-H_2O

Temp/°C	25	50	75	100	200	400	600
α	3.62	3.05	2.77	2.55	1.98	1.52	1.30

(b) *The Barr tower in the Canadian plant at Trail*

Murphy [6] has given an account of the large scale heavy water plant set up at Trail in Canada in 1942-54 employing the Barr towers. These consist of catalytic beds and bubble-cap absorption plates stacked alternately (Fig. 9.4).

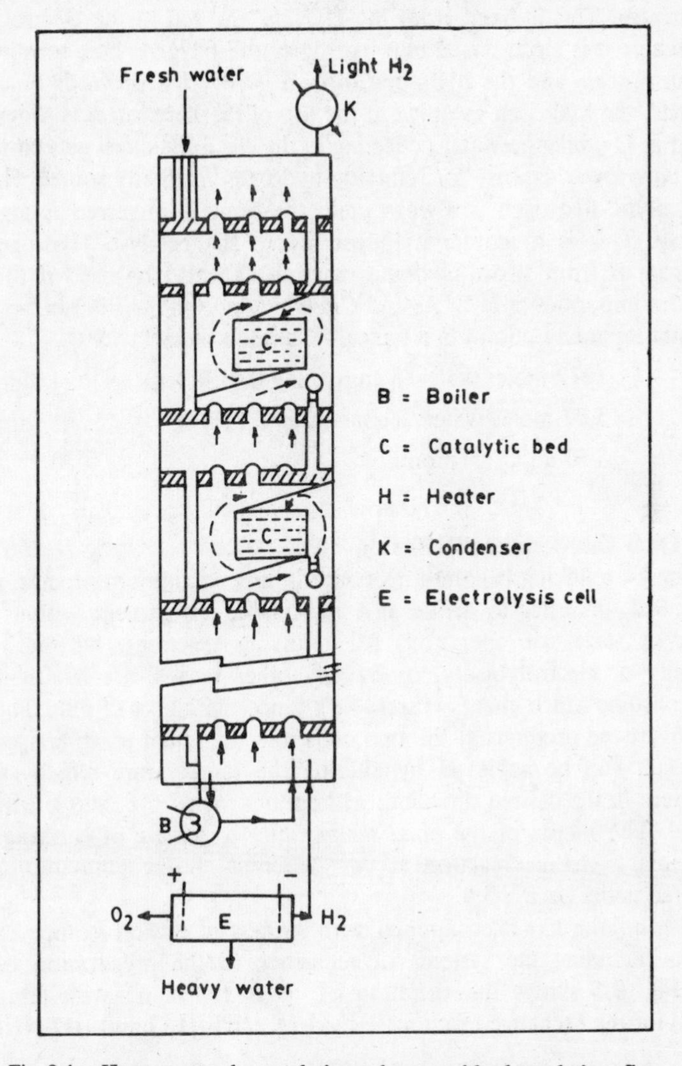

Fig. 9.4: **Heavy water by catalytic exchange with electrolytic reflux** (Reprinted with permission from *La Séparation des Isotopes* by M. Chemla and J. Périé p. 222 © Presses Universitaires de France, 1974)

Water fed at the top plate flows down bypassing the catalyst beds, but meeting the up-coming stream of HD-water vapour mixture. The latter mixture passes through successively in each stage a pair of absorption plates, a superheater which prevents the condensation of water or spray formation, and a catalyst bed which consists of finely divided platinum deposited on carbon, or of nickel on chromium trioxide. In passing through each catalyst bed, some of the deuterium is transferred from the HD to the water vapour by isotopic exchange. The HDO formed is condensed by the downcoming water stream. This is freed from the HD gas and led to the electrolysis cell below where it is electrolysed into hydrogen and oxygen. This constitutes the refluxing process and the hydrogen formed is recycled upwards once again. In the end, the hydrogen escaping at the top of the Barr tower is progressively depleted in D while the water collecting in the electrolytic cell gets enriched.

The Barr tower can use for refluxing hydrogen from any source. However, as electrolytic hydrogen is always pure, the same is preferred though more expensive. This is to confer a longer life to the catalyst. Hydrogen from steam-iron or from steam-methane reaction can also be used if the gas is freed from impurities of H_2S, As and CO which poison the catalyst.

The net input and output in a typical separation were:

Input : 1972 moles water/h containing 0.0138% D

Output : 5.07 moles water/h containing 2.14% D

Net output : 5.50 kg D_2O/month

9.1.6 **Dual Temperature Refluxing**

Recycling by refluxing becomes necessary in any equilibrium process, physical or chemical, in order to arrive at a reasonable enrichment within a given number of stages of operation. All refluxing reactions, whether effected chemically or electrolytically, or by any other means, are irreversible and hence consume much energy (Sect. 7.3). Since the object of refluxing is only to reconvert the products at the two ends into the initial reacting species, the same result can be achieved by altering the temperature which shifts the equilibrium in the desired direction, without destroying the isotope enrichment achieved. The merits of this *dual temperature technique* of carrying out an equilibrium exchange reaction at two different chosen temperatures, were considered under Sect. 7.3.5.

This technique has been applied with success in several isotope exchange reactions including the systems of relevance to the preparation of heavy water. Fig. 9.5 shows the variation of $\ln K$ ($= \ln \alpha$) with temperature (Eq. 7.8) for the exchange reactions HD-H_2O, HDS-H_2O and HD-NH_3

(a) *Dual temperature HD-H_2O gas phase exchange :*
The process of Harteck and Suess

The dual temperature process developed by Harteck and Suess[9], consists of a cascade of alternate hot (600°C) and cold (80°C) catalytic reactor columns. The feed water and the depleted gas enter the hot column (*H1*) of stage 1 (Fig. 9.6) where some of the deuterium from the water is transferred to *the gas* corresponding to the *slight shift* in the equilibrium to the left due to the lowering of the equilibrium constant from 2.8 at 80°C to 1.3 at 600°C. The reacted mixture is condensed and diverted along with the depleted water to the cold column of the preceding stage (or to waste if there is no lower stage). Rest of the depleted feed water and the enriched HD gas are heated and the resulting mixture admitted to the cold column

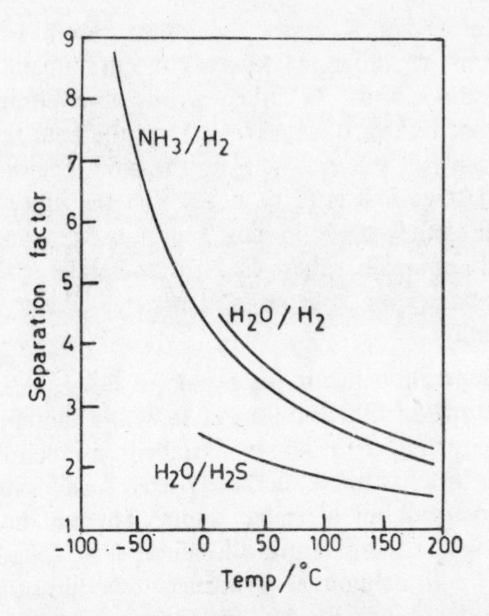

Fig. 9.5: Variation of separation factor with temperature for the exchange reactions HD-H_2O, HDS-H_2O and HD-NH_3.
(Reprinted with permission from *Isotope Separation* by S. Villani p. 351 © American Nuclear Society 1976)

(*C1*). Here the *water gets enriched* corresponding to the higher equilibrium constant at the lower temperature. The output of *C1* (enriched water) is again condensed and sent to the hot column (*H2*) of the next forward stage. The cycle of operations are repeated over several stages till water of the desired enrichment is obtained from the last cold stage.

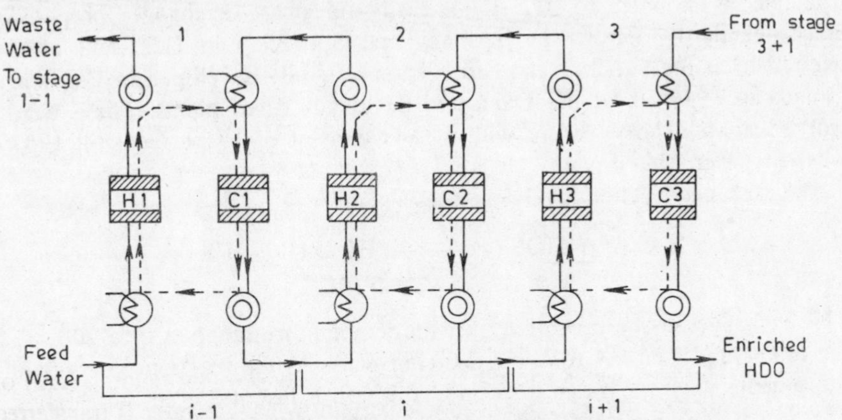

Fig. 9.6: Dual temperature HD-H_2O gas phase exchange:
The process of Harteck and Suess H: Hot tower; C: Cold tower; 1,2,3: Stages in series; O: Condenser; ⓢ Evaporator; —: Water; liquid or vapour; --→ Hydrogen
(from *Separation of Isotopes,* Ed. By H. London George Newnes Ltd., 1961, p. 173; reprinted by permission from William Heinemann Ltd.).

As in all cascade processes of isotope separation, care has to be taken that no mixing of streams of different degrees of enrichment occurs *i.e.* of different isotopic compositions. In other words, the composition of the downflow from the hot reactor of stage $(i + 1)$ to the cold reactor of the ith stage should be the same as the upflow from the cold reactor of stage $(i-1)$ to the hot reactor of stage i. It is to be noted that the amounts of gas (and water) are conserved as they move in closed circuits from hot (cold) to cold (hot) columns and back again. There is no irreversible refluxing, only *the result of enrichment passes on from stage to stage*, which is indeed a clever way of isotope separation.

The single stage separation factor for a pair of hot-cold columns is 1.48 for the dual temperatures of 600 and 80 °C. It would therefore need a large number of stages to arrive at a highly enriched sample of heavy water. Again, at every stage water is to be alternately heated and vapour condensed, which involves a large amount of energy input. Though the efficiency will improve by increasing the temperature difference, it is found that 80°C is about the limit of the cold column, as on further lowering of temperature the vapour tends to condense heavily, and as mentioned earlier, the catalyst cannot function in the presence of liquid water.

(b) *Dual temperature H_2S-H_2O gas-liquid phase exchange*:
Though efficient in principle, the dual temperature exchange as applied to separation of heavy water by the H_2-H_2O system, has not made a break-through, mainly because the hydrogenation catalyst used functions only with water in the vapour phase, necessitating the evaporation of large amounts of water at every stage and this last process is expensive. However, the same principle as applied to H_2S-H_2O gas-liquid phase exchange, proposed independently by Girdler in Germany and Spevack[10] in U.S., and often referred to as *the G-S process*, had been successfully developed in U.S. and Canada in 1947-50, at the Dana and Savannah River plants. These were worked to yield heavy water with a D_2O content of 15 per cent till 1957 when they were shut down*.

The exchange reaction involved is

$$H_2O\,(l) + HDS\,(g) \rightleftharpoons HDO\,(l) + H_2S\,(g)$$
$$K = 2.22\ (25\,°C)$$

This exchange has been studied over the temperature range of 0 to 200 °C, the K varying from 2.4 to about 1.6 (Fig. 9.5). As the reaction proceeds in the aqueous medium via an ionic mechanism, it is fast and needs no catalyst. This last is the important feature of the G-S method. The exchange occurs between the H^+_{aq} and D^+_{aq} ions.

* Around 1957, the American interest shifted more towards ^{235}U-enriched light water reactors and consequently the demand for heavy water fell.

$$H^+_{aq} + HDS \rightleftharpoons D^+_{aq} + H_2S$$

$$D^+_{aq} + OH^- \rightleftharpoons HDO$$

The optimum temperatures of the cold and hot towers are arrived at from a consideration of the basic H_2S-H_2O phase-equilibrium diagram (Fig. 9.7) studied by Selleck, Carmichael and Sage[11]. Since the G-S process involves isotopic exchange between liquid water and gaseous H_2S, the operations have to be confined to the region of the coexistence of these two phases (region 3, Fig. 9.7). The deposition of the solid H_2S hydrate has to be avoided. Hence the temperature of the cold tower has to be above the quadruple point which is 29° C. The cold tower is usually worked at 30° C. The main consideration for the hot tower is that the water remains in the liquid state. A value of 130° C and a pressure of around 20 atmospheres are chosen for the hot tower to be within reasonable limits of power consumption.

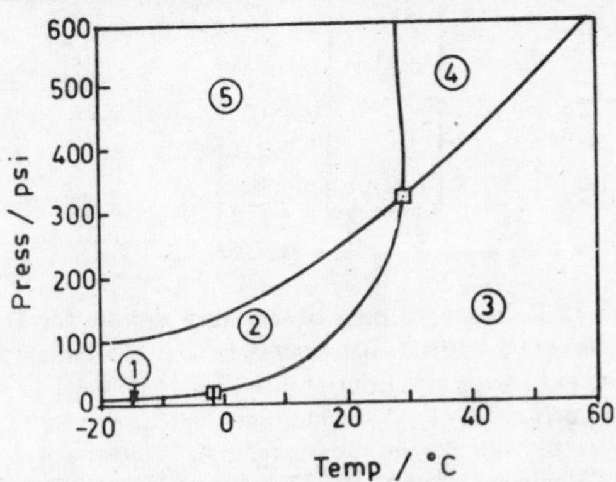

Fig. 9.7: **Phase-equilibrium diagram of the water**—H_2S system. Phases 1—Ice + gas; 2—Solid H_2S hydrate + gas; 3—Aqueous liquid + gas; 4—Aqueous liquid + H_2S liquid; 5—Solid H_2S hydrate + H_2S liquid; □ Quadrupule points (after F.J. Selleck, L.T. Carmichael and B.H. Sage, *Ind. Engng. Chem.*, 1952, *44*, 2219)

The Girdler-Spevack process

Figure 9.8 depicts schematically a single stage of the G-S process. Feed water from $(i - 1)$ stage is admitted at the top of the cold tower of the ith stage. It flows down meeting countercurrently H_2S gas from the bottom of the hot tower. A part of the enriched water arriving at the bottom of the cold tower is passed on to the next stage $(i + 1)$ for further enrichment, while the rest of the water enters the hot tower below and thereon to waste. The depleted H_2S gas leaving the top of the cold tower is heated and readmitted into the base of the hot tower. Since in the hot tower, the α $(= K_{130°})$ is around 1.83, the tendency is for inverse enrichment, *i.e. the gas gets relatively enriched*. This makes it well suited, after it is cooled to 30° C, to enter the cold tower as reflux, where the α $(= K_{30°})$ is around 2.17. Thus, by successive stages highly enriched water containing around 15 per cent D_2O is obtained which is suitable for use in a finishing process as distillation or electrolysis, whereby D_2O of above 99 per cent purity results.

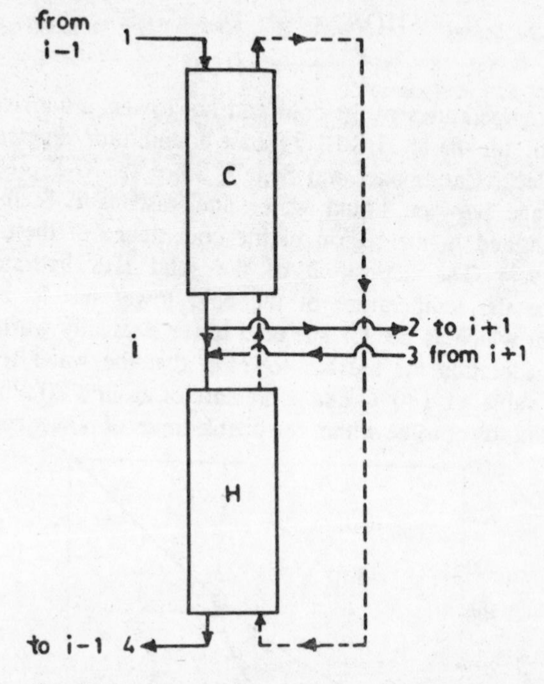

Fig. 9.8: **The Geib—Spevack dual temperature process for H_2O—H_2S exchange :** 1—Feed water from stage $i-1$ 2—Enriched water to sgate $i+1$ 3—Water from $i+1$; 4—Depleted water to $i-1$; H—Hot tower at 130° C; C—Cold tower at 30° C ---- water; –H_2 (Adopted with permission from *La Se'paration des Isotopes* by M. Chemla and J. Périé, P. 227 © Presses Universitaires de France, 1974)

A major item of expense is in providing the heat necessary for raising the temperature of the water and the gas entering the hot tower and for the humidification of the gas. Only a part of the heat is recoverable by heat exchangers before the streams are cooled to enter the cold tower. The provision of a humidifying section at the base of the hot tower and a dehumidifier at the bottom of the cold tower with arrangement for circulation of water round each of these adds to the efficiency by lowering the power consumption.

Details of calculations of the degree of extraction of deuterium from the feed water, the process of series-parallel coupling of stages, and relevant design equations are discussed by London[3].

Usually around 18 000 kg of water as steam are needed for a kg of D_2O, the power consumed being 800 kWh.

(c) *Hydrogen-ammonia gas-liquid phase exchange*

On account of the very low proportion of deuterium in natural hydrogen (0.015 atom per cent) and on account of poor degree of extraction by any process, some 10 000-20 000 times more hydrogen gas (or water) are thrown out for every part of D_2 collected. This makes no heavy water plant viable unless the vast amount of hydrogen, the main product, is usefully and economically utilised, and the heavy water is treated as a by-product. The synthetic ammonia industry would suggest itself in this context, since the demand for combined nitrogen is ever on the increase.

A second factor favouring the coupling of the heavy water production with the ammonia industry is the exceptionally high separation factor for the HD-NH_3 exchange:

$$HD(g) + NH_3(l) \rightleftharpoons H_2(g) + NH_2D(l)$$

$$\begin{aligned} K &= 8.5\ (-70\ ^\circ C) \\ &= 3.6\ (+25\ ^\circ C) \\ &= 2.8\ (+100\ ^\circ C) \end{aligned}$$

According to Bigeleisen and Perlman the variation of K ($\simeq \alpha$) for the HD-NH_3 exchange is given by

$$\log \alpha = \frac{237}{T} - 0.2428$$

for the temperature range -73 to $+227\ ^\circ C$.

The wide temperature range over which the separation factor ($\simeq K$) remains high (Fig. 9.5) makes the reaction suitable for dual temperature exchange as well. A third and an important factor in this context has been the discovery by Claeys, Dayton and Wilmarth[12a] that potassamide (NH_2K) functions as an efficient catalyst for the above exchange. Being soluble in liquid ammonia, it serves as a homogeneous catalyst, and the problem of separating the solid catalyst, as in the HD-H_2O vapour exchange, is obviated. The French Atomic Energy Commission were the first to build a heavy water plant based on HD-NH_3 exchange attached to a synthetic ammonia unit (the Mazingarbe plant). The successful functioning of the plant had enabled them to design more such plants for some other countries, including the one for Baroda, India[7]. As the catalyst is susceptible to decomposition by H_2O, O_2, CO and CO_2, the process gas has to be carefully purified so that the amounts of H_2O and O_2 are <1 ppm.

Dual temperature HD-NH_3 exchange

In view of the conveniently large separation factor of around 9 at $-70\ ^\circ C$ and around 3.2 at $+50\ ^\circ C$ (Fig. 9.5), the HD(gas)-NH_3 (liquid) exchange can be developed as a dual temperature process along the same lines as the G-S process for the H_2S (gas)-H_2O (liquid) exchange described in the preceding section. Liquid ammonia containing dissolved potassamide, the catalyst, is circulated in a closed cycle and this helps reaching the exchange equilibrium rapidly, without there being any necessity of catalyst recovery. In spite of these advantages there is a major problem associated with the efficient contacting of the gas and liquid phases at such low temperatures as -40 to $-70\ ^\circ C$. This difficulty makes normal distillation or exchange towers inefficient and powerful mechanical gas dispersers have to be developed to make the process a success. In the face of this problem, classical chemical refluxing is resorted to. The French plant at Mazingarbe based on these principles is described below.

The Mazingarbe plant

This consists principally of two towers: the first for isotopic exchange and the second for ammonia synthesis. The exchange tower is split into an upper and a lower half. Natural hydrogen and liquid ammonia enter the upper part of the exchange tower which also receives the catalyst potassamide dissolved in liquid ammonia (Fig. 9.9). Here most of the exchange takes place, the hydrogen being depleted of deuterium.

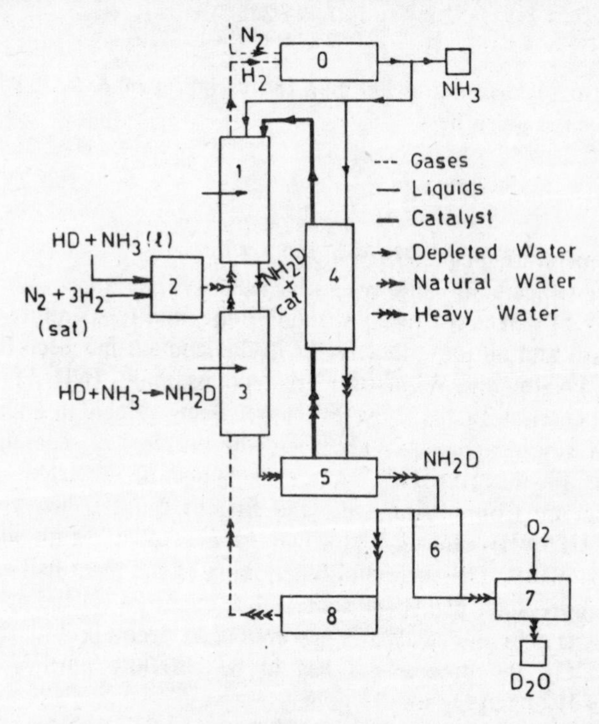

Fig. 9.9 : **Scheme of the Mazingarbe heavy water plant:** 0 — Ammonia synthesis; 1 — Depletion tower; 2— Saturation tower; 3 — Enrichment tower; 4 — Catalyst washer; 5 — Catalyst extraction; 6 — Extraction of ND_3; 7 & 8 — Cracking chamber
(Adopted with permission from *La Sèparation deș Isotopes* by M. Chemla and J. Périé, p 232 © Presses Universitaires de France, 1974)

$$HD(g) + NH_3(l) \rightleftharpoons H_2(g) + NH_2D(l)$$

The deuterium content of ammonia rises by a factor of about five. The hydrogen stripped of deuterium then enters the synthesis tower together with nitrogen and the main product ammonia is synthesised.

$$N_2 + 3H_2 \rightleftharpoons 2NH_3$$

The above constitutes a single stage. In practice, it is necessary to have a cascade of multiple stages. This necessitates chemical refluxing at the two ends. Of these, the synthesis of ammonia, occurring at the top end, constitutes one of the refluxing processes, while for the other some of the liquid ammonia is heated to cracking point so that it breaks back into nitrogen and enriched hydrogen

$$2NH_3(+D) \rightleftharpoons N_2 + 3H_2(+D)$$

This mixture of nitrogen and enriched hydrogen enters the bottom part of the lower half of the exchange tower and ascends up, meeting countercurrently liquid ammonia containing the catalyst, dripping from the upper half of the exchange tower. Here the isotope exchange is carried to near completion, yielding highly enriched ammonia with a 100-fold higher deuterium content compared to normal. Even so this is equivalent to only three per cent of deuterium.

The importance of the need for a powerful mechanical disperser was mentioned earlier. This is because gases, unlike liquids, offer a very low viscosity at such low temperatures*. In order that the isotope exchange may proceed to a maximum extent, even counting on the high separation factor available at the low temperature, it is absolutely necessary to ensure as large a surface of contact as possible between the gas and the liquid phases. This is realized in the Mazingarbe plant by a series of emulsifier-ejector units, one for each plate of the column. In this device, the hydrogen gas under pressure and liquid ammonia are dispersed into each other to form an emulsion of microfine particles of one in the other, approximating a homogeneous phase (Fig. 9.10). The isotope exchange proceeds with maximum efficiency over the greatly expanded surface of contact. This is followed by a mechanical separator which breaks the emulsion into pure gas and liquid phases. The hydrogen thus purged of the liquid particles and stripped of deuterium goes to the ammonia synthesis tower, while the liquid ammonia enriched in deuterium flows to the chamber where it is washed to have the dissolved catalyst extracted, which is used in the next cycle of operations as indicated in the figure.

The rest of the operations are straight-forward. The highly enriched liquid ammonia having around three per cent deuterium is fractionally distilled in a conventional way to obtain pure ND_3 of over 99.8 per cent purity. This last is finally burnt in oxygen and practically 100 per cent pure heavy water collected

$$3O_2 + 4ND_3 \longrightarrow 2N_2 + 6D_2O$$

* Gas viscosity is a function of the mean velocity of the gas molecules which in its turn is a function of $\sqrt{T}$.

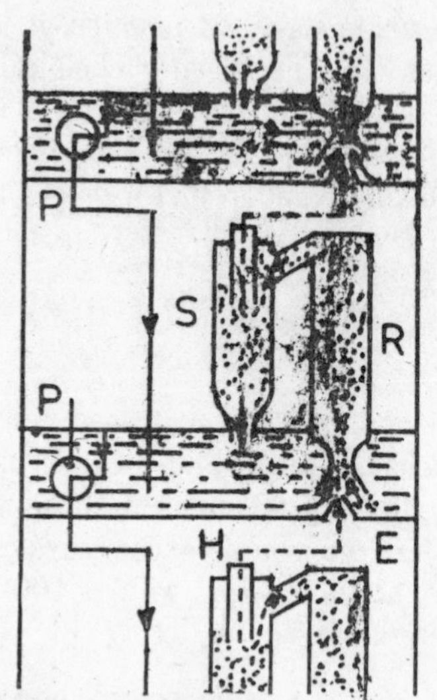

Fig.9.10 : **Emulsifier—ejection unit** : P — pump; S — separator; R — reaction chamber; E — ejector; H — high pressure gas.
(Reprinted with permission from *La Se'paration des Isotopes* by M. Chemla and J. Périé, P 233 © Presses Universitaires de France, (1974).

Methylamine-hydrogen bithermal exchange

Some of the major limitations of the ammonia-hydrogen exchange are:

(i) The exchange rate in the cold column at −25° C is very slow, which makes it necessary to employ a column of very large volume (1000 m^3),
(ii) The vapour pressure of ammonia at the temperature of the hot column (+ 60° C) is very high which also necessitates a column of very large volume, and
(iii) The hydrogen has to be under a very high pressure (~ 300 atm) to make it dissolve in liquid ammonia.

To avoid these difficulties the use of methylamine (CH_3NH_2) in place of ammonia has been suggested. On account of its lower vapour pressure, the cold and hot columns could be worked at −40° C and 70° C respectively, the catalyst remaining the same. Energy-wise, however, the amine-HD exchange is costlier as the following data show.

Table 9.3: HD exchange with NH_3 and CH_3NH_2

	Ammonia-HD	Methylamine-HD
Cold column/° C	−25	−40
Hot column/° C	+ 60	+ 70
pH_2/atm	300	65
Column/m^3	1000	600
D_2O yield/%	80	85
Power/kwh (kg D_2O)$^{-1}$	560	750

From *Isotope Separation* by S. Villani (American Nuclear Society, 1976.)

9.1.7 Deuterium Enrichment by Other Methods

Laser enrichment of deuterium has been recently reported by van der Leeden[13]. Laser induced pre-dissociation spectrum of formaldehyde showed one HDCO molecule to be dissociated for 3500 molecules of H_2CO which corresponds to a separation factor of nearly two.

The fractionation of isotopes of hydrogen by algae and by microbial action were described in Sec. 8.5.2 and 8.5.5. Most of these refer to a significantly faster metabolism of molecules with the H atom as compared to those with the D atom in the corresponding position. The enrichmemts being all on a micro level, these methods are not available for a large scale separation of deuterium.

9.2 LITHIUM ISOTOPES

Natural lithium consists of two stable isotopes ^{6}Li and ^{7}Li in the proportion of 7.5 and 92.5 atom per cent. Though their physical and chemical behaviour is nearly identical, displaying only feeble isotopic effects, their nuclear characteristics vary enormously. ^{6}Li is one of the only four stable nuclides in which both the proton and neutron numbers are odd*. What has made the isotope ^{6}Li so important in the atomic age is its very high cross section for capturing a neutron to form tritium by the reaction

$$^6\text{Li} + n \longrightarrow {}^4\text{He} + {}^3\text{T}$$

the product tritium being of paramount importance to thermonuclear reactions both for peaceful and military uses of nuclear energy. The cross section of ^{6}Li for the above reaction is 940 barns.

The isotope ^{7}Li, freed from ^{6}Li, is also of importance in another context. Because of its very low cross section for neutron capture (*viz.* 37 mb), it can serve in the molten state as an ideal heat exchanger in special nuclear reactors, as the fast breeder one, to transmit the heat from the reactor core to the outside to be used for steam generation needed for power production. The high thermal capacity of metal lithium is an additional advantage for this use**

These factors have made the separation of lithium isotopes a matter of strategic importance to the nations who consider themselves as nuclear powers. The result has been, as may be expected, most lithium compounds in the market have been secretly and illegally depleted of their ^{6}Li content.

We shall recapitulate briefly the different methods available for separating the isotopes of lithium, described in the text earlier. These are :

---

* The only stable nuclides having both the proton and neutron numbers odd are : ^{2}H, ^{6}Li, ^{10}B and ^{14}N.

** However, because of the high cost of the separated ^{7}Li, the natural monoisotopic sodium is preferred as a reactor coolant, despite the fact of its forming radioactive ^{24}Na by (n, γ) reaction. The molten metal sodium is all contained in a closed loop.

1. Chromatography,
2. Molecular distillation,
3. Ionic migration.

9.2.1 Chromatographic Separation

It was by ion exchange chromatography that Urey and Taylor[14] first effected a partial enrichment of the isotopes of lithium, besides those of potassium in 1938. Using a 30 m high column loaded with zeolite, they observed a significant depletion of ^{6}Li in the first fraction of the element, showing thereby that ^{6}Li is held up by the zeolite in preference to ^{7}Li. They concluded a value of 1.022 for the separation factor for the exchange

$$^7\mathrm{Li}_z^+ + {}^6\mathrm{Li}_{aq}^+ \rightleftharpoons {}^7\mathrm{Li}_{aq}^+ + {}^6\mathrm{Li}_z^+$$

Later workers did not confirm this high value.

It was not before another 12 years had lapsed that some progress may be said to have been achieved in the separation of isotopes of lithium. By then the theory of ion exchange chromatography and the use of organic ion exchanger resins had been well developed. Also the vital role of the isotope ^{6}Li in the new defence strategy began to be realized and the race for the separation of this and some other isotopes by chromatography was on. The first breakthrough in this was made by Glueckauf, Barker and Kitt[15] when they used an organic ion exchanger, zeokarb-HI, sulphonated coal designed to be equivalent to 1000 plates per cm. The equilibrium reaction is virtually irreversible.

$$\mathrm{LiAc}^{**} + \mathrm{HR} \underset{K=60\,000}{\rightleftharpoons} \mathrm{LiR} + \mathrm{HAc}$$

With K of such a magnitude, the lithium is all retained by the resin. When the column (LiR) was eluted, the first fraction (0-0.5 ml) of the eluant was found to be nearly wholly depleted of ^{6}Li, (Fig. 9.11). This gave for the equilibrium

$$^7\mathrm{Li}_r^+ + {}^6\mathrm{Li}_{aa}^+ \rightleftharpoons {}^7\mathrm{Li}_{aq}^+ + {}^6\mathrm{Li}_r^+$$

$$\alpha = \left[\frac{^7\mathrm{Li}^+}{^6\mathrm{Li}^+}\right]_{aa} \Bigg/ \left[\frac{^7\mathrm{Li}^+}{^6\mathrm{Li}^+}\right]_{resin} = 1.0025$$

This value of α was confirmed by other workers, using other sulphonated exchangers of a similar (Dowex-50) type.

By using ion exchanger Dowex-50 with increasing extents of cross linking with sulphonated divinyl benzene, Lee and Begun[16] showed that the separation factor can be boosted upto 1.003 8 with 24 per cent of cross linking.

---

* Here Ac stands for the acetate ion and R for the organic resin ion exchanger.

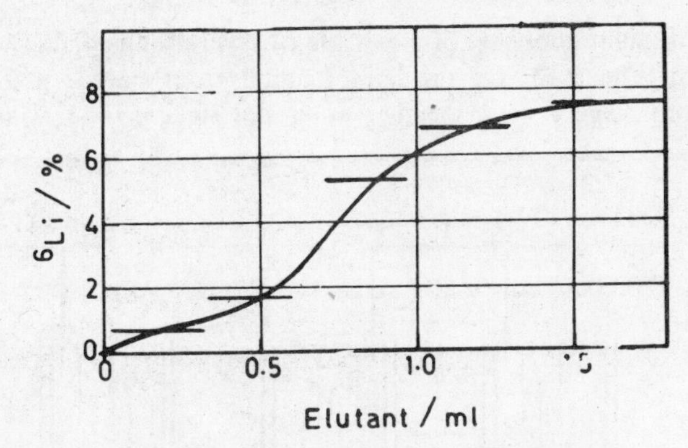

Fig. 9.11 : **Chromatographic separation of lithium isotopes** (after T.I. Taylor and H.C. Urey, *J. Chem. Phys.*, 1938, *6*, 429)

9.2.2 Molecular Distillation

The Brewer-Madorsky still for the separation of mercury isotopes by molecular distillation (Sec. 8.1.3) has been continuously improved to adopt the same for the separation of lithium isotopes. Lang *et al*[16] developed two types of stills : one a circular one with condensate receivers arranged in spiral stages and the other a large six stage one, each of 60 cm^2, which is described below. It consisted in the main of two plates: a stainless steel base on which metallic lithium was distilled between 500 and 600 °C and a condensation plate held 5 cm above it, the two plates being joined by convolution bellows, the space inside being under vacuum. The condensate was led from the ceiling by drainage channels 12 mm wide and spaced 5 cm apart into collection troughs suspended from the top plate at a level above that of the liquid metal on the basal pan, (Fig. 9.12). Measured distillation rates varied from 0.11 to 0.68 g h^{-1} cm^{-2}. Because of its low density and high surface tension, liquid lithium did not tend to form droplets, but collected simply in the receiving troughs.

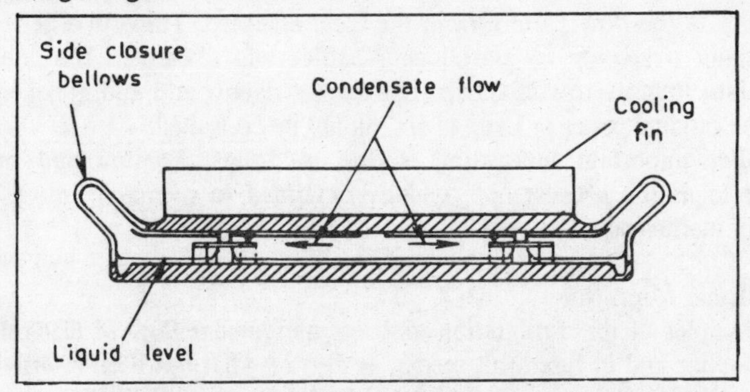

Fig. 9.12: **Lang's still for the molecular distillation of lithium isotopes** (after D.B. Trauger, J.J. Keyes, G.A. Kuipers and D.M. Lang, *Proc. Intern. Symp. on Isotope Separation*, North-Holland Publishing Co., 1958)

The entire plant consisted of a cascade of several units connected together as shown in Fig. 9.13, *i.e.* the light fraction from stage i and the heavy fraction from stage $i + 2$ together enter the stage $i + 1$. Under correct

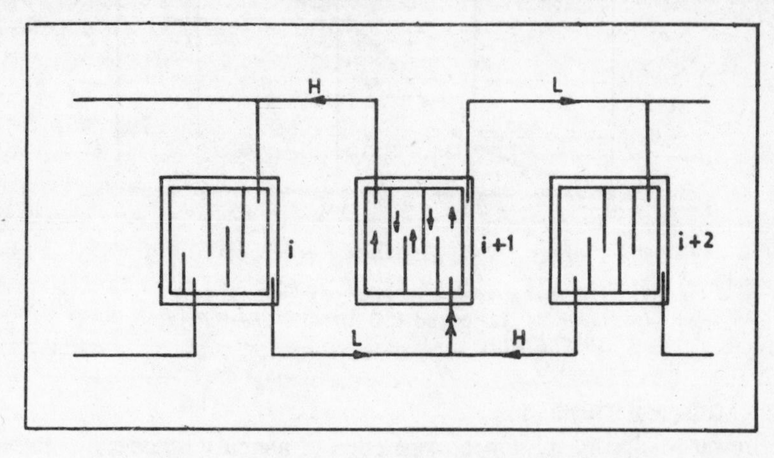

Fig. 9.13: **Molecular distillation in cascade : H — heavy fraction; L — light fraction**

working conditions, the compositions of these two streams should be the same. In a cascade where all the stages are of the same size, the enrichment factor per stage is enhanced by a factor of 2 ln 2 or 1.4, thus

$$\epsilon_{(effective)} = 1.4$$

Expressing the enrichment factor in terms of the separation factor and noting that the limiting value of the latter in the case of lithium isotopes is $\sqrt{7/6}$, we have

$$\epsilon_{(effective)} = 1.4\left(\sqrt{7/6} - 1\right) = 0.113$$

In practice, however, this high value is not reached, mainly because of some reevaporation of the light isotope from the condensate on the ceiling and deviations in the flow pattern from the ideal assumed. These two defects can be partially overcome by introducing baffles which convert the collection trough into long narrow channels. For further details and additional innovations, the original paper of Lang *et al*[16] should be consulted.

Another important innovation is due to Johns, London and others[16] referred to as the *process box*, which is claimed to overcome most of the causes of inefficiency.

9.2.3 Ionic Migration

The principles of ionic migration with countercurrent flow of electrolyte in both aqueous and in fused salt media, as also of what has been referred to as *simple ionic migration*[28] without counterflow, or zone migration, in the two media, were described in Chapter 8. The chief point in favour of the fused salt medium, as first pointed out by Klemm[17], is the absence of ion solvation resulting in larger mass effects in the migration of isotopic ions. Insofar as the separation of isotopes of lithium is concerned, results available are the contributions of two groups namely, of (*i*) Klemm and collaborators in Germany on ionic migration with countercurrent flow in fused salt media, and (*ii*) Chemla and collaborators in France on simple ionic migration, or zone migration, without countercurrent flow in both aqueous (agar gel) and in fused salt media.*

The principal results of these studies are briefly reported here without repeating the details presented earlier (§ 8.3.7, 8 and 9).

9.2.3.1 *Countercurrent migration in fused salt medium*

Under a programme spread well over a decade (1947-1958), Klemm and collaborators worked on the separation of isotopes of some 15 elements including lithium by countercurrent electromigration of their ions in a medium of molten electrolyte. On the basis of these extensive studies Klemm was able to arrive at a new concept *viz.* relative mass effect (§ 8.3.3),

$$-\mu = \frac{\Delta v/v}{\Delta m/m}$$

which he later reinterpreted in terms of the frequencies of two types of jumps the ion makes in electromigration *viz.* the spontaneous ones and those induced by the other ions of the medium (Eq. 8.14 *a*). This important concept was later found acceptable as a generalization valid in electromigration both in aqueous and fused salt media.

Lunden's cell

The cells earlier used by Klemm, Hintenberger and Hoernes[17] for the enrichment of ^{41}K and ^{7}Li isotopes by ionic migration in fused LiCl (KCl), was modified by Lunden[18,19] to include the refluxing of lithium forming at the cathode by bromine vapour. The cell used by Lunden is sketched in Fig. 9.14. The essential part of the cell consists of a long tube of Supremax glass the middle part of which holds fused LiBr into which dip carbon rods serving as cathode around which a flow of bromine vapour is maintained. The lower half of the tube is packed with properly rounded glass granules and frits ending in a porous diaphragm at the bottom. A coaxial tube surrounding the LiBr and the packed column contain fused $PbBr_2$ whose top end forms the anode compartment holding arc lamp carbon rod anode. The entire cell is held in a vertical furnace at the controlled temperature of 500 °C.

The lithium discharging at the cathode is immediately reconverted into LiBr by the bromine present there and refluxed back into the cell. This refluxing of the lithium not only provides countercurrent flow but it ensures that no metallic lithium is allowed to form in the cell as it would attack and destroy the glass cell at that temperature (above 400 °C). The boundary

*Subsequently Chemla too adopted countercurrent electromigration.

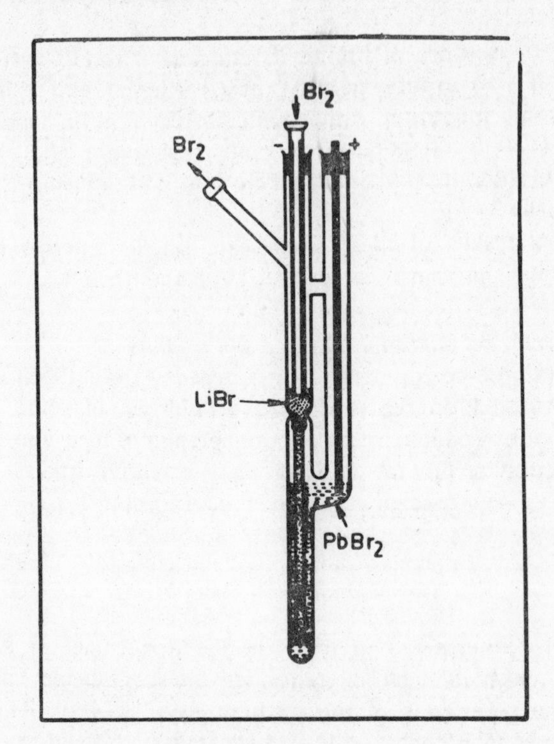

Fig. 9.14: **Lunden's cell for the separation of lithium isotopes by countercurrent electromigration in fused lithium bromide**[18,19] (after A. Lunden, *Z. Naturf.*, 1956, *11 a*, 590)

formed between the molten LiBr and $PbBr_2$ phases serves to monitor the countercurrent flow as a self-regulatory device. Also the hold up of enriched ^{7}Li in this region is reduced which enables the completion of the process in a shorter period. The process yielded highly enriched ^{7}Li at the $PbBr_2$ anodic border corresponding to a mass effect $(-\mu)$ of 148×10^{-3}, the highest value reported in the separation of lithium isotopes.

The method of countercurrent electromigration in fused salts involves several difficult and delicate techniques, such as filling the cell with the molten electrolytes in vacuum, maintaining the proper bromine pressure for refluxing, and avoiding the formation of air bubbles. In any case, the cell life is very limited. Sometimes it can be used only once.

Klemm's Plant for ^{7}Li

Overcoming the difficulties, Klemm[20] succeeded in setting up a commercial plant for producing ^{7}Li by countercurrent electromigration in fused LiCl. The unit consists of several refractory tubes each 22.8 cm long packed with high temperature non-corrosive material and closed at the bottom with a porous diaphragm. The tubes are held in a bath of fused LiCl. The anodes are carbon rods in contact with the upper end of the cell (Fig. 9.15).

Chloride ions migrate from the cathode to the anode, while gaseous chlorine returns to the cathode which refluxes the lithium collecting there

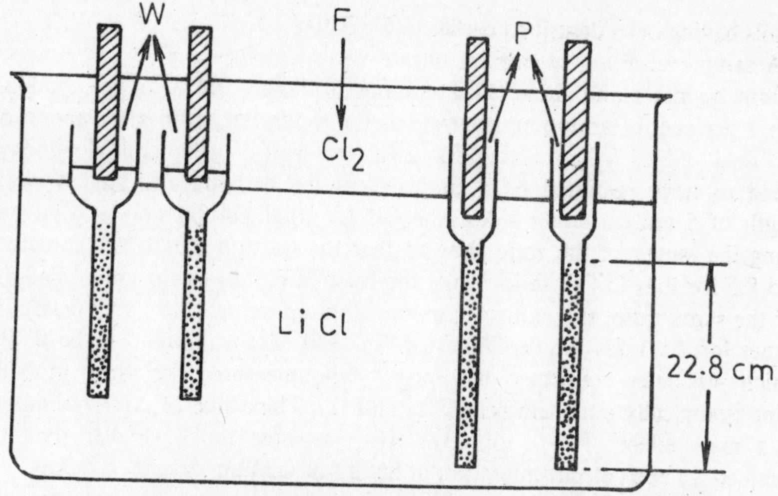

Fig. 9.15 : **Klemm's mini—plant for ^{7}Li from molten LiCl by counter-current electromigration.** F—feed; P— product; W — waste (after A. Klemm, *Proc. Intern. Symp. on Isotope Separation.* North Holland Publishing Co., 1958; and *Proc. Intern. Symp. on Molten Salt Chemistry and Technology, Kyoto, 1883)*

back into the ionic form. Relative to the Cl^- ions, $^7Li^+$ migrates 2.2 per cent slower than $^6Li^+$ ions and thereby ^{7}Li gets enriched around the anode. Under a direct current of 9 A/cm^2, the plant is stated to operate with a separation factor of 300, yielding 99.974 per cent pure ^{7}Li. It is claimed that the separation factor with this "mini-plant" is same as with any other large sized plant[-21]. The power consumed is stated to be 11 kWh/g of lithium.

Recently, Okado[22] has modified the plant to enrich ^{6}Li by including molten ammonium nitrate (m.p. 170°C) on the cathode side, as the mobility of the NH_4^+ ion is greater than that of the Li^+ ion. The temperature of the column and the cathode compartment are maintained at 300 and 180°C respectively. The decomposition products of NH_4NO_3, prevent the electro-deposition of metal lithium, and the Li^+ ions are thus constantly regenerated and flushed back, thus maintaining countercurrent electromigration.

9.2.3.2 *Simple Zone Migration in Agar Gel*

As already pointed out (§ 8.3.6), the Chemla-Arnikar method of simple ionic migration without countercurrent flow has the merit of simplicity and quickness, providing at the same time data in respect of relative isotopic ion mobilities and the mass effect of the same precision as obtained in the more elaborate countercurrent electromigration. However, the amount of the substance that can be treated in any one experiment is only a small fraction of a gram, usually of the order of 10-20 mg, and the enriched samples attainable are in fractions of this. Obviously, the method of zone migration, as it is, is not capable of being scaled up beyond, probably an order of magnitude, and hence is unsuited for large scale production. We summarize here the results observed in the separation of lithium isotopes by this method. The procedural details having been described earlier[23] (§ 8.3.10).

A sample of 5 mg of lithium nitrate was localized in a 10 mm zone in a 10 mm diameter and metre long column of 1 per cent agar gel containing also 1 per cent of ammonium nitrate as the electrolyte. Under the action of a d.c. field of 1.5 V/cm acting for 24 h, the centre of the initial zone was found to have migrated 67.5 cm towards the cathode and spread over a length of 5 cm on either side. Analysis for total lithium and isotopic ratio along the length of the zone showed that the fraction closest to the cathode had 8.5 for the $^7Li/^6Li$ ratio, while the fraction closest to the anode had 15.0 for the same ratio, the value for natural lithium being 11.5 (Fig. 9.16). The values for $\Delta v/v$ for the mobilities of $^6Li^+$ and $^7Li^+$ was found to be 0.0036, which was later confirmed by Fuoss[24] who measured the same mobilities using isotopically pure samples of ^6Li and ^7Li. This value of $\Delta v/v$ corresponds to a mass effect of $-\mu^+$ of 23×10^{-3}, a value much smaller than that obtained for zone electromigration in fused salt medium described below.

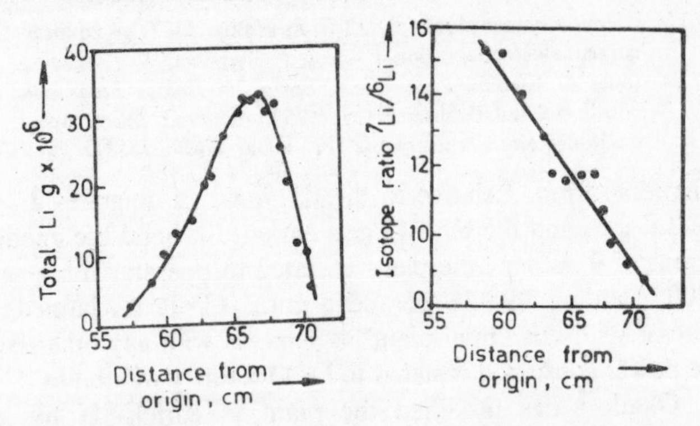

Fig. 9.16 : Separation of Li isotopes by electromigration in agar gel medium (Arnikar[23])

9.2.3.3 *Simple Zone Migration in Fused Alkali Nitrate Medium*

So far as isotope separations are concerned, the merits of the fused salt medium over the aqueous hold good for the technique of simple ionic migration as for the elaborate migration against countercurrent flow of electrolyte. Similarly, the limitations of the simpler technique apply equally to both media.

The procedure adopted by us[25,26] for the enrichment of ^6Li isotope by electromigration in fused alkali nitrate spread on an asbestos strip was the same as described earlier for the separation of the isotopes of potassium and rubidium (§ 8.3.10). At the end of 4 hours of electromigration under constant field, the current varying between 250 to 325 mA, the strip was withdrawn from the furnace and cut up into 5 mm long segments. The total lithium content and the isotopic ratio in each of these segments were determined by flame photometry and mass spectrometry respectively. The results for electromigration at 250 °C are shown in Fig. 9.17.

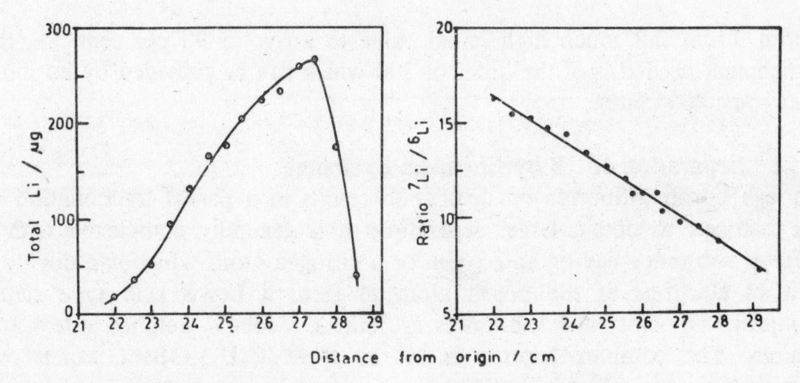

Fig. 9.17 : Separation of Li isotopes by electromigration in fused (Na+K) NO_3 (Anikar and Chemla[26])

Results show that the lithium isotopes had been progressively fractionated during electromigration, such that the fractions closest to the cathode and anode had 7.2 and 16.20 for the ratio of $^7Li/^6Li$, the value for the natural element being 11.5. The value for the ratio of ionic mobilities expressed as $\Delta v/v$ is 0.014 implying a separation factor of 1.41. The magnitude of the mass effect comes out to be $-\mu^+ = 89 \times 10^{-3}$ while the value obtained in aqueous agar medium was 23×10^{-3}. If the Li^+ ions be assumed to be monohydrated in aqueous solution the relative mass difference for $(^6LiH_2O)^+$ and $(^7LiH_2O)^+$ would be 1/24.5 (= 0.041) as against 1/6.5 (= 0.154) the corresponding value for the "dry" ions in the fused state. These values are in fair agreement with the four fold higher mass effect observed for electromigration of these ions in the fused medium compared to in aqueous medium, assuming a nearly constant value for $\Delta v/v$ for the two media.

9.3 BORON ISOTOPES

Natural boron consists of 20 per cent of ^{10}B and 80 per cent of ^{11}B. The importance of this element in the atomic age is due to its use as a control rod to regulate the neutron flux in a reactor within preset limits for the steady operation of the same. This becomes possible on account of the high neutron capture cross section that ^{10}B offers, *viz.* 3836 b, as against 5 mb of ^{11}B. The only other nuclide rivalling this is ^{113}Cd with a cross section of 19 910 b, and abundance of 12.4 per cent. The detector most frequently used for measuring slow neutron fluxes consists of a proportional counter, or an ionization chamber, filled with BF_3 gas enriched in ^{10}B. The overall reactions with neutrons are represented by

$$^{10}B\,(n, \alpha)\,^7Li \longrightarrow \alpha + T, \quad \text{or} \quad ^{10}B\,(n, 2\alpha)\,T$$

which show that ^{10}B may compete with the more rare ^6Li as a starting material for obtaining tritium on a large scale. This constitutes an additional demand on the ^{10}B isotope, besides its use as reactor control rod.

A special advantage with ^{10}B is its high abundance of 20 per cent in natural boron. From this much high initial value to arrive at 98 per cent ^{10}B, the enrichment needed is of the order of 200 which can be provided by no more than some 400 plates.

9.3.1 Separation of ^{10}B by Chemical Exchange

Though boron trifluoride on distillation results in a partial fractionation of the isotopes, to obtain larger separations it is generally complexed with a suitable substance having an oxygen or a nitrogen atom which can donate a pair of electrons to the boron atom to form a Lewis acid type stable complex. Amongst such substances are ethers, alcohols, ketones, esters and amines. The complex formed is of the type $(C_2H_5)_2OBF_3$, which can participate in an exchange equilibrium

$$^{10}BF_3(g) + {}^{11}BF_3O(C_2H_5)_2(l) \rightleftharpoons {}^{11}BF_3(g) + {}^{10}BF_3O(C_2H_5)_2(l)$$

$$K = 1.034\ (20\ °C)$$

The ^{10}B tends to concentrate in the liquid phase. Table 9.4 lists the equilibrium constants of some exchange reactions involving boron trifluoride.

Table 9.4: Equilibrium constants for exchange reactions with BF_3

Complexing agent	Temp/°C	*K*	Ref
Methyl ether	100	1.027	27
Ethyl ether	20	1.034	28
	60	1.020	27
	70	1.026	28
Butyl ether	20	1.029	29
Anisol (Phenyl methyl ehter)	20	1.013	30
	25	1.032	31
Phenol	25	1.027	31
Water	20	1.025	29

If reflux can be obtained at both ends, these would be efficient processes for the separation of ^{10}B. While at the top end of the column this is possible via the spontaneous combination,

$$BF_3 + (C_2H_5)_2O \rightarrow (C_2H_5)_2OBF_3$$

it is not easy to break the complex to provide the reflux needed at the bottom end, what is desirable is the dissociation involving the fission of the B—O bond and not the decomposition by the rupture of the B—F bond:

(a) *The use of methyl and ethyl etherates*
The methyl and ethyl ethers fulfil this requirement. Two plants using these exchange reactions have been set up, one in Britain and the other in U.S. around 1957.

The British plant is based on the BF_3-ethyl ethyrate exchange. A 5 m high column packed with stainless steel Dixon rings (§ 7.1.5) is operated at 75 °C. A plate height of 1 cm obtains providing a separation factor of 1.026. The vapour pressures employed are 20 and 50 mm at the top and bottom of the column. The yield is 300 g/year of 98 per cent ^{10}B. Later, the plant was adopted to work with methyl ester complex with wider diameter tubes when the yield rose to kg quantities per year of 96 per cent pure ^{10}B.

The American plant was a much larger one, using Monel metal columns of total length of about 100 m with a more sophisticated packing. The gauze used was dipped in a slurry of nickel particles of 200 mesh, dried and sintered at 950° C. This resulted in a self-wetting packing not requiring preflooding. The annual yield was of the order of 3 tons of ^{10}B of 95 per cent purity.

The decomposition of the complex providing the reflux at the bottom end is accompanied by side products as methyl borate and methyl fluoride

$$3(CH_3)_2OBF_3 \rightleftharpoons (CH_3)_3B\ 2BF_3 + 3CH_3F$$

This reduces the reflux effect but this side reaction can be considerably suppressed by maintaining an excess of methyl fluoride. Though the presence of water has a positive catalytic effect, the hydrofluoroboric acid, one of the decomposition products, has a severe corrosive action.

$$(CH_3)_2O.BF_3 + 2H_2O \rightleftharpoons 3HF.HBO_2 + (CH_3)_2O$$

(b) *The use of anisole etharate*
The special advantage of complexing BF_3 with anisole (phenyl methyl ether) is that the complex is sufficiently stable at its melting point (−37.8°C) but is fully dissociated at a temperature well below its boiling point (155°C). This permits the normal bottom reflux to be effected by simply heating the complex to drive off the BF_3 at atmospheric pressure

$$C_6H_5.OCH_3.BF_3 \rightleftharpoons BF_3 + C_6H_5O.CH_3$$

$$K = 1.013\ (20°C)^{3,30}$$
$$= 1.032\ (25°C)^{3,31}$$

These greatly simplifying factors are offset by some operational difficulties due to a tendency for the deposition of solid impurities during above dissociation. These solid particles can however be washed out by water and the anisole rendered free from them, provided the same is thoroughly dried before recycling, the upper limit of water permissible is stated to be below 100 ppm. Till the early sixties the process was mostly in an experimental stage of development.

9.3.2 Separation of ^{10}B by Laser Irradiation of Boron Trichloride

There is an overlap between the emission of a CO_2 laser and the absorption of BCl_3 vapour in the region around 10.6 μm. Laser-specific product formation is possible availing of the significant isotope shift in the absorption of $^{10}BCl_3$ and $^{11}BCl_3$ in this region. The absorption spectrum of natural boron tricholoride has two sharp peaks around the frequencies of 956 and 995 cm^{-1} for the two isotopic molecules $^{11}BCl_3$ and $^{10}BCl_3$ respectively. Fig. 9.18 also shows the relative proportions of the two isotopic molecules remaining after natural BCl_3 is irradiated by the two laser frequencies separately. Laser irradiation at the frequency of one of the selected isotopic molecules excites an asymmetric stretching vibration in that species alone and the same suffers decomposition in the presence of excess of oxygen (or air) and this can be separated chemically from the other unaffected isotopic species. Soviet scientists have described details of separating ^{10}B isotope by the laser irradiation of BCl_3 vapour in the presence of 25 times excess air[32,33]

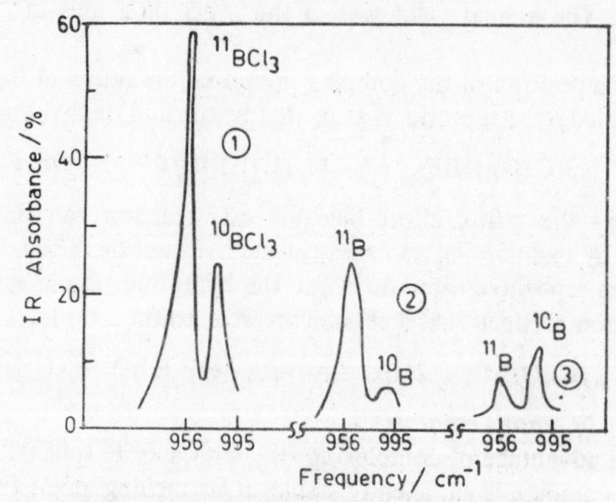

Fig. 9.18 : **Separation of boron isotopes by laser irradiation**

9.4 TITANIUM ISOTOPES

The separation of ^{50}Ti was described under § 8.4.9.

References

1. H.C. Urey, F.G. Brichwedde and G.M. Murphy, *Phys. Rev.*, 1932, *40*, 1.
2. J. Bigeleisen and E.C. Kerr, *J. Chem. Phys.* 1955, *23*, 2442.
3. H. London, *Separation of Isotopes*, George Newnes Ltd., London, 1961.
4. K. Clusius and K. Starke, *Z. Naturf.*, 1939, *4A*, 549.
5. D.C. Gani, D. Gupta, N.B. Prasad and K.C. Sharma, II UN Conf. on Peaceful Uses of Atomic Energy.
6. G.M. Murphy, *Production of Heavy Water*, McGraw-Hill Book Co., New York, 1955.

7. M. Chemla and J. Perié, *La Séparation des Isotopes*, Presses Universitaires de France 1974.
8. P.T. Walker, *AERE Report*, 1860 (1956).
9. Harteck and Suess, (See Ref.3).
10. Geib and J. Spevack (see Ref. 3).
11. F.J. Selleck, L.T. Carmichael and B.H. Sage, *Ind. Engng. Chem.*, 1952, *44*, 2219.
12. S. Villani, *Isotope Separation* (American Nuclear Society, 1976).
12(*a*) Y. Claeys, J. Dayton and W.K. Wilmarth, *J. Chem. Phys.*, 1950, *18*, 759.
13. J.C. van der Leeden, *Laser Focus*, 1977.
14. T.I. Taylor and H.C. Urey, *J. Chem. Phys.*, 1938, *6*, 429.
15. E. Gluekauff, K.H. Barker, and G.P. Kitt, *Faraday Soc. Disc.*, 1947, *7*, 199.
16. D.B. Trauger, J.J. Keyes, G.A. Kuipers and D.M. Lang., *Proc. on Internat. Symp. on Isotop Separation*, N. Holland Publ. Co. 1958.
17. A. Klemm, H. Hintenberger and P. Hoernes, *Z. Naturf. 1947, 2 a*, 245.
18. A. Lunden, *Thesis*, Univ. Goetberg, 1956.
19. A. Lunden, *Z. Naturf.*, 1956, *11 a*, 590.
20. A. Klemm, *Proc. Intern. Symp. on Isotopes*, 1957. 1958.
21. A. Klemm, *Proc. Intern. Symp. on Molten Salt Chemistry & Technology*, Kyoto, 1983.
22. I. Okado, *et al. ibid.*
23. H.J. Arnikar, *J. Inorg. and Nucl. Chem.*, 1959, *11*, 249.
24. R. Fuoss, (*private communication*).
25. H.J. Arnikar Thése, University of Paris 1958. *Annal de Phys*, 1959, *13* (4), 1291.
26. H.J. Arnikar, and M. Chemla, *Radioisotopes in Scientific Research*, Vol. 2. (Proc. UNESCO Conf. on Radioisotopes, Paris, 1958), Pergamon Press, London, 1958, p.421.
27. I. Kirshenbaum, N. Sabi and P.W. Schutz, *The Separation of Boron Isotopes*, Nuclear Energy Series, Vol. III-5.
28. R.W. McIlroy and F.C.W. Pummery, *Proc. Intern. Symp. on Isotope Separation*, (Amsterdam, 1957), Ch. 11.
29. S.V. Ribnikar, *ibid.*, Ch.14.
30. G.M. Panchenkov, V.D. Moiseyevc and A.V. Makarov, *Proc. Acad. Sci., USSR*, 1957, *112*, 659.
31. R.M. Healy and A.A. Palko, *J. Chem. Phys.*, 1958, *28*, 211.
32. R.V. Ambartzumian and V.S. Latokhov, *Laser Focus*, 1975.
33. R.V. Ambartzumian, N.V. Chakalin, Yu. A. Gorokhov, V.L. Letokhov and G. N Makaro, *Soviet J, Quant, Electronics*, 1975, *5*, 1196.

Chapter 10

REACTOR FUEL ISOTOPES

The heaviest element occurring in nature, uranium, had been looked upon with interest to know if any element of higher mass or atomic number could be prepared by making it capture some nucleons. Apart from this and apart from its natural radioactivity, there was little further interest in uranium. The only use of uranium salts before 1940 was in the colouration of glass and ceramics. The chemistry of the metal and its compounds were known to be consistent with the place allotted to it earlier in the Periodic Table, namely in Group VIA as the analogue of tungsten, and subsequently as the 5*f* analogue of neodymium. Interest in the element suddenly mounted up in 1939 when it was found to undergo nuclear fission, an altogether new phenomenon, on capturing a neutron and releasing in the process an amount of energy some 50 to 100 times more than in most nuclear reactions known till then. Till some years before that, even the density and the melting point of the metal were not known with precision[1]. It is now known to be the heaviest metal with a density of 19.04 g/cm^3 and a melting point of 1132°C. The metal is obtained by the reduction of UF_4 by Mg

$$UF_4 + 2Mg \rightarrow U + 2MgF_2$$

When Nier[2] perfected an improved version of the mass spectrometer involving a two stage acceleration of the ions (§ 5.2.2), a mass resolution ($M/\Delta M$) of 10^6 became possible and Nier separated the first few micrograms of the isotope ^{235}U in 1940. When Nier further found that it was this rare isotope of uranium, occurring to 0.72 per cent in the natural element, that is responsible for the fission observed in the element due to the capture of a slow neutron, a momentus discovery may be said to have been made, whose full implications were hardly realized by any one at that time. With America entering the War, events moved very fast and the race for harnessing atomic energy and the bomb project received an additional dimension. The separation of ^{235}U on a large scale became a vital necessity to win the War. Almost any method suggested, for its separation was accepted for detailed examination, under the cryptic code name of the *Manhattan Project* totally unmindful of the costs involved. The secrecy was absolute and it was only after first three atom bombs (two using ^{235}U and one ^{239}Pu), had been exploded between July and August 1945 that some data came to be declassified and appeared as the famous Smyth's Report [1]*.

Since then remarkable progress has been made in harnessing atomic energy for peaceful uses through nuclear reactors. Unfortunately it has also resulted in stock piling of atom bombs and other nuclear weapons on a scale enough to destroy the entire world population many times over the moment the delicate divine capacity for right thinking ceases (may it never happen !). While very large amounts (scores of tonnes) of uranium lightly enriched in ^{235}U (3-4 per cent) are needed for fuelling power reactors of the swimming pool type, employing light water as the moderator, much smaller amounts but very heavily enriched in ^{235}U (90-95 per cent) are needed for weapon-grade uranium (*i.e.* for making atom bombs).

10.1 CHARACTERISTICS OF REACTOR FUEL ISOTOPES

Natural uranium consists of three isotopes ^{234}U, ^{235}U and ^{238}U. All are radioactive and decay by alpha emission with long half-lives. Their characteristics and data relevant to their use as a reactor fuel are summarized in Table 10.1. Included in the table are also the plutonium isotope ^{239}Pu, which is used as a fuel in certain special reactors, as the fast breeder, and the isotope ^{233}U, which has a potential as a reactor fuel when natural thorium comes to be considered as a suitable substitute. Neither ^{233}U nor ^{239}Pu occur in nature. They are wholly man-made. We shall consider here the saparation of the isotopes ^{235}U. ^{239}Pu and ^{233}U.

Table 10.1: Characteristics of Reactor Fuel Isotopes[3-6]

	^{233}U	^{234}U	^{235}U	^{238}U	^{239}Pu
Abundance	0	0.00 55	0.72	99.28	0
α decay half-life/y	1.59×10^5	2.44×10^5	7.04×10^8	4.47×10^9	2.44×10^4
Total n capture σ/b	47.7	100.2	98.6	2.7	268.8
Fission n capture σ/b	531.5	0.65	582.2		742.5
Activation energy for fission by thermal neutrons/MeV	4.6		5.3	5.5	4.0
Excitation energy following thermal neutron capture/MeV	6.6		6.4	4.75	6.4

10.1.1 Natural Uranium

Of all the 274 stable and some 20 radionuclides present on the earth, it is strange, but fortunately so, that only one *viz.* the nuclide ^{235}U, with an

* One of the US army generals was stated to have been in great distress at this publication and wanted it to be withdrawn, but some one pointed out that "the egg cannot be put back into the hen".

isotopic abundance of less than one per cent should be capable of fissioning on capturing a slow neutron and release in the process on the average 2.3 fresh neutrons besides a large amount of energy. The excess 1.3 neutrons (over the one used) can cause similar fissions in other ^{235}U nuclei and thus set up chain reactions with anticipatable consequences. The other and more abundant isotope ^{238}U also plays a key role in sustaining these chain reactions in a reactor and in governing the neutron multiplication factor of critical importance. It is therefore proper that we consider the processing of natural uranium, from its ores to the metal or oxide form, before taking up the separation of the isotope ^{235}U.

10.1.2 Uranium—from Ore to Metal

The chief ores of uranium, with the weight per cent of the element in parentheses, are

Pitchblende or Uraninite (U_3O_8) (40-50)
Carnotite ($K_2O.U_2O_3.2VO_4.6H_2O$) (10).

The ore is first treated to rid it of the gangue by differential leaching and froth floatation process. This consists in adding an oil-water mixture to the ore and bubbling an inert gas when the mineral fraction floats up with the oil layer and the earthy matter settles down in the water layer.

The ore concentrate thus obtained is subjected to following analytical operations.

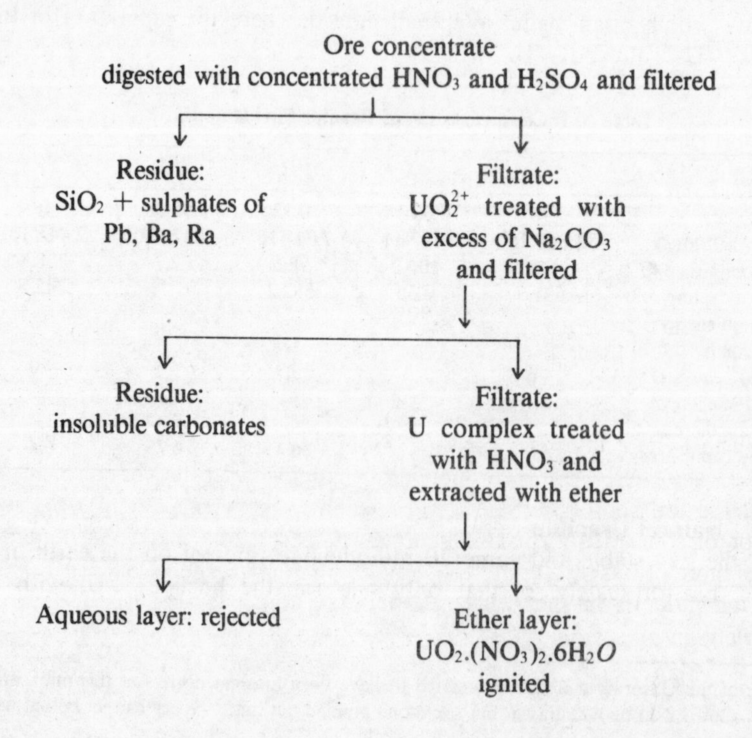

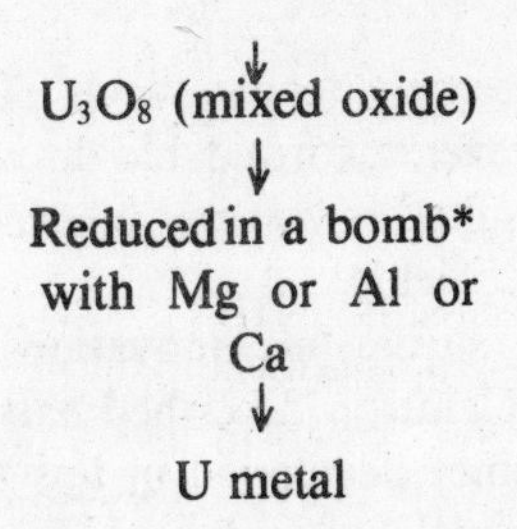

Alternatively, the mixed oxide is reduced by hydrogen when UO_2 results. This last treatment with hot anhydrous HF (or HCl) yields UF_4 (or UCl_4), which is used in one or the other of following ways:

a. $UF_4 + F_2 \rightarrow UF_6$ used for isotope separation.

b. $UF_4 \xrightarrow{\text{Na or Ca}}$ U metal : used in slow neutron reactor as a source of neutrons.

c. $UF_4 + 2KF \rightarrow K_2UF_6$: on electrolysis yields very pure metal U.

10.2 SEPARATION OF ISOTOPE ^{235}U

In chapters 6-9 wherein different methods of isotope separations were described, several references had been made to the separation of the isotopes of uranium. The more important of the methods of which some have been adopted for a large scale production of ^{235}U are:

1. Electromagnetic separation,
2. Gaseous diffusion across a membrane,
3. Nozzle separation,
4. Ultracentrifugation,
5. Thermal diffusion, and
6. Laser induced separation.

A summary of the results achieved by these methods is presented below.

10.2.1 Electromagnetic Separators

In fact, the first experiments on isotopes and their separation became possible only by the use of a kind of electromagnetic set up, the positive ion analyser of Sir J.J. Thomson, and later the mass spectrometer of Aston (§ 1.1.1 and 5.2.1). The principles of an electromagnetic isotope separator were discussed in § 5.2.3 and the essential components beginning with the charge material (UCl_4), the ion source, the electrode assembly, the magnet(s) and the ion collector of some of the machines designed to separate macro amounts of isotopes more or less on a commercial scale were described in § 5.2.4-7.

The chief merit of the electromagnetic separator is that it can resolve completely, in one single operation, all the isotopes present in a state of total

* A thick walled closed steel vessel.

purity. The extremely feeble yields of the order of a microgram per hour constitutes, however, a formidable drawback of the process. Efforts to combat this difficulty resulted in several ingeneous innovations of employing inhomogeneous magnetic fields[7].

Amongst the successful innovations employing inhomogeneous magnetic fields[7-9(a)], the Calutron described earlier (§ 5.2.8), has had no less than 68 prototype machines developed in less than two years' time (1942-43) at the Oak Ridge National Laboratory, U.S.A., mainly for the separation of ^{235}U in kilogram quantities. There have been other separators designed and developed in Britain, France and USSR, each with some specific points of merit. These latter include separators with a cylindrically symmetrical field and those with two stage analysers, magnetic and electrostatic or both magnetic. These have been described earlier in § 5.2.8-10. Using the calutron, Lawrence and his team succeeded at first in obtaining centigram amounts of ^{235}U per hour with centiampere ion currents. Later, with a battery of 68 calutrons, kilogram amounts of ^{235}U were obtained whose stockpiling together with the output of the diffusion plants (see below), sufficed for exploding two atom bombs of this isotope between July and August of 1945.

10.2.2 Separation by Diffusion

Diffusion is one of the earliest methods used for the separation of isotopes. However, in the case of uranium isotopes, interest in which was suddenly generated in about 1940, when it became imperative for the countries at War to obtain large amounts of ^{235}U and in as short a time as possible, separation by both methods, electromagnetic and diffusion across a porous membrane, were developed side by side in America. The theory of the process was discussed in § 6.1.1. It was shown that the maximum enrichment factor is: $\epsilon = \Delta M/2\bar{M}$ (= 0.004) theoretically, while in practice, the value is lower due to imperfect mixing, back diffusion and other factors. To arrive at meaningful enrichments, the process is worked in a cascade of a large number of stages (Fig. 6.1), which involves many problems some of which are discussed below.

(a) *Uranium Hexafluoride—the Diffusant*

The only form in which uranium can be obtained as a gas or vapour at low temperatures is its hexafluoride (UF_6), subliming at 56.5° C under normal pressure, when the vapour has a mean free path of 0.02 μm. To obtain pure uranium hexafluoride, the starting material preferred is UO_3 which is converted in several stages to UF_6

$$UO_3 \xrightarrow{H_2} UO_2 \xrightarrow{HF} UF_4 \xrightarrow{F_2} UF_6$$

The reactions involved, which are all exothermic, are:

(i) The reduction of UO_3 to UO_2 by hydrogen at 540-575° C, or by ammonia which by decomposition provides the hydrogen,

$$UO_3 + H_2 \rightarrow UO_2 + H_2O; \Delta H = 105 \text{ kJ/mole},$$

(ii) The fluorination of UO_2 to UF_6 at 450° C

$$UO_2 + 4HF \rightarrow UF_4 + 2H_2O; \quad \Delta H = 189 \text{ kJ/mole},$$
$$UF_4 + F_2 \rightarrow UF_6; \quad \Delta H = 84 \text{ kJ/mole}.$$

After the $^{235}UF_6$ is separated, it is reconverted by hydrolysis to UO_2F_2 in the presence of an alkali

$$UF_6 + 2H_2O \rightarrow UO_2F_2 + 4HF$$

UF_6 forms colourless crystals. Chemically, the substance presents many problems as it tends to dissociate into corrosive constituents

$$UF_6 \rightarrow UF_4 + F_2$$

The vapour is highly reactive to most common substances. Other products of reaction with air and moisture are HF, other lower fluorides and oxyfluorides which are even more corrosive. This necessitates the use of special materials as teflon (tetrafluroethylene $CF_2{=}CF_2$) for the pipes, gaskets, pump linings and other components coming into contact. Even the lubricants offer a problem. An additional difficulty arises when some of the solid decomposition products tend to block the fine membrane pores.

The need to keep UF_6 vapour out of contact with air and at the same time maintain its circulation from vessel to vessel and from stage to stage, creates a problem. Shafts and vanes revolving at high speeds of the order of a thousand times per minute have their ends exposed on one side to the atmosphere and on the other to the vapour of UF_6. To prevent contact with air, an additional vessel is included in between through which oxygen-free nitrogen is kept flowing at a pressure only slightly negative to that of the UF_6 (Fig. 10.1)[9].

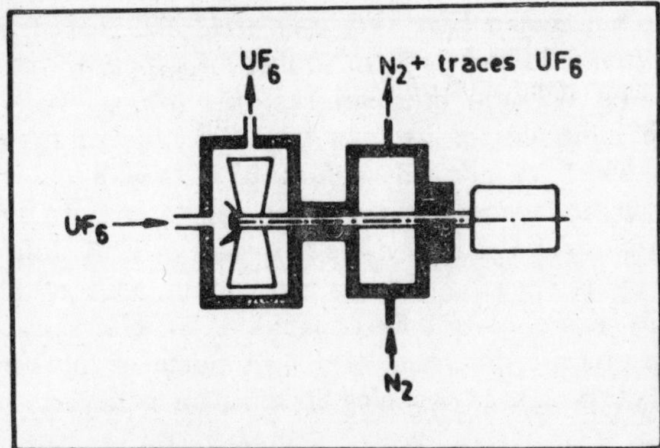

Fig. 10.1: **The circulation of UF_6 out of contact with air**
(Reprinted with permission from *La Séparation des Isotopes* by M. Chemla and J. Périé p. 118 © Presses Universitaires de France. 1974).

The operative pressures and temperatures have to be strictly within the limits dictated by the phase-equilibrium diagram of UF_6 (Fig. 10.2). The pressure has to be low enough to ensure that the gas mean free path is several times longer than the mean pore diameter of the membrane (~ 0.04 μm). At the same time, the pressure has to be high enough to let the vapour diffuse across to the other side of the barrier. A pressure range between 0.2 to 1.0 atm is permissible with the membranes in use. This limits the temperature to values between slightly below the atmospheric sublimation point of UF_6 of 56.5° C and the triple point at 64.05° C, 1.5 atm.

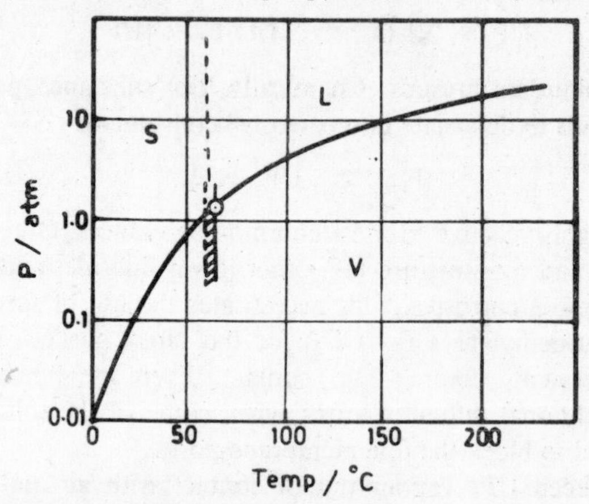

Fig. 10.2: **Phase-equilibrium of UF_6 S—liquid; V—vapour phase**
(Reprinted with permission from *Isotope Séparation* by Stelio Villani p. 169 © American Nuclear Society 1976)

(b) The Porous Membrane

The membrane across the pores of which the UF_6 has to diffuse should not only be chemically resistant to the action of the vapour and its decomposition products, but should have very even sized and shaped pores of mean diameter between 0.03 and 0.05 μm, so that the vapour of meanfree path of about 0.1 μm (at 0.2 atm pressure) can pass through before suffering a collision with molecules of its own kind. In addition to these stringent specifications, which are normally difficult to meet with, the membrane as a whole has to be mechanically strong to withstand a pressure difference of the order of an atmosphere acting on the two sides of it. A diffusion barrier is characterized by its P/r ratio, where r is the mean radius of the pores and P its permeability, expressed in units of moles of air diffusing across a m^2 per hour under a pressure difference of 1 torr. Some of the barrier materials suggested and the process of obtaining the same are listed below.

1. Alloy of Au(40) + Ag(60), etched by HNO_3 of density 36° Be,
2. Alloy of Ag(66) + Zn(34) etched by dilute HCl,
3. Al sheet anodically oxidized by 5% H_2SO_4 at 25 °C,
4. Fine Al dust got in a nitrogen arc forced into a mesh of the same metal,
5. Sintered Al or Ni powders,
6. Teflon granules pressed into a grid,
7. Ceramic or fritted membranes.

The entire membrane sheet has to be absolutely free from fissures and microcracks. No wonder the details of the membrane material and its fabrication are guarded as absolute secrets.*

(c) The Number of Cascade Stages
The theoretical single stage separation factor in the diffusion of $^{238}UF_6$ and $^{235}UF_6$ is 1.004 29**. As pointed out earlier, the factor obtaining in practice is closer to 1.001 4. Accepting this latter value, it can be seen that for a tenfold enrichment (*i.e.* to raise the content of ^{235}U from 0.72 to 7 per cent), the number of stages needed to be operated in series is close on 1650***, which is a very large number indeed. However, if weapon-grade uranium is the object (*i.e.* a ^{235}U content around 90 per cent) the number of stages needed is of the order of 3500. Though these numbers are not accurate, they show the order of the magnitude of the number of stages involved in realizing the above enrichments.

(d) Housing the Plant
The above gives an idea of the gigantic size of the overall plant. It is reported that a typical plant at Oak Ridge (US) uses several thousand kilometers of piping and hectares of diffusion barriers. The nature of the operation is such that all this has to be under one roof, for the functioning of the cascades in series. This last is not a happy factor in war time, despite the strictest security arrangements.

(e) The Hazard of the Blue Flash
^{235}U being a fissile material, it has to be stored in small, subcritical amounts, each lump being separated from the other. Once a stray neutron induces fission in one of the ^{235}U nuclei, the neutrons released in the process lead to a chain of fissions in neighbour nuclei, releasing an increasing number of neutrons and the chains multiply; but in a subcritical mass, the chains converge and die out due to the escape of some of the neutrons out of the uranium mass. As the mass of the latter increases, the fraction of neutrons escaping diminishes till a limiting or critical mass is reached when no neutron escapes, when the chains become diverging resulting in an explosion. This is nothing but a veritable miniature atomic bomb explosion, referred to as the *blue flash*[9]. The danger is from the high intensity of the radiations due to gammas and neutrons, and radioactivity of the fission fragments. The overall result can be fatal to the people around. Such blue flashes must have occurred in the early plants. These days, precise knowledge of the critical mass, efficient monitoring of its approach by sophisticated instruments, and the exercise of regulations in regard to radiation protection in the strictest

---

* The secrecy is so complete that it is often considered that any detail given out in a paper or in a conference is just such as would not work if adopted.

** $(^{238}UF_6/^{235}UF_6)^{1/2} = (352/349)^{1/2} = 1.004\ 29$

*** To a first approximation : $(1.0014)^n = 10$, whence $n = 1650$.

manner, verging on the draconian, ensure against the reaching of the blue flash stage. The necessity of these safe-guards apply to all methods of isolation of ^{235}U in a high state of isotopic purity.

(f) Some of the ^{235}U Diffusion Plants

By 1942, the theory of ^{235}U separation by gaseous diffusion was well worked out and the dimensions involved in a plant to produce kilogram quantities of the isotope involved were recognized. Work on the first such plant began at Oak Ridge (USA) in the summer of 1943, though it was not before the lapse of two years that production had reached the capacity planned. Considering the scale of operations involved in each cascade of over 1000 stages and the associated technical difficulties and health hazards, it is aptly described in the words of Smyth[1] that it was a "monument to the courage and persistence as well as the scientific and technical ability" of the scientists and engineers who built some three such plants. On one estimate the cost of a large size diffusion plant well exceeds one billion US dollars (1980 prices).

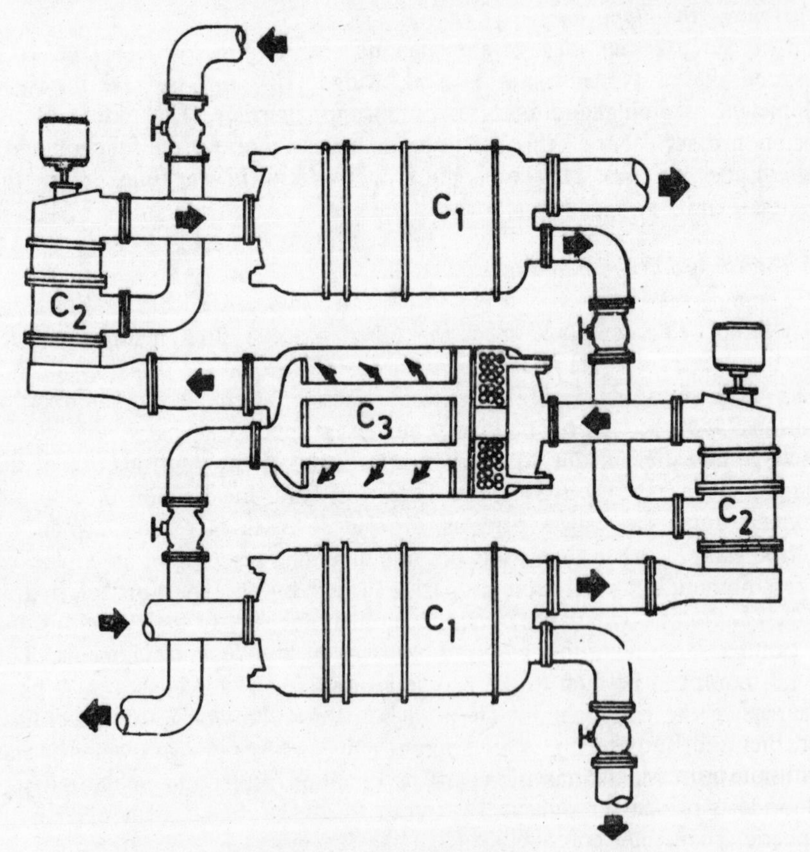

Fig. 10.3(*a*): **Arrangement of diffusion stages in series** C_1: Converter; C_2: Compressor; C_3: Coolant
(Reproduced by permission from *Isotope Separation* by S. Villani, American Nuclear Society, 1976, p. 198)

It is known that there exist similar diffusion plants in some other countries as Great Britain, USSR, France, and possibly China, though the information available about them is close to nil, as each country has been guarding the details in the strictest secrecy. A plant of the size comparable to the one at Oak Ridge was constructed at Pierrelatte in France with meticulous planning, as evidenced by its performance at peak capacity for over a decade. It was, however, ironical that even before the infrastructure of this plant was ready, diffusion as a process for the enrichment of ^{235}U became obsolete, as it was by then established with certitude that ultracentrifugation was a far superior process to diffusion. However, as the investments made in the Pierrelatte plant were so heavy and the construction had progressed so far that the point of no return had been reached and the French government had to sustain the plant. Other countries as Germany and Netherlands and possibly Italy, opted for the technically more efficient and economically relatively cheaper centrifuge for obtaining enriched uranium in commercial quantities.

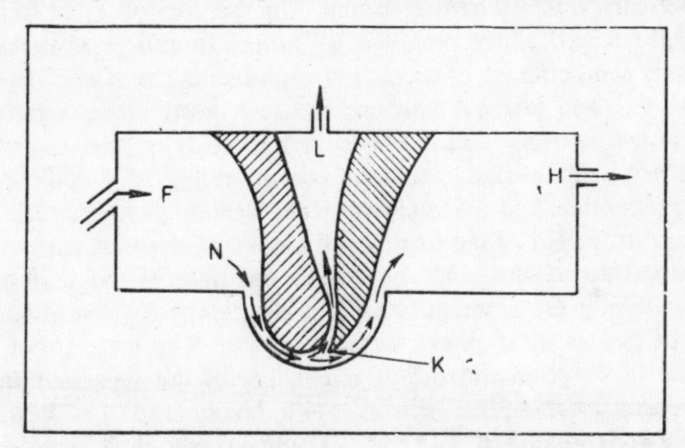

Fig. 10.3(*b*): **The double blade nozzle for separating the isotopes of uranium.** F—feed gas UF_6 (5%) + He (95%); H—heavy fraction; L—light fraction; K—knife edge.
(after E.W. Becker, K. Bier and H. Burghoff, *Z. Naturf.*, 1954, *9a*, 975 and 1955, *10a*, 565)

The merits and limitations of the diffusion and centrifugation processes were discussed in § 6.1.3. and § 6.5.5.

10.2.3 Nozzle Separation

Isotope separation by the nozzle or the jet method, developed in 1955 by Becker, Bier and Burghoff[10] is an innovation based on the same principles as the diffusion process. The gaseous mixture of isotopic molecules of uranium hexafluoride is passed through an expanding air nozzle at a pressure below atmospheric. The lighter component $^{235}UF_6$ having more of the faster moving molecules, tends to concentrate on the outside of the gas stream with the heavier component $^{238}UF_6$ in the inside. A "pairing tube" fixed opposite the centre of the nozzle collects the heavier component. The principle of the nozzle or jet separation illustrated in Fig. 6.4, was discussed under § 6.3. In a freely expanding jet there exist concentration, temperature and pressure gradients leading to corresponding diffusions, though it is the pressure diffusion which dominates[11].

(*a*) *Becker's Slit Type Nozzle*

The slit-type ceramic nozzle depicted in Fig. 6.4 was first developed by Becker *et al*[10], for the separation of the uranium isotopes. The slit and diaphragm widths, their separation and geometry were described in § 6.3. The variation of the elementary separation factor with the dimensions of the slit (0.085-0.045 mm) and the diaphragm (0.29-0.19 mm). their separation (0.11-0.081 mm) and the optimum flow rates, for different feeding pressures, upto 30 mmHg, were evaluated by the authors[10].

(*b*) *Double Blade Nozzle*

In the version developed in the Kernforschungszentrum, Karlsruhe (West Germany)[12, 12a], an aluminium pipe of 10 cm diameter and 2 m length was divided internally into 10 radial sectors. On the top of each sector and running all the length of the pipe was a "gutter" (a narrow channel) of an approximately semi-circular cross section of radius 0.2 mm and depth 0.03 mm. Above this and just not touching are two metal blades rigidly fixed. These are of special shapes as indicated in Fig. 10.3(*b*), one is curved and the second has a knife-edge. The feed gas, consisting of 5 mole per cent uranium hexafluoride and 95 mole per cent helium at a pressure of 600 mmHg enters to the left of the first curved blade and expands partly into the region between the blades, and the rest to the right of the knife-edge. A partial separation of the uranium isotopes is effected by the centrifugal forces set up in the gas jet as it strikes the curved wall. The presence of helium increases the flow speed and delays remingling of the separated fractions. The component escaping between the two blades and that beyond the second blade are enriched in ^{235}U and ^{238}U respectively, their pressures being around 150 mmHg each (Fig. 10.3 *b, c*).

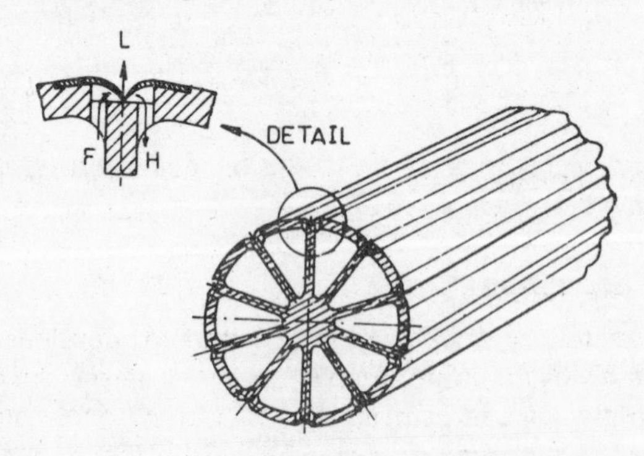

Fig 10.3(*c*): **Separation nozzle arrangement**
F: Feed gas; H: Heavy fraction; L: Light fraction
(Reproduced, with permission, from *Isotope Separation.* by S. Villani, American Nuclear Society, 1976, p. 267)

The separation factor in the nozzle separation is of the order of 1.01. This process also lends itself to multiplication of the overall separation by operating in a cascade of multiple stages. The Karlsruhe unit had 81 tubes of the type described above.

(c) Merits of the Nozzle Separator

The greatest merit of the nozzle separator is the avoidance of the very cumbersome porous membrane barriers of a tremendous total area of a few hectares, together with the associated pumps needed in the diffusion process. The single stage separation factor is also higher which means far fewer number of cascade stages are needed to arrive at a given overall enrichment. This is brought out by the approximate figures presented in Table 10.2.

Table 10.2: A comparison of the physical process of separation of ^{235}U isotope of purity > 90 per cent

Process	Separation factor	Number of stages
Diffusion	1.0014	4000
Nozzle	1.01-1.015	100
Ultracentrifuge	1.2-1.5	50

The major drawback of the nozzle separator at the present time is the need of high suction volume pumps on account of the low operation pressure. No doubt improved versions would be forthcoming if the world needs of the ^{235}U isotope are not met by other methods.

10.2.4 Ultracentrifugation

The most attractive feature of isotope separation by the centrifuge is that the separation factor here depends simple on the mass difference ΔM $(= M_2 - M_1)$, unlike in gaseous diffusion where it depends on $\sqrt{(M_2/M_1)}$, or as $\Delta M/(M_1 + M_2)$ as in thermal diffusion. Hence the efficiency of the method in separating the uranium isotopes is same as that in separating the isotopes of light elements. The avoidance of the vast porous membrane barriers and associated pumps is another great advantage. The practical interest in the method was further heightened with the availability of modern vacuum centrifuges spinning at some 100 000 times per minute generating accelerations equivalent to some 10^5 times earth's gravity. Enrichments resulting in some 50-100 stages of gaseous diffusion can be realized in a single stage of centrifugation. Hence, today the gas centrifuge stands unrivalled as the most efficient, relatively economical and technically the simplest means of separating the uranium isotopes.

(a) The Late Start

However, it was not before the late fifties that the use of the gas centrifuge became an important competitor in producing enriched uranium on a commercial scale*. A complete theory of isotope separation by centrifugation was first available through the important work of Martin and Kuhn[13]. It was some ten years later that a comprehensive theory, specially applicable to the large scale production of ^{235}U isotopes was published by Cohen[14]. A total concentration of interest and resources on the problem led to rapid developments in the designing and successful operation of the ultracentrifuge capable of delivering weapon-grade ^{235}U in Germany, Netherlands and possibly in some other countries by the late fifties.

(*b*) *Technical Difficulties*

The design of an ultracentrifuge for obtaining enriched uranium is, however, beset with many technologically difficult problems. These include the selection of special materials as maraging steel for the rotor and other components which are resistant to the action of the vapour of UF_6 and its decomposition products, the delicate mechanism of suspending the rotor revolving at 40 000 to 90 000 times per minute, and fixing shafts which allow sufficient flow of gas and also are elastic enough to provide necessary mechanical disconnection of the rotor from the bearings at the top and bottom ends of the rotating system. The nature and type of the material used in making the rotor, its length and the length to diameter ratio, the maximum revolution rate *vis- à-vis* its tensile strength and density, its very shape, straight or curved, hollow or solid, etc., are known to affect the overall performance and the separation factor. Presence or otherwise of radial attachments to serve as vanes or blades also is important. All such data and studies on these variables are closely guarded secrets. Information and data on minor but strategically important components as bearings, damping bearings, gaskets and packings, and lubricants etc. are all classified. Some gas centrifuges, with generally known details were described earlier in § 6.5.2.

(*c*) *The Countercurrent Centrifugation*

Though the centrifuge permits a high single stage separation factor of the order of 1.2, thereby needing fewer stages in the cascade, the output per stage is low compared to the diffusion process. This necessitates the provision for countercurrent centrifugation which raises the separation potential. For this, Martin and Kuhn[13] had suggested dividing the rotor into a number of chambers and leading the process gas (*i.e.* the gas mixture to be separated), through these chambers in countercurrent. Thermopumps were to be inserted for maintaining a continuous circulation. A temperature gradient between the

---

* The reason is not far to seek. The immediate needs of the US dominated War (1941-45), and their heavy investments in the diffusion factories, which were yielding needed results, would not allow a reinvestment in a new technology irrespective of the merits of the latter. The development of the centrifuge technology was thus left to other countries. Also, materials with special properties, needed for the modern gas centrifuge, became available only after progress in other areas, as space research.

hot axis and the outer cold wall would sustain a flow of the lighter isotope from the lower to the upper chamber and of the heavier isotope in the reverse direction (Fig. 10.4). Each compartment, it is obvious, serves the same purpose as a plate in a distillation column (§ 7.1.4). The separation factor gets multiplied and the cylindrical column constitutes a cascade of centrifuges, the light isotope escaping at the top and the heavy isotope at the bottom.

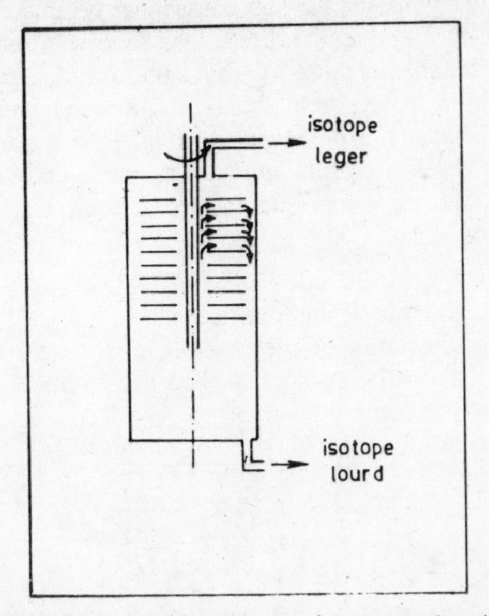

Fig. 10.4: **Martin's idea of dividing the rotor into compartments for effecting counter centrifugation**
(Reprinted with permission from *La Séparation des Isotopes* by M. Chemla and J. Périé p. 134 © Presses Universitaires de France, 1974).

It was soon discovered that even without subdividing the rotor into chambers, the same multiplication effect by countercurrent circulation can be achieved with any centrifuge with a simple cylindrical rotor, provided the latter was sufficiently long. It was observed that rotors of length about a meter, narrow diameter about 15-20 cm, and revolving at high speeds generate thermogravimetric currents necessary for effecting the cascade multiplication effect. Each single stage centrifuge is the equivalent of some 100 stages of diffusion (see Table 10.2).

(d) A Zippe-type Ultracentrifuge for the Separation of Uranium Isotopes

The centrifuge of the type known as the Zippe has served as a prototype which has been adopted with modifications by most countries engaged in obtaining enriched uranium by centrifugation. It consists of an aluminium cylinder rotor about 1 m long and about 20 cm in diameter*. The lower end

* Various lengths between 66 to 120 cm and diameters between 13 and 20 cm had been used in the models ZG 3 and UZ III B.

of the rotor shaft rests on a flexible needle pivot in a hardened concave metal seat, while the upper end is held in place by a magnetic coupling. An induction motor drives a steel disc fixed concentrically with the rotor at its bottom. The stator coils are also at the bottom. The version of the Zippe centrifuge used by Beyerle and Groth[14a] is depicted in Fig. 10.5. Friction is reduced to a minimum by elastic shock absorbers even when the rotor attains maximum speeds of the order of 90 000 revolutions per minute. To eliminate friction with outside air and consequent heating effects, the centrifuge is housed in an outer rigid cylinder, the space between the two being evacuated. This vacuum is realized without the use of a pump by the high speed rotation of the rotor itself, whose motion squeezes out the air in between through helicoidal grooves cut in its wall, the action being much the same as in a molecular diffusion pump. The rotor enclosed in the vessel under vacuum is all one piece upto the lower shaft including the flexible needle resting in the socket.

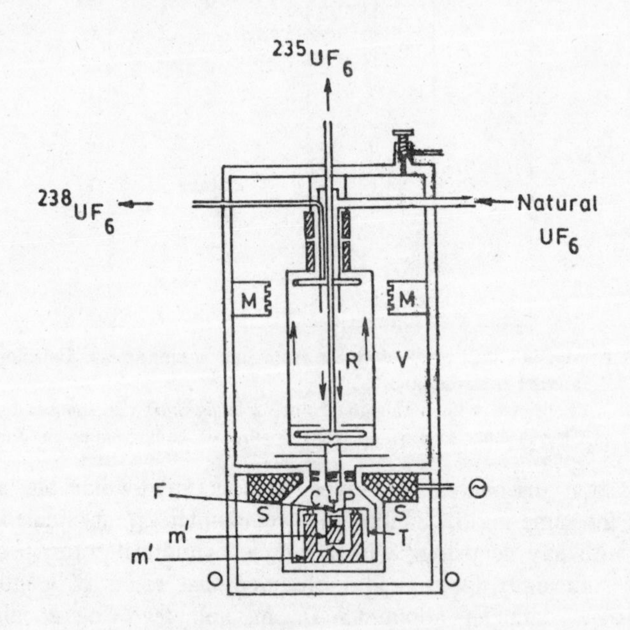

Fig. 10.5: **Zippe type ultracentrifuge for separating uranium isotopes.**
R : rotor upper and end held by a magnet M; P : lower end; F : a flexible needle resting in a hollowed metal block C; V : vessel under vaccum; S : stator; T : footing held in place by magnets m'. m'.
(after K. Beyerle and N. Groth, *Proc. Intern. Symp. on Isotope Separation,* North Holland Publishing Co., 1958).

The isotopic gas mixture of UF_6 enters through the middle co-axial inlet passage at the top surrounding the axis of the rotor. During centrifugation, the heavier isotopic molecules are pushed towards the outer wall while the lighter species collects nearer the axis of rotation. The accompanying inner circulation of the gas parallel to the axis under a pressure gradient of the order of 3000 between the axis and the wall, brings about the cascade action described above. The heavier species is pushed downwards and the lighter species pushed upwards, without any pump being necessary. This results in an overall multiplication by nearly five times the separation factor of 1.19-1.23, even with a medium peripheral velocity of 250-280 m/s.

(e) A Unit for Experimental Study

A flow sheet for the experimental study for obtaining and verifying relevant data in the separation of the isotopes of uranium by ultracentrifugation is described in Fig. 10.6. Purified UF_6 is stored in a container (B) maintained at a constant temperature in a thermostat above the sublimation point of 56.5° C. By controlling flow rates by valves (D) a steady amount of the vapour enters at A the centrifuge C operating under steady conditions. Samples enriched in ^{235}U and ^{238}U leave via exits E_1 and E_2 as shown in the figure, which are controlled by valves and flow resistances W_1, W_2. Finally they are condensed and collected in liquid nitrogen cooled receivers K_1 and K_2. At any instant, samples E_1 and E_2, or their mixture, can be tapped for analysis by mass spectrometer

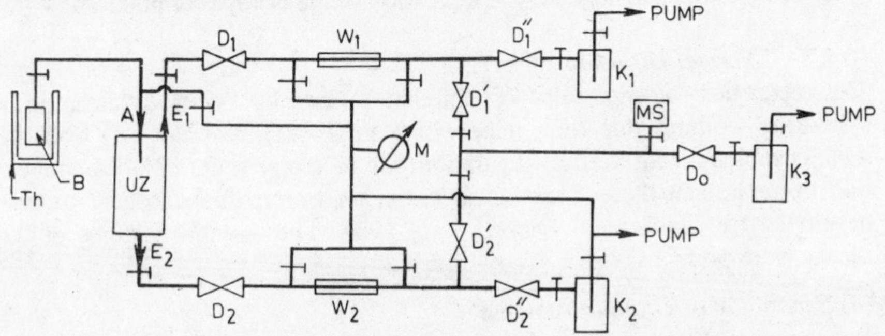

Fig. 10.6 : **Separation of uranium isotope by ultracentrifugation of UF_6**
Th: Thermostat; B: Stock of pure UF_6; A: Inlet of UF_6 to UZ; E_1: Outlet for enriched UF_6; E_2: Outlet for depleted UF_6; D_1, D_2...: Flow rate valves; W_1, W_2: Flow resistances; K_1, K_2, K_3: Liquid nitrogen cooled receivers; M: Membrane manometer; MS: Mass spectrometer.
(from *Separation of Isotopes*, Ed. by H. London, George Newnes Ltd., 1961, p. 283; reprinted by permission from William Heinemann ltd.)

With a test unit of this kind, it had been possible to obtain necessary valuable data for arriving at the technically and economically optimum conditions for separating enriched uranium needed not only for light water moderated reactors but for military purposes. Though most of the vitally important data are classified and unavailable, London[7] and Villani[12] have given valuable information in regard to the influence of experimental parameters as the dimensions of the rotor, its revolution rate, temperature, rate of throughput, *etc.* on the enrichment and multiplication factors and the separation potential. Some of these values are given in Tables 10.3 and 10.4 obtained with UZ III B and ZG3 centrifuges.

Table 10.3: Variation of the separation factors (single stage a_0, optimum A_{opt})*, the multiplication factor (K) and the separative power (δU) with the rotor peripheral velocity in the countercurrent centrifugation of UF_6

Peripheral velocity $\omega r_a/\mathrm{ms}^{-1}$	a_0	A_{opt}	K	δU/ g $UF_6 h^{-1}$
252	1.038	1.188	4.65	0.081
280	1.047	1.225	4.45	0.132

Table 10.4 : Energy inputs for the separation of uranium isotopes by ultracentrifugation[7]

Centrifuge length/cm	63.5		120		350	
Peripheral vel./m s^{-1}	300	400	300	400	300	400
MWh/kg U treated	6.8	3.8	3.6	2.0	1.23	0.69

The corresponding energy input for treating one kg of uranium to arrive at the same degree of enrichment by the diffusion of UF_6 vapour is 9.0 MWh. This *inter alia* other technical factors described above, establishes the superiority of the ultracentrifuge for the separation of the isotopes of uranium.

10.2.5 *Thermal Diffusion*

The separation of molecules of different masses by radial diffusion in a cylindrical column due to a temperature gradient across the axis and the wall, combined with vertical separation due to temperature induced convection currents, constitutes *thermal diffusion*, an irreversible statistical process developed by Clusius and Dickel[15,16] in 1938. The essential features of the theory were presented in § 6.4.1 and § 6.4.2.

(a) Separation of Uranium Isotopes

Recognizing the attractive features of the process, thermal diffusion was included under the Manhattan Project for obtaining enriched uranium on a large scale. No less than 2 100 Clusius-Dickel columns for treating uranium hexafluoride in the liquid state were set up at Oak Ridge since June 1940 with the limited object of obtaining therefrom large amounts of samples enriched in ^{235}U to the extent of about one per cent. This was to serve subsequently as the starting material for the isolation of weapon-grade uranium by electromagnetic separation in the calutrons.

For this, the column consisted of an inner nickel cylinder of 15 m height and maintained at 280° C, while the outer cylinder was of copper cooled to 64 °C by circulating water. The separation between the two tubes was extremely narrow being less than one mm. Maintaining strict parallelism between the two cylinders with such a narrow separation all along the 15 m length involves much technical perfection. The outer wall temperature being close to the triple point, the uranium hexafluoride could be maintained in the liquid state in the column. For attaining the desired enrichment, the columns

* These quantities are defined in Eqs. 6.26 and 6.27 (6.5.1).

were joined on the principle of a square cascade (§ 4.4.4), *i.e.* in a pyramid of series-parallel arrangement. For instance, a four-stage cascade had 22 columns in parallel in the first stage, 14 in the second, eight in the third and three in the last stage.

After the attainment of a state of equilibrium, which needed a time of the order of several weeks, the unit could treat 50 g of uranium per day with a separation factor close to two.

Despite the advantages of simplicity of installation and high performance efficiency, the Oak Ridge authorities gave up thermal diffusion as a process for obtaining enriched uranium on the sole ground of very high expenses involved in meeting the energy required in maintaining the high temperature gradient over the prolonged periods involved. Taking into account all factors, the diffusion process appeared to be more acceptable compared to thermal diffusion in the separation of isotopes of uranium during the middle and late forties, *i.e.* before the development of the gas centrifuge.

10.2.6 Laser Induced Separations

1. *Principle of photoseparation*

Isotopic atoms and molecules display a small shift in their peak absorption spectra. The methods of isotope separation, based on the shifts in their absorption peaks, were described in § 8.4.

Attractive as the photochemical process is for the separation of isotopes, no success had been reported for long. In fact, Urey[7] was one of the earliest to study the photochemical separation of ^{235}U by irradiating UF_6 and UCl_6 with the hyperfine structure lines of uranium. The result was negative, due, in all probability, to side reactions leading to remixing of the separated components.

2. *Essential requirements*

Conditions essential for isotope separation by photon absorption were outlined under § 8.4.3, in respect of the system and the light source. Evidently these requirements are best met only by a tunable laser, which can have the necessary narrow line-widths and high power. No wonder therefore that progress in isotope separation by photon absorption became possible only after the development of laser technology.

3. *Principal photon-induced reactions*

The principal photon-induced reactions relevant to isotope separation are (i) capture of two photons leading to excitation and ionization; (ii) excitation and dissociation of the molecule; (iii) capture of a single photon leading to predissociation; (iv) chemical reaction: and (v) capture of multiple photons leading to dissociation. These processes and the spectral data relevant to uranium isotopes are described below.

A. Two-photon atomic process

(i) *The lasers*

Following spectral regions are found to be convenient for the preferential excitation of the ^{235}U isotopic species, in the first stage in the two-photon excitation-ionization process of laser induced separation.

(i) *Visible violet-blue region* The absorption spectra of atomic uranium in the vapour state in the violet-blue region has the following characteristics[17].

(*a*) Absorption peak : $\lambda = 424.63$ nm. $\nu = 2.355 \times 10^4$ cm^{-1}
Isotope shift $\bar{\nu}_{238} - \bar{\nu}_{235} = 0.28$ cm^{-1} (= 8.4 GHz)
Doppler line-width = 0.055 cm^{-1}: (= 1.65 GHz)
(*b*) Absorption peak : $\lambda = 426.62$ 66 nm. (or $\nu = 2.344 \times 10^4$ cm^{-1})
Isotope shift : $\Delta \bar{\nu} = 0.32$ cm^{-1} : (= 9.6 GHz)

The absorption spectrum of atomic uranium in the vapour state in the 400 nm region is depicted in Fig. 10.7[17-20]. It may be seen that while ^{235}U has a hyperfine structure with eight peaks, that of ^{238}U consists of a single sharp peak, well separated from the former by a frequency shift of the order of 7-8 GHz*

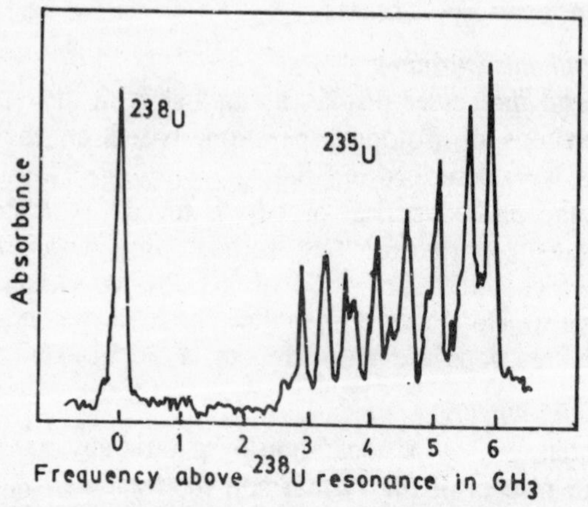

Fig. 10.7: **Absorption spectrum of atomic uranium vapour in the 400 nm region.**
(after B.B. Snavely, with permission from F.S. Becker and K.L. Kompa, *Nuc. Tech.*, 1982, *58*, 329).

(ii) *The separation unit operating in the green-yellow region* The unit using the green-yellow laser to obtain enriched uranium is sketched in Figs. 10.8 and 10.9. A collimated beam of uranium atoms is got by volatalizing a uranium-rhenium alloy or by evaporating the metal by an electron beam. This is focussed by weak magnetic fields into a water cooled copper crucible from where it expands with a velocity of about 400 m/s. The atomic beam is simultaneously irradiated by two light sources; (i) a dye laser** of wavelength 591.54 nm which raises only the ^{235}U species to the excited state leaving unaltered the

* 1 GHz = 10^9 Hz.

** The dye laser consists of an active dye dissolved in a liquid and pumped by flash lamps. The efficiency is usually low, of the order of 0.2 per cent.

^{238}U species, and (ii) a 2.5 kW mercury arc lamp together with a filter which cuts out all radiation of wavelength shorter than 210 nm (*i.e.* of energy greater than 6 eV). The ionization threshold of $^{235}U^{+}$ being 6.187 eV above the ground state, the filtered mercury radiation ($210 < \lambda < 310$ nm)

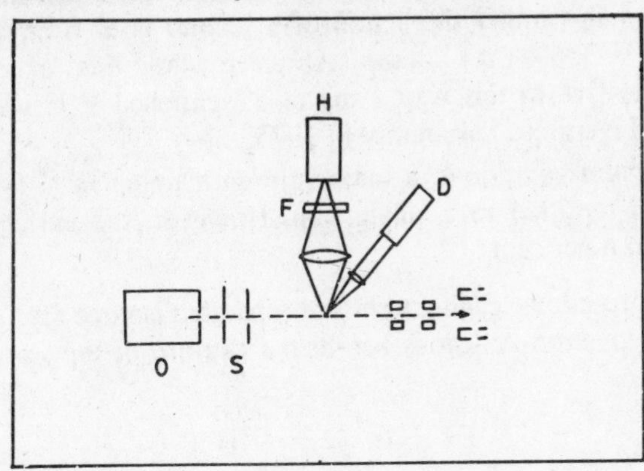

Fig. 10.8: **Laser separation of uranium isotopes:** O : oven production beam of uranium atoms: S : slits; D : dye laser (591.54 nm); H : 2.5 kW mercury lamps; F : filter to cut out $\lambda < 210$ nm.
(Reprinted with the permission from *Industrial Applications of Lasers* by J.F. Ready p. 504 © Academic Press, 1978)

causes the ionization of only $^{235}U^{*}$ and not the unexcited species (Fig 10.9). The ionized species is deflected out by an electric or a magnetic field.

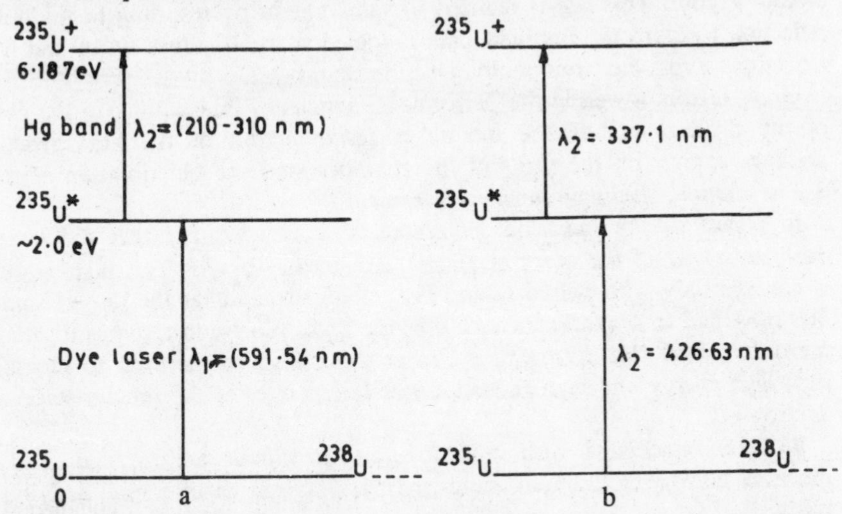

Fig. 10.9 : **Energy levels of ^{235}U and ^{238}U in the (a) 591.54 nm region and (b) 426.63 nm region.**

Alternatively[19], the two-step ionization is also possible using a dye laser precisely tuned to 426.626 6 nm (λ_1) which excites ^{235}U bypassing ^{238}U whose excitation level is just 0.32 cm^{-1} below it (Fig. 10.10) by a *uv* nitrogen laser of 337.1 nm (λ_2) which ionizes $^{235}U^{*}$. This procedure is reported to have yielded 50 per cent enriched ^{235}U samples in one single stage.

As yet another variation of the two-step process, the uranium atomic beam is irradiated with a xenon ion laser (378.1 nm) when the ^{235}U excites selectively, and this under the action of a second laser from a krypton ion source (350.7 / 356.4 nm) ionizes. An electrostatic field extracts the $^{235}U^+$ ions free from ^{238}U. In this way 4 mg of 3% enriched ^{235}U was obtained in the Lawrence Livermore Laboratory in 1975.

Under optimum conditions, a maximum enrichment in ^{235}U of 60-70 per cent had been reported in a single step. However, the enrichment rapidly falls to close on 6 per cent.

Important side reactions leading to loss of separated isotope are:

(a) Deactivation collisions before the capture of the second (ionizing) photon

$$^{235}U^* + U \rightarrow \, ^{235}U + U$$

(b) Charge transfer collisions before removal by the deflecting field

$$^{235}U^+ + \, ^{238}U \rightarrow \, ^{235}U + \, ^{238}U^+$$

To minimize these wasteful collisions, the uranium atom density should be low, around $10^{10}/cm^3$, corresponding to a low pressure of the order of 10^{-6} torr, but lowering the particle density affects adversely the atom-photon interaction. Lower the pressure, longer has to be the photon path in the atomic vapour. This last is realized to some extent by resorting to multiple reflection by mirrors. The light path is folded some 10 times by to and fro reflections across the atom-photon reaction chamber.

Other factors lowering the performance efficiency and adding to the cost of the process include the intense corrosive action of the hot metallic uranium vapour on the walls of the chamber, and the deterioration of the dye in solution, needing frequent replacement.

In regard to the radiation requirements, it is necessary that the lasers dissipate power of the order of 10 mJ/cm^3 in the vapour. This high power is possible only with pulsed lasers. The vapour expands at the rate of some 400 m/s, and to synchronize with this, the pulse repetition rate has to be of the order of 10^4 Hz. It strains dye laser technology to the limit to provide such high energy and high repetition rate laser pulses at the present stage of development.

Problems associated with metallic uranium vapour are avoided in the processes employing uranium compounds, as uranium hexafluoride, described below.

B. Multiphoton molecular decomposition process

Amongst the uranium compounds, the most convenient is naturally UF_6. It has the lowest sublimation point (56.5 °C), and fluorine is monoisotopic and does not add to the number of isotopic species. Again UF_6 is prepared in large amounts for use in other isotope separation processes. Also detailed spectroscopic data of the compound are now available. The early impetus for using UF_6 for effecting laser induced separation of ^{235}U came from the success achieved in separating the isotopes of sulphur (^{32}S and ^{34}S) by the laser irradiation of chemically analogous SF_6 vapour (§ 8.4.8).

The spectral absorption characteristics of UF_6 in the *ir* region at 300 K are as follows[17,20]

Absorption peak $\lambda = 16.05\ \mu m$ ($\bar{\nu} = 625\ cm^{-1}$)
Isotopic shift $\bar{\nu}_{235} - \bar{\nu}_{238} = 0.65\ cm^{-1}$
Rotational band width = $16\ cm^{-1}$

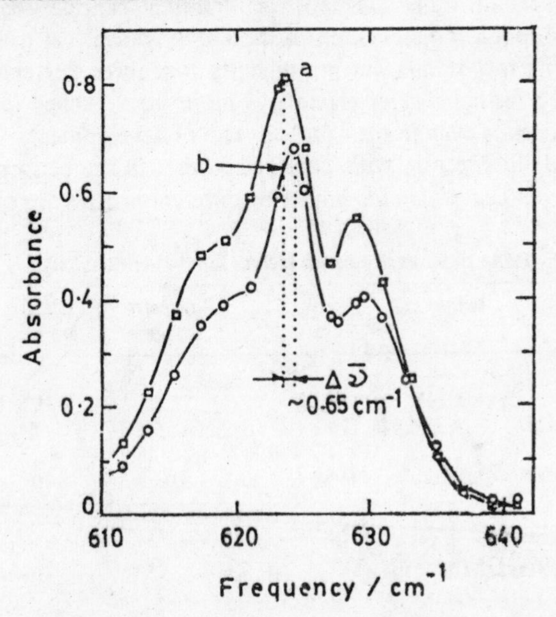

Fig. 10.10: **Absorption spectra of (*a*) $^{235}UF_6$ and (*b*) $^{238}UF_6$**
(after B.B. Snavely, Separation of uranium isotopes by laser Photochemistry, VIII *Intern. Conference on Quantum Electronics,* San Francisco, 1974) (with permission from F.S. Becker and K.L. Kompa, *Nuc. Tech.,* 1982, *58,* 329).

The actual absorption spectra of $^{235}UF_6$ and $^{238}UF_6$ at 300 K are shown in Fig. 10.10[18,20]. The complexity of the *ir* spectrum at room temperature is considerably simplified by flow cooling the UF_6 vapour mixed with nitrogen in an expanding supersonic jet stream.

Other *ir* laser frequencies which may possibly be used for the separation of the uranium isotopes are 823, 1160 and 1294 cm^{-1}. As only *ir* lasers are involved, the energy input per ^{235}U atom separated would be less for the molecular process compared to the atomic process needing visible and *uv* lasers.

Basically, the multiphoton decomposition selectively of one of the isotopic molecules of uranium hexafluoride is similar to the separation of ^{34}S from sulphur hexafluoride described under § 8.4.8. The absorption of a number of successive infra-red photons, say from a carbon dioxide laser, or a pumped CF_4 laser, pushes the selected isotopic molecule up the ladder of approximately equally spaced excited levels, until the molecule dissociates eventually. The dissociation products can be chemically separated from the unaltered UF_6 species. The multiphoton process of dissociating UF_6 is, however, still under development.

10.2.7 Performance Factors for Different Processes of ^{235}U Enrichment

From what has been described in the previous section, laser techniques offer advantages over all other methods of uranium isotope separation. The molecular multiphoton dissociation method is, however, yet to be developed technically. The fact stands out prominently that three per cent enrichment of ^{235}U, needed for light water reactors, is more than reached in a single step by all laser processes even in their present state of development.

We conclude this section with Table 10.5 wherein the performance factors of about three per cent ^{235}U enrichment by different process are compared.

Table 10.5 : Performance factors for ^{235}U enrichment[17]

Process	Diffusion	Nozzle	Ultracentri-trifuge	Laser-atomic	Laser-molecular
Separation factor	1.0014*	1.011 8	1.25	10	33
No. of stages needed	335; 4000 ‡	120	6; 50 ‡	1	1
Energy needed /kWh per SWU**	2100	3500	210	170	151

10.3 PLUTONIUM

Plutonium is an entirely man-made element, if the existence of one part in 10^{14} parts of uranium ores be neglected. If any of its isotopes had existed they must all have been radioactive and decayed practically completely in the past. Some 15 isotopes from ^{232}Pu to ^{246}Pu have been synthesised to date most of which decay by α emission and fission. The long-lived Pu isotopes ($\tau > 1$ year) are listed in Table 10.6.

Table 10.6: Long-lived plutonium isotopes

Mass No.	236	238	239	240	241	242	244
Half-life/y	2.85	87.75	2.44×10^4	6.5×10^3	14.89	3.87×10^5	8.26×10^7

Of these ^{238}Pu and ^{239}Pu are better known, the former in 5-10 curie

* This is the value realizable in practice, though the theoretical value is somewhat higher, 1.0043

** ‡ for weapon grade SWU is energy needed in kilowatt hours per *separative work unit.*

amounts mixed with beryllium as a convenient laboratory neutron source of fluxes of the order of 10^6n/s, and the latter as a nuclear fuel and as a strategic bomb material. That this species of matter, ^{239}Pu should serve man in his day to day energy needs and also be his potential destroyer, is ironical indeed. At any rate, its synthesis atom by atom, into tens, or perhaps hundreds of kilograms per year, is an achievement of the atomic age.

10.3.1 Creation of ^{239}Pu

The interest in the isotope ^{239}Pu arises from its nuclear characteristics which make it an eminently suitable nuclear fuel element, as fission can be readily induced in it by neutrons of thermal energy with a high cross section of 742.5 barns (Table 10.1). The nuclide has its birth as a waste (!) product in a nuclear reactor employing natural uranium as the fuel element. First, a ^{238}U nucleus captures a neutron of specific resonance energy. The immediate product is ^{239}U in an excited state which by two successive β^- decays yields ^{239}Pu

$$^{238}\mathrm{U} + \mathrm{n(res)} \longrightarrow {}^{239}\mathrm{U}^* \xrightarrow[23.5\ \mathrm{min}]{\beta^-} {}^{239}\mathrm{N}_p \xrightarrow[2.355\ \mathrm{d}]{\beta^-} {}^{239}\mathrm{Pu} \xrightarrow[24\,400\ \mathrm{y}]{\alpha}$$

The very long life of ^{239}Pu permits its separation, purification and stockpiling in kilograms, in *sub-critical* amounts no doubt.

Once the optimum amount of the fuel element is consumed, the problem of the recovery of ^{239}Pu and the unused ^{238}U in the spent fuel begins. Both constituents are very precious and regulations require their total input-output accountability to within a fraction of one per cent. The separation and recovery of the two species ^{238}U and ^{239}Pu are considered in the next section.

10.3.2 Recovery of Plutonium and Uranium from the Spent Fuel

As no isotope separation is involved, there is no difficulty in the separation and recovery of plutonium and uranium each in a pure form and free from accompanying fission fragments. The only problem is of analysis and isolation of very small amount of plutonium mixed up with a large excess of uranium. In addition, there is the problem of serious health hazard arising from the intense radioactivity of the spent fuel rods, estimated to be equivalent to a few tons of radium! Besides its high chemical toxicity, the specific α activity of plutonium makes it an internal radiation poison. It is therefore imperative to adopt only remote control operations from behind heavily shielded walls, with lead glass windows for viewing. Precautions include the total trapping of air-borne traces of plutonium whose inhalation or ingestion can be fatal.

General principles of U-Pu separation

The chemistry of the separation of uranium and plutonium and the fission fragments is based on simple differences in the properties of the chemical groups to which they belong. These include:

(*a*) *The relative stability of the MO_2^{2+} state,*
UO_2^{2+} is more stable than PuO_2^{2+}.

(*b*) *Solvent extractability*

(i) MO_2^{2+} ions from a nitrate solution are readily extractable into an organic solvent.

(*ii*) M^{4+} ions from a 6M HNO_3 solution can be extracted into tributyl phosphate (TBP) dissolved in kerosene.

(*iii*) M^{3+} ions can be similarly extracted into TBP in kerosene, but from a much stronger (10-16 M) nitric acid solution.

(*c*) *Precipitation as fluorides or phosphates*

Only the M^{3+} and M^{4+} yield insoluble fluoride or phosphate precipitates, specially in the presence of sulphate ions, with which the ions of higher oxidation states form soluble complexes.

The final purification is effected by special ion exchangers.

U-Pu separation programmes

To start with, the intensely active spent fuel rods are "cooled" for an aging period of a few months under water in deep shielded wells when most of the fission fragments decay and the general level of radiation comes down. Sometimes the radiation energy released during the cooling process, which is considerable, corresponding to doses of the order of 20-30 kGy* per hour, which would in any case go waste, is availed of in effecting certain radiation induced reactions as the synthesis of ethyl bromide, gammexance, ethylene glycol, egosterol from yeast, silicone lubricants *etc.* (*vide* books on *Radiation Chemistry*[6,21,22]). This aging also helps in reducing subsequent radiation-induced corrosions and other reactions during the separation process.

The broad stages involved in the reprocessing of the spent fuel are indicated in the following two processes which are typical.

(*i*) *The tributyl phosphate (TBP) process*

The first step is to dissolve out the external canning material, aluminium or zircalloy (Zr + Al), in hot caustic soda **

The alkali digested material is filtered

↓

Filtrate : $NaAlO_2$

Residue: U + Pu digested with 13M HNO_3 under reflux

↓

Concentratre of U + Pu subjected to mechanically pulsed

↓

*1 gray (Gy) = 100 rad.

** Mechanical decanning of the uranium slugs is dangerous and hence the chemical process is resorted to.

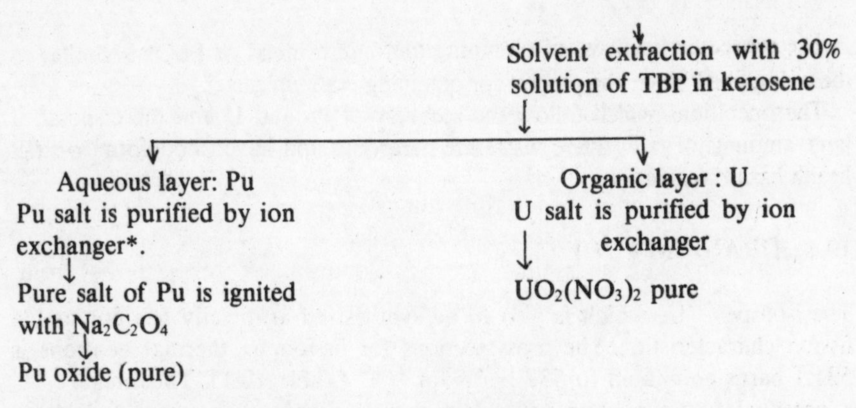

(ii) *Methyl isobutyl ketone (hexone) + LaF_3 Process*

This process was originally developed by McMillan and Abelson[23] for the isolation of neptunium from the products of uranium-neutron interaction. The earlier stages are same as in the TBP process described above.

The concentrate of UO_2^{2+}, Pu^{4+} and fission fragments (*ff*) extracted into methyl isobutyl ketone (hexone)

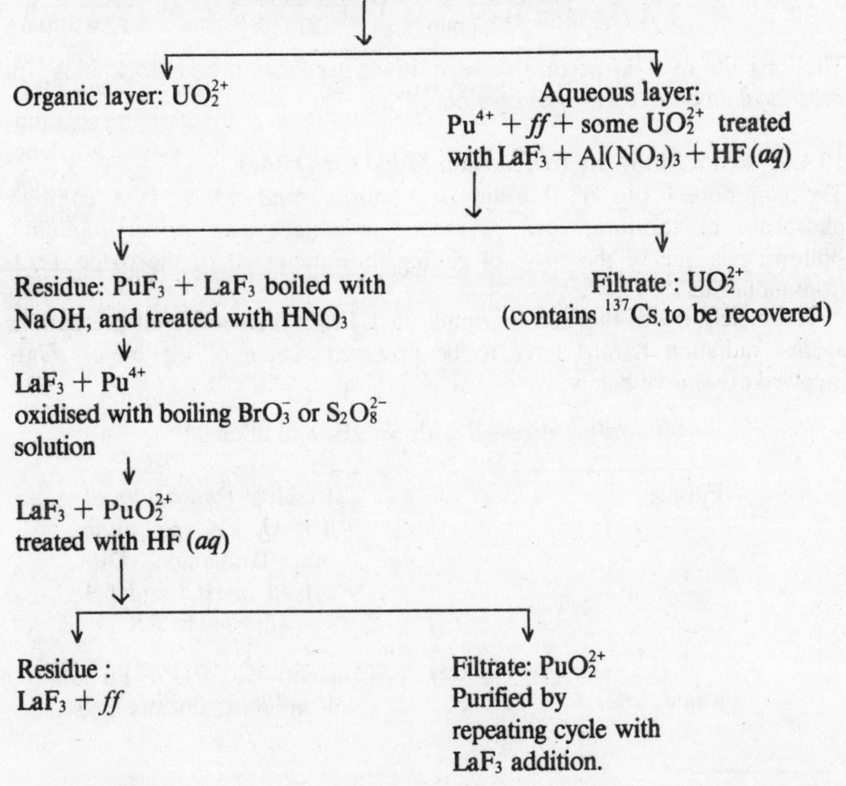

* The waste aqueous solution contains $NaUO_3$ + radioactive Cs. The latter is recovered as ^{137}Cs (see § 11.4).

The subsequent process of obtaining plutonium metal or PuO_2 is similar to those described under the process for obtaining uranium metal.

The operations which follow the recovery of Pu and U and the disposal of large amounts of radioactive waste are hazardous and have to conform to strict health hazard regulations.

10.4 URANIUM (^{233}U)

The isotope ^{233}U, which is also to be synthesized artificially has favourable fission characteristics. The cross sections for fission by thermal neutrons is 531.1 barns compared to 582 barns for ^{235}U (Table 10.1). This makes ^{233}U a potential future nuclear fuel when natural uranium sources get depleted and even when adequate plutonium fuel ^{239}Pu may be bred. This hopeful expectation is on account of the vastly greater abundance of thorium, the parent of ^{233}U, in nature compared to uranium sources. The nuclide ^{233}U is prepared from natural ^{232}Th (monoisotopic) in the same way as ^{239}Pu is obtained from natural ^{238}U.

$$^{232}Th + n(res) \longrightarrow {}^{233}Th^* \xrightarrow[22.3\ min]{\beta^-} {}^{233}Pa \xrightarrow[27.0\ d]{\beta^-} {}^{233}U \xrightarrow[1.59 \times 10^5\ y]{\alpha}$$

The long life of ^{233}U permits its separation, purification and stockpiling (in *subcritical* amounts) as in the case of ^{239}Pu.

10.4.1 From Monozite to Thorium Metal (or Oxide)

The commonest ore of thorium is monozite sand which is a complex phosphate of thorium, cerium, other lanthanides and some uranium*. Following is one of the ways of getting thorium metal or the oxide ThO_2 from monazite.

The metallurgy of thorium is complicated. Being radioactive all precautions against radiation hazard have to be observed. Some of the major stages involved are shown below:

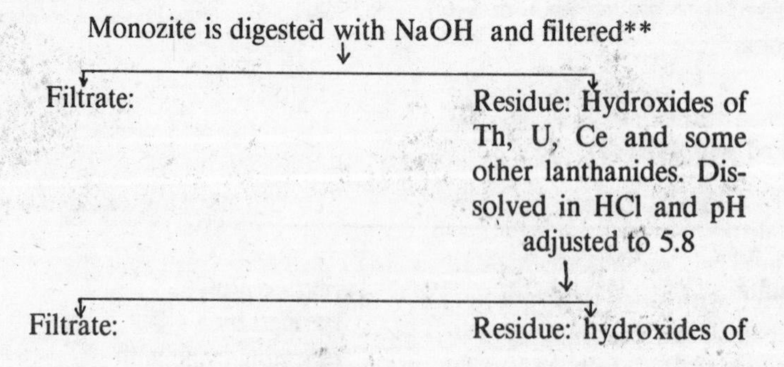

* India has vast deposits of monozite on the coast of Kerala.

** When monozite is heated the occluded helium is liberated at the rate of approximately one cm^3 at S.T.P. per gram. This is an important source of helium.

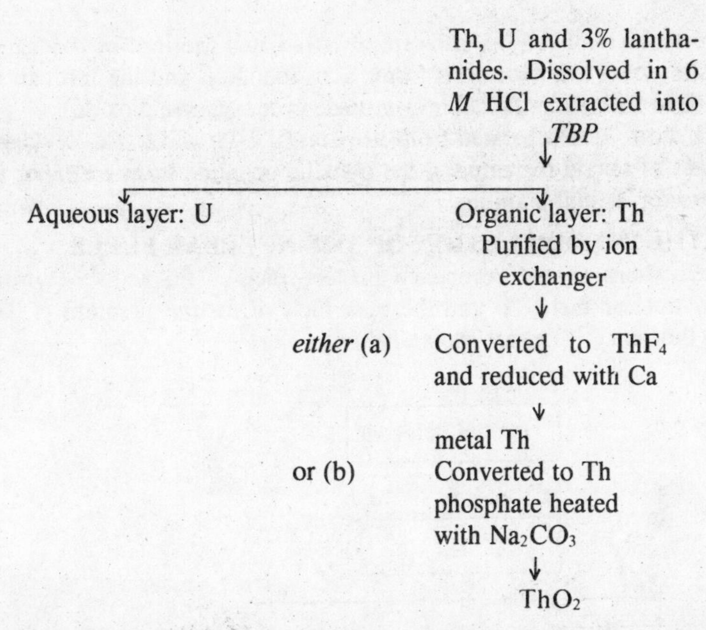

It is of the utmost importance to ensure that the thorium is all free from traces of uranium (^{238}U) whose presence would isotopically dilute the ^{233}U to be prepared and render it less efficient as a nuclear fuel.

10.4.2 From Thorium to ^{233}U

The thorium metal, or oxide, or the basic carbonate, is subjected to bombardment by thermal neutrons when the reactions described above take place leading to the formation of ^{233}U.

The separation of ^{233}U from thorium

The problem of separating very small amounts of ^{233}U from a large excess of thorium and associated fission fragments is similar to that encountered earlier in the recovery of ^{239}Pu from the spent uranium fuel rods. ^{232}Th exposed to neutrons of flux 10^{10} n cm^{-2} s^{-1} would be in equilibrium with the daughter elements ^{233}Pa, some 5 kCi per tonne of Th, and about 200 mg of ^{233}U. Here too the procedure followed is similar to the one used for the recovery of ^{239}Pu. First, the irradiated thorium is "cooled" to an aging period of 5-6 half-lives of ^{233}Pa (*i.e.* about as many months). This also decays down the fission fragments and brings down the general radiation level.

The Th-U separation, specially when the latter is in a minute proportion, is beset with analytical problems. While Th and U resemble completely in the + 4 oxidation state, there is no convenient compound of uranium in + 6 state which is insoluble under conditions when Th is in solution. Solvent extraction is the only practical way out of the difficulty. The "aged" Th + ^{233}U together with the fission fragments is dissolved in aqueous nitric acid and ether added when $^{233}UO_2.(NO_3)_2$ salts out and is thus separated from the thorium which goes into the organic layer. Alternatively, most of

^{233}Pa may be removed from the freshly irradiated thorium by oxidising it to Pa^{5+} state which on hydrolysis forms a radiocolloid and the last can be co-precipitated with either tantalum pentoxide or manganese dioxide.

Once the ^{233}U is separated from its parents ^{232}Th and ^{233}Pa, its subsequent treatment to obtain the same in the metallic or oxide form is similar to that described for natural uranium.

10.5 THE INTERRELATION OF THE NUCLEAR FUELS

The derivations of the secondary nuclear fuels ^{239}Pu and ^{233}U from the primary nuclear fuel ^{235}U and the bulk fuels of natural uranium (^{238}U) and natural thorium (^{232}Th) are shown below.

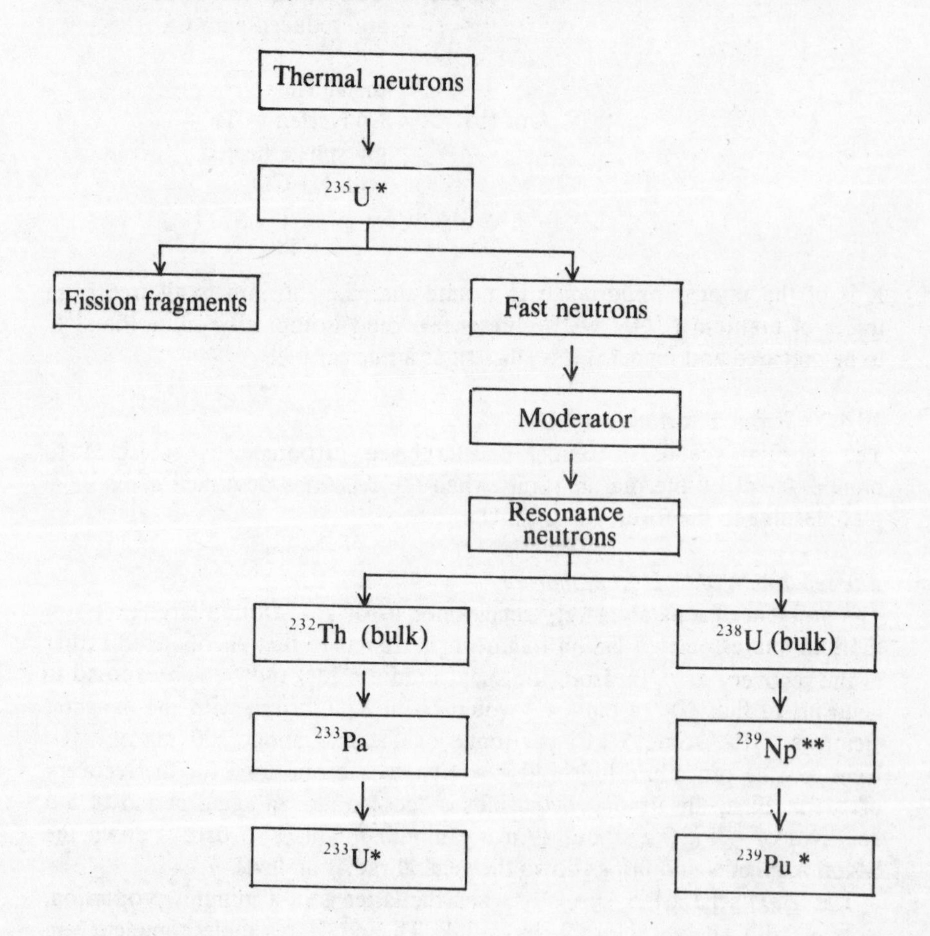

Scheme showing interrelation of nuclear fuel nuclides

References

1. H.D Smyth *Atomic Energy for Military Purposes,* Princeton University Press, Princeton, N.J. 1945; *Rev, Mod. Phys.,* 1945, 17, *351.*
2. A.O. Nier. *Rev. Sci. Instruments,* 1940, *11,* 212
3. Karlsruher Nuklidekarte, 4 Auflage. 1974.
4. I. Kaplan, *Nuclear Physics,* Addision Wesley, Reading MA, 1963
5. R.W. Lamphere, *Phys. Rev.,* 1956, *104,*1654.
6. H.J. Arnikar, *Essentials of Nuclear Chemistry,* Wiley Eastern Ltd., New Delhi II Ed, 1987.
7. H. London, *Separation of Isotopes,* G. Newnes, London, 1961.
8. C. Cassignol, *Proc. Internat. Symp. on Isotope Separation,* N. Holland Pub. Co., Amsterdam, 1957.
9. M. Chemla et J. Périé, "*La Separation des Isotopes*", Presses Universitaires de France, Paris 1974.

9.*a* R. Bernas, *J. Phys. Radium,* 1960, *21,* 191 A.

10. E-W. Becker, K. Bier and H. Burzhoff, *Z. Naturf.* 1955, *10a* 565.
11. S. Chapman and T.G. Cowling, *The Mathematical Theory of Non-uniform Gases,* Cambridge Univ. Press, 1939.
12. S. Villani, *Isotope Separation,* American Nucl. Science Soc. Monograph 1976.

12.a E.W. Becker *et al. Z. Naturf.,* 1954, *9a,* 975; 1955, *10a,* 565.

13. H. Martin and W. Khun, *Z. Phys. Chem.,* 1940, *189,* 219.
14. K. Cohen, *The Theory of Isotope Separation as Applied to Large Scale Production of* ^{235}U, Nat. Nucl Energy Series, Manhatten Project, Vol IB., McGraw-Hill Book Co., New York, 1951.

14.a K. Beyerle and W. Groth, *Proc. Intern. Symp on Isotope Separation,* (N. Holland Publications 1958.)

15. K. Clusius and G. Dickel, *Naturwiss,* 1938 *26,* 546; *Z. Phys. Chem.,* 1939, *B44,* 397, 451; 1942, *B52,* 348; 1944, *193,* 274.
16. K.E. Grew and T.L. Ibbs, *Thermal Diffusion in Gases,* Cambridge Univ. Pres, 1952.
17. J.F. Ready, *Industrial Applications of Lasers,* Academic Press, New York, 1978.
18. B.B. Snavely, *Proc. Conference on Laser-75 Opto-Electronics,* Munich, 1975.
19. G.S. James *er al.,* I.E.E.E. *J. Quantum Electronics,* 1976, *12,* 111.
20. F.S. Becker, and K.L. Kompa, *Nucl. Tech.,* 1982, *58,* 329.
21. J.W.T. Spinks and R.J. Woods, *An Introduction to Radiation Chemistry,* John Wiley & Sons, Inc., New York, 1964.
22. A.J. Swallow, *Radiation Chemistry of Organic Compounds,* Pergamon Press, Oxford, 1960.
23. E.M. McMillan and P.H. Abelson, Phys. Rev., 1940, *57,* 1185 L.

Chapter 11

RADIOISOTOPES

11.1 NATURAL AND MAN-MADE RADIOISOTOPES

We have on the earth 81 elemental forms of matter which together with their stable isotopes constitute some 274 stable nuclides. On the present understanding, a nuclide is considered stable if its half-life is longer than 10^{12} years.* As against this, some 2000 nuclides of widely varying half-lives have been reported, all but some 60 of which are man-made.

Till 1934, the only radionuclides available were the long-lived naturally occuring ones as ^{3}H, ^{14}C, ^{40}K, ^{87}Rb and the initiators of the three disintegration series ^{232}Th, ^{235}U and ^{238}U together with some 45 of their descendents with half-lives varying from 0.3 μs (^{212}Po) to 14 billion years (^{232}Th). This collection started growing suddenly in 1934 when Frédéric and Irène Joliot-Curie created the 3 minute radiophosphorus (^{30}P), the 5 s radiosilicon (^{27}Si) and the 10 minute radionitrogen (^{13}N) by bombarding with α particles common elements aluminium, magnesium and boron respectively. This was the birth of *artificial radioactivity.* Soon Fermi added to the list 37 radioisotopes of common elements by bombarding them with neutrons. Today, one can prepare a radioactive isotope of almost any element, identify it and put it to the beneficial use of mankind as excellent examples of the peaceful uses of atomic energy.

Besides bombardment by neutrons and α particles obtainable from natural radioelements, a variety of alternative and more powerful techniques have been developed for the synthesis of radioisotopes. High voltage particle accelerators as the cyclotron and synchrotron are available in specialized forms capable of providing a variety of projectiles of wide ranges of mass and energy. These accelerated heavy ions are being successfully employed in synthesizing a host of new nuclides with large excess, or deficiency, of neutrons, the *exotic* nuclides.

With the discovery of nuclear fission by Hahn and Strassmann in 1939 yet another avenue was opened for obtaining radionuclides, mainly in the form of short-lived fission fragments of middle mass elements. A sequel to

* It is possible that this limit may be further pushed up in the future with the perfection of techniques to detect feebler and slower decay processes.

the studies of fission was the construction of the nuclear reactor which serves both as a power generator and as a facility for irradiating selected target substances to furnish corresponding radioisotopes. Today, the reactor is the principal mode of obtaining on a commercial scale radioisotopes used in research, medicine and industry.

These developments of isotope production have been well matched by progress in the techniques of their analysis and identification. Sophisticated techniques exist for the on-line analysis and characterization of extremely short-lived species, as they are created amongst the fission fragments, or the recoils in the spallation due to high energy particles, or by radiation emerging from accelerating machines. These studies of high energy photonuclear reactions, which simulate fission on one hand and spallation on the other, of direct target reactions and of deep inelastic collisions, are providing new results leading to a better understanding of the nuclear processes.

11.2 UTILITY OF RADIOISOTOPES

Radioisotopes can be used to serve as probes at the atomic level and reveal how a reaction proceeds, be it in *vitro* or *vivo*[1]. If to a substance participating in a reaction, a very small amount of a radioisotope of the element in the same chemical form is added, the two isotopic molecules function identically*. The only difference which the radioisotope makes is in the emission of radiation signals $\beta^{\pm}$ and/or γ which can be detected with a great precision. Thus, the movements of the reactant atom get monitored from the initial state, through the intermediates, to the end product. Examples of the use of *radioisotopes as tracers,* will be presented in the next chapter.

11.3 THE MORE COMMONLY USED RADIOISOTOPES

It is only about 100-150 radioisotopes which find common applications in scientific research, medicine, industry and agriculture. These are shown in the form of a Periodic Table (Fig. 11.1). The principal modes of their preparations, and their radioactive characteristics are presented in Appendix 1.

11.3.1 Sources of Radionuclides

Following are the principal sources of radionuclides:

(i) *Natural nuclear reactions caused mainly by neutrons of cosmic origin.* The natural formation of radionuclides ^{3}H and ^{14}C by neutrons with atmospheric nitrogen are examples of this type.

$$^{14}N + n \rightarrow {}^{12}C + {}^{3}H$$
$$^{14}N + n \rightarrow {}^{14}C + {}^{1}H$$

* This is strictly true only under conditions that there is no significant isotope effect (§ 1.4.2).

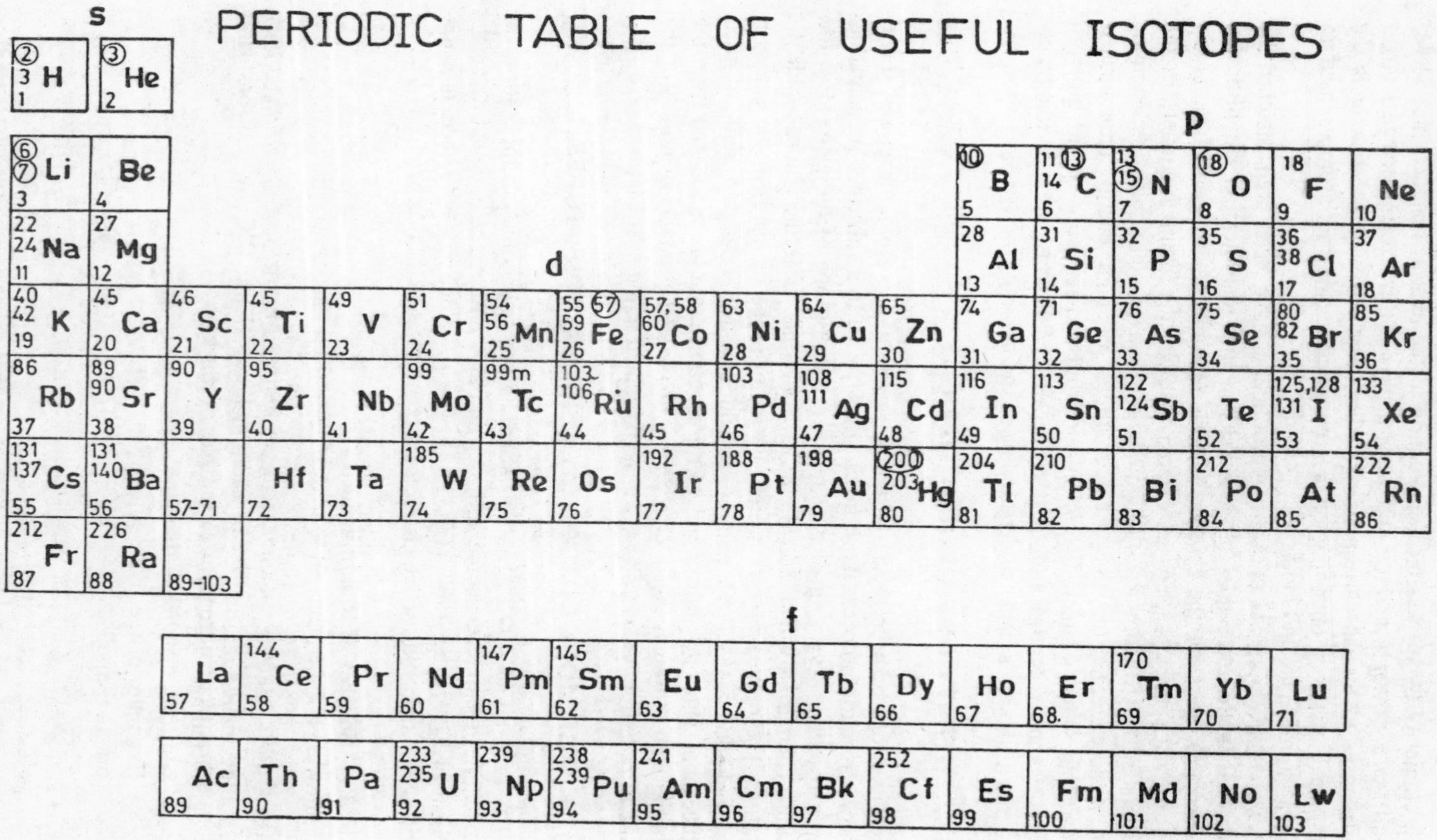

PERIODIC TABLE OF USEFUL ISOTOPES

s		d										p					
H 1: (2), 3																	He 2: (3)
Li 3: (6), (7)	Be 4											B 5: (10)	C 6: 11, (13), 14	N 7: 13, (15)	O 8: (18)	F 9: 18	Ne 10
Na 11: 22, 24	Mg 12: 27											Al 13: 28	Si 14: 31	P 15: 32	S 16: 35	Cl 17: 36, 38	Ar 18: 37
K 19: 40, 42	Ca 20: 45	Sc 21: 46	Ti 22: 45	V 23: 49	Cr 24: 51	Mn 25: 54, 56	Fe 26: 55, (57), 59	Co 27: 57, 58, 60	Ni 28: 63	Cu 29: 64	Zn 30: 65	Ga 31: 74	Ge 32: 71	As 33: 76	Se 34: 75	Br 35: 80, 82	Kr 36: 85
Rb 37: 86	Sr 38: 89, 90	Y 39: 90	Zr 40: 95	Nb 41	Mo 42: 99	Tc 43: 99m	Ru 44: 103, 106	Rh 45	Pd 46: 103	Ag 47: 108, 111	Cd 48: 115	In 49: 116	Sn 50: 113	Sb 51: 122, 124	Te 52	I 53: 125, 128, 131	Xe 54: 133
Cs 55: 131, 137	Ba 56: 131, 140	57-71	Hf 72	Ta 73	W 74: 185	Re 75	Os 76	Ir 77: 192	Pt 78: 188	Au 79: 198	Hg 80: (200), 203	Tl 81: 204	Pb 82: 210	Bi 83	Po 84: 212	At 85	Rn 86: 222
Fr 87: 212	Ra 88: 226	89-103															

f

La 57	Ce 58: 144	Pr 59	Nd 60	Pm 61: 147	Sm 62: 145	Eu 63	Gd 64	Tb 65	Dy 66	Ho 67	Er 68	Tm 69: 170	Yb 70	Lu 71
Ac 89	Th 90	Pa 91	U 92: 233, 235	Np 93: 239	Pu 94: 238, 239	Am 95: 241	Cm 96	Bk 97	Cf 98: 252	Es 99	Fm 100	Md 101	No 102	Lw 103

Fig. 11.1 : More commonly used isotopes of elements with atomic numbers at bottom corners. Stable isotopes are circled.

(ii) *Decay of Natural Radioelements of the 4n* (^{232}Th), *4n* + 2 (^{238}U) and 4*n* + 3 (^{235}U) series. Of about 45 radionuclides produced in the decay of these natural radioelements only some 10-15 are important. These are:

(*a*) *from* ^{232}Th *series:* ^{232} Ra (*Ms Th*). ^{212}Po (*Thc′*) *and* ^{208}Tl (*Thc″*)
(*b*) *from* ^{238}U seıies: ^{238}U. ^{234}Pa, ^{226}Ra, ^{222}Rn, ^{210}Bi (*RaE*), and ^{210}Po (*RaF*) and ^{210}Pb (*RaD*).
(*c*) *from* ^{235}U *series:* ^{235}U (*Ac-U*), ^{223}Fr and ^{219}At

(iii) *Fission Products*
While some 200 fission products are known, only some as ^{85}Kr, ^{90}Sr (+ Y) ^{99}Mo (+ Tc), ^{131}I, ^{133}Xe, ^{131}Cs (+ Ba), ^{144}Ce and ^{147}Pm are isolated on account of their use in research and other areas.

(iv) *Bombardment of substances by nuclear particles of natural origin.* The celebrated Nobel prize winning experiments of Irène and Frédéric Joliot-Curie (1934) of preparing the first artificial radioisotopes using α from ^{212}Po come under this category*. Some of these reactions are:

$$\alpha + {}^{27}\text{Al} \rightarrow n + {}^{30}\text{P} \xrightarrow[3\text{ min}]{\beta^+} {}^{30}\text{Si}$$

$$\alpha + {}^{24}\text{Mg} \rightarrow n + {}^{27}\text{Si} \xrightarrow[5\text{ s}]{\beta^+} {}^{27}\text{Al}$$

$$\alpha + {}^{10}\text{B} \rightarrow n + {}^{13}\text{N} \xrightarrow[10\text{ min}]{\beta^+} {}^{13}\text{C}$$

Neutrons from laboratory sources as (Ra + Be) and (Pu + Be) are routinely used for obtaining short-lived radioisotopes for research, mainly by (n, γ) reactions, *e.g.* ^{23}Na (n, γ) ^{24}Na, similarly ^{32}P, ^{38}Cl, ^{56}Mn, ^{76}As, ^{128}I, and many others.

(v) *Bombardment of substances by high energy accelerated particles and radiation.* A variety of high voltage machines exist which can accelerate charged particles as H^+, $^{2}H^+$, $^{7}Li^{3+}$, $^{12}C^{6+}$, $^{14}N^{6+}$, $^{20}Ne^{6+}$ etc. to fixed energies of 4 or 10 MeV (linear acceleators), or 100 MeV and beyond (cyclotrons, van de Graaff generators, synchrotrons). In addition, neutron generators based on $D + T$ reaction provide monoergic (14 MeV) neutrons. One can also avail of short pulses (μs to ns) of radiation of 10-20 MeV. All these sources have been used in effecting a large variety of nuclear reactions resulting in the formtion of many new radionuclides including some *exotic* nuclides, bearing either a vary large excess of neutrons, or a heavy deficit of neutrons (e.g. ^{20}Na to ^{33}Na, ^{32}Cl to ^{42}Cl) Examples of the production of radiosiotopes by these means and bearing practical applications are:

* In the earlier experiment of Rutherford (1919), of bombarding nitrogen with α yielded a *stable* and not a radioisotope.

$$\alpha + {}^{14}\text{N} \rightarrow {}^{17}\text{O} + p$$

^{18}F, ^{22}Na, ^{55}Fe and some others. The reactions leading to their production are:

$$^{19}F\,(n,\,2n)\,^{18}F \xrightarrow[110\text{ min}]{\beta^+} {}^{18}O$$

$$^{24}Mg\,(d,\,\alpha)\,^{22}Na \xrightarrow[2.6\text{ y}]{\beta^+} {}^{22}Ne$$

$$^{55}Mn\,(n,\,p)\,^{55}Fe \xrightarrow[2.7\text{ y}]{\beta^+} {}^{55}Mn$$

Other examples include the formation of high Z elements, the transcurides, as:

$$^{238}U\,(^{12}C,\,4n)^{245}Cf; \qquad ^{238}U\,(^{14}N,\,6n)^{246}Es;$$

$$^{238}U\,(^{16}O,\,xn)^{254-x}Fm; \qquad ^{246}Cm\,(^{12}C,\,6n)^{252}No;$$

$$^{252}Cf\,(^{11}B,\,xn)^{263-x}\,Lw;$$

and several photonuclear and spallation reactions as

$$^{19}F\,(\gamma,\,2n)^{17}F;$$

$$^{75}As\,(\gamma,\,\alpha p2n)^{68}Zn \text{ (spallation)}$$

(vi) *"Cow-milking" systems*

Sometimes a radioactive nuclide is in equilibrium with her daughter nuclide which is also radioactive but with a shorter life-time than the parent, ($\tau_P > \tau_d$; or $\lambda_p < \lambda_d$)*. It can be shown that on the attainment of equilibrium, the parent and the daughter decay at the same rate (equal to that of the parent) and the ratio of the concentrations N_d/N_p, or the amounts of daughter to parent, remains constant, i.e.

$$\frac{N_d}{N_p} = \frac{\lambda_p}{\lambda_d - \lambda_p} \qquad (11.1)$$

This is known as *transient equilibrium.* In the special case where $\lambda_d \gg \lambda_p$, Eq 11.1 simplifies to

$$\frac{N_d}{N_p} = \frac{\lambda_p}{\lambda_d} \qquad (11.2)$$

The equilibrium is then referred to as a *secular* one.

In both cases of equilibrium, if the daughter species be separated from the parent by physical or chemical means, the same will start growing again in

*The subscripts p and d stand for parent and daughter, used with τ (half-life period). λ (decay constant) or N (the number atoms).

the parent species to a maximum concentration *i.e.* till the two are once again in equilibrium, in about one or two *mean life-times* $\bar{\tau}$

$$\bar{\tau} = \frac{1}{\lambda_d} = \frac{\tau_d}{0.693} = 1.443\ \tau_d$$

At this stage, the cow (the parent) can be 'milked' again to collect a second quantity of the daughter species. This 'milking' process can be repeated many times as long as the parent activity is high enough. That is why these are referred to as *cow-milking systems.* Normally, some 90 per cent of the daughter species can be separated each time without altering the activity of the parent in anyway. Following are common examples of cow-milking systems.

(1) *Radium-Radon* The secular equilibrium is between

$$^{226}\text{Ra} \xrightarrow[1600\ y]{\alpha} {}^{222}\text{Rn} \xrightarrow[3.82\ d]{\alpha} {}^{218}\text{Po} \xrightarrow{\alpha}$$

In this case, radon being a gas, the milking operation is simple. One has only to heat the radium compound and collect the radon once every 4 or 5 days. Also in this case the cow can supply milk for about 2000 years or more.

(2) *Molybdenum-Technecium* The transient equilibrium is between

$$^{99}\text{Mo} \xrightarrow[66.0\ h]{\beta^-} {}^{99m}\text{Tc} \xrightarrow[6.0\ h]{\text{I.T.}} {}^{99}\text{Tc}$$

It may be recalled that Tc is a wholly man-made element. The isomer ^{99m}Tc in the form of sodium pertechnatate ($NaTeO_4$) is a valuable radiopharmaceutical widely used in various forms for scanning cancerous condition of different organs. The milking system as such is marketed (vide § 11.4.4). The parent ^{99}Mo is loaded on an alumina or a resin column, and eluting the same with 0.9 per cent NaCl solution once every day the isomer ^{99m}Tc can be milked out, while the parent recreates the same at the same rate as shown in Fig. 11.2.

(3) *Strontium-Yttrium* The secular equilibrium is between

$$^{90}\text{Sr} \xrightarrow[28.5\ y]{\beta^-} {}^{90}\text{Y} \xrightarrow[61.7\ h]{\beta^-} {}^{90}\text{Zr}$$

Simple chromatography on asbestos paper with dilute HCl as solvent, separates the two species, the R_F values of Sr and Y being 0.76 and 0.18 respectively[1a].

11.3.2 **Recovery of Radioisotopes**

For the recovery of the radioisotopes, resulting in any of the above processes, from the target and other substances mixed with it, one or the other of following physico-chemical techniques is made use of.

(1) *The Recoil Effect: The Szilard-Chalmers' Reaction*

In a nuclear reaction wherein a particle or a photon escapes with a momentum p, the residual product atom must recoil with an equal and opposite momentum such that

$$mv + p = 0 \tag{11.0}$$

where m and v are the mass and velocity of the recoil atom. In the case where the ejectile is a photon of energy $E\gamma$ its momentum is $E\gamma/c$, where c

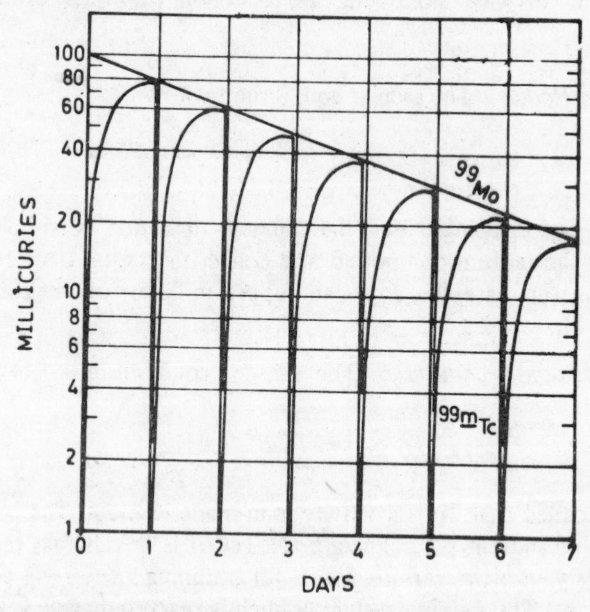

Fig. 11.2: **The Cow milking system:** Transient equilibrium between ^{99}Mo (66h) and ^{99m}Tc (6h)
(from *Radionuclides in Medicine and Biology,* by E.H. Quimby, S. Feitelberg and W. Gross; Lea and Febiger, Philadelphia, PA, 1970. Reproduced with permission.)

is the velocity of light. This leads to a simple expression for the energy of recoil in eV

$$E_r = \frac{1}{2} mv^2 = \frac{P^2}{2m} = \frac{E\gamma^2}{2mc^2} \tag{11.4}$$

or

$$E_r = 536\ E\gamma^2/m \text{ eV*} \tag{11.5}$$

---

* However, if two or more photons are emitted in different directions, the recoil energy would be much less. In the case of 2 photons of energies $E\gamma_1$ and $E\gamma_2$ going at an angle θ

$$E_r = \frac{E_{r1}^2 + E_{r2}^2 + 2E_{r1}\ E_{r2} \cos\theta}{2\ mc^2}$$

where $E\gamma$ is in MeV and m in atomic mass unit u. As $E\gamma$ in most cases is of the order of 1-5 MeV, the recoil energy of a medium mass element ($50 < m < 100$) comes to 5-250 eV. This energy is many times greater than the bond energy holding the atom in the molecule which rarely exceeds 5 eV. Hence the bond invariably breaks. This is the principle of the Szilard-Chalmers' reaction[2] of separating the radioisotope formed in an (n, γ) reaction. Irradiation of ethyl iodide by slow neutrons results in the formation of free iodine ^{128}I atoms by the rupture of the C-I bond, and if water be added most of the radioiodine passes into the aqueous layer.

Substances best suited for this mode of separation are organic chloro-, bromo- and iodo-compounds, stable complex ions, several oxyanions and organo-metallic compounds as As $(C_6H_5)_3$, Sn $(C_6H_5)_4$ and Pb $(C_6H_5)_4^1$.

The principle of separation by the recoil effect can be availed of in other nuclear reactions as well where α or β is emitted, or in isomeric transitions involving the emission of conversion electrons. The net momentum of the ejected particles has to be properly calculated. Nesmeynov[3] describes these recoil effects in detail.

(2) *Solvent Extraction*

On the addition of a mixed solvent (usually water + an immiscible organic liquid), the radioactive species enters preferentially into the water or the organic phase *e.g.* the separation of Pu from U by TBP (§ 10.3.2)

(3) *Coprecipitation with a Stable Isotopic Species*

To separate for example trace amounts of ^{128}I formed, one adds a few mg of stable iodide (KI) as isotopic carrier and some lead nitrate when the radioisotope coprecipitates with I as PbI_2. The method was first used by F and I. Joliot-Curie. The product of this method is not carrier-free.

(4) *Coprecipitation with an Isomorphous Non-Isotopic Species*

Sometimes a stable isotope may not be available, or the radioisotopic species may be required to be carrier-free. In such a case, an analogous element forming a similar compound is added and precipitated, *e.g.* trace amount of Ra can be co-precipitated in the presence of a barium salt by adding a sulphate solution. This method was first used by Madame Curie in her classical work.

(5) *Absorption-Coprecipitation*

Certain substances when freshly formed come down as a non-crystalline gelatinous phase with a large surface *e.g.* $Fe(OH)_3$ $Al(OH)_3$. If a radionuclide be present the same gets strongly adsorbed and coprecipitates. The precipitate can be dissolved in dilute acid and the iron extracted into an ether layer. The radioisotope remains carrier-free in the water layer.

(6) *Adsorption*

In some cases strong adsorbents, as activated charcoal, silica gel, ion exchanger resins etc. adsorb strongly the radionuclide. Thus freed from other materials, the adsorbate may be recovered by subsequent elution. The ^{32}P formed by *(n,p)* reaction on sulphur of carbon disulphide can be recovered by adsorption on charcoal. Sometimes the active species may form a radiocolloid and held up on a filter paper, *e.g.* the ^{27}Mg formed by *(d, 2p)* reaction on Al.

(7) *Distillation or Expulsion*

Where the radioisotopic product happens to be a gas or a volatile substance and the target is involatile, the latter after irradiation is melted or dissolved and a carrier gas bubbled through when the radionuclide gets expelled into the gas phase. Radioiodine and radiobromine, as well as radiokrypton and radioxenon formed in uranium fission, are thus recovered from the uranium target.

(8) *The Use of Charged Plates*

Pioneer workers on natural radioactivity had used charged plates for isolating certain radioelements. This method is based on the fact that radioactive atoms at the *moment of their formation* carry a large charge and can be separated to some extent from their parent substrate. Fermi, Rasetti *et al* [4]. Paneth[5] and others[6,7], showed that radioiodine can be collected in a wire-in-cylinder type discharge vessel containing methyl iodide vapour under neutron irradiation. Similarly, Saha[8] was able to collect ^{32}P forming in carbon disulphide by *(n,p)* reaction by immersing a pair of electrodes in the liquid. The technique was improved by Arnikar[1,9] who used two silver electrodes immersed in an organic halogen compound as ethyl bromide or iodobenzene and connected to a *dc* potential of 100-500 V while the system was under slow neutron irradiation. Washing the plates in the end with sodium thiosulphate solution yielded the radiohalogen in a state of high specific activity.

(9) *Leaching*

This consists in leaching (dissolving) out the active product by a suitable solvent in which the target is insoluble. This is the reverse of the precipitation process. As an example may be cited the leaching out of ^{22}Na formed on an MgO target by (d,α) reaction by an extremely dilute solution of sodium hydroxide. The latter substance also serves as a carrier.

(10) *Other Methods*

Several other common analytical techniques as paper chromatography, thin layer and ion exchange chromatography[12] have been successfully used for isolating radionuclides formed in any of the processes described under § 11.3.1.

11.4 PREPARATION OF COMMONLY USED RADIONUCLIDES

We shall now describe the preparation of some of the more commonly used radionuclides in the order of their atomic numbers. Besides the methods described here alternative methods are also adopted in some cases.

(1) ^{3}H (tritium) (decay : $^{3}H \xrightarrow[12.33\ y]{\beta^-} {}^{3}He$)

(a) *Formation in Nature*
Significant amounts of ^{3}H are formed continually in nature by the action of fast neutrons of cosmic origin (flux 2.6 n cm^{-2}s^{-1}) with the nitrogen of the atmosphere. The reaction $^{14}N\ (n,t)^{12}C$ was described under § 11.3.1. In due course the tritium gets oxidised to HTO and through rains reaches the seas, rivers and other large surfaces of water. The equilibrium ratio T/H in such open waters is $\sim 10^{-18}$. Once a finite amount of the water is withdrawn and its contact with the atmosphere is cut off, its T content decreases with its characteristic half-life.

(b) *By Neutron Irradiation of ^{6}Li*
The high neutron capture cross section ^{6}Li of the order of 940 barns is made use of in obtaining large quantities of tritium needed for studies of nuclear fusion reactions and for defence purposes, the reaction involved is $^{6}Li\ (n, \alpha)T$. ^{6}Li as metal or as ^{6}LiF is irradiated by neutrons in a reactor and the tritium is expelled by heating the target lithium either in vacuum or in the presence of small amount of hydrogen. It is purified through absorption in palladium and finally oxidised to water, T_2O or HTO, by passing over heated copper oxide.

^{3}H is marketed either as gas enriched to 95 per cent in pyrex ampoules, or as HTO (water) of specific activity 0.2-5 Ci/ml.

Uses
The major amount of tritium produced goes for defence purposes, for the making of hydrogen bombs. Small amounts in the form of *D-T* pellets, adsorbed on zinc or copper, are used in researches on thermonuclear fusion reactions. Tritium adsorbed targets are also used in neutron generators by $D + T$ reaction. Tritium labelled compounds are routinely used in research for studying reaction mechanisms. Microgram amounts of ^{3}H labelled compounds can be readily detected by autoradiography or by a liquid scintillation counter.

(2) ^{7}Be (decay : $\xrightarrow[53.4\ d]{\beta^+} {}^{7}Li$)

(a) The (n, α) reaction on ^{6}Li is followed by the capture of some of the tritium by the target giving $^{6}Li(t, 2n)^{7}Be$ reaction. In the end the target is dissolved in hot water and the ^{7}Be recovered by adsorption- coprecipitation with ferric hydroxide.

(b) Alternatively, ^{6}Li may be bombarded with deuterons when the reaction $^{6}Li(d,n)^{7}Be$ occurs. The lithium is dissolved out in dilute HCl and the pH raised to 9 when $Be(OH)_2$ separates as a radiocolloid which is retained on a glass filter

(3) ^{14}C (decay $^{14}C \xrightarrow[5730\,y]{\beta^-} {}^{14}N$)

(a) *Formation in Nature*

Large amounts of ^{14}C are formed continually in nature by an alternative reaction of fast neutrons of cosmic origin on the nitrogen of the atmosphere $^{14}N(n,p)^{14}C$ as described in § 11.3.1. In due course the ^{14}C changes into $^{14}CO_2$ which together with the ordinary CO_2 of the atmosphere is taken up by living plants to provide the food for their growth by a photosynthetic process. The $^{14}C/^{12}C$ in any part of a *living* plant or animal is the equilibrium mass ratio, corresponding to 1.6×10^{-12}, yielding 16.1 ± 0.3 disintegrations $min^{-1}g^{-1}$ (total C)[10]. However, once the plant or animal is dead, the ^{14}C in it decays with its characteristic half-life. Thus, the actual $^{14}C/^{12}C$ ratio in a specimen of plant or animal origin can be used to determine its 'age', *i.e.* the period it has been dead. This is the principle of Willard Libby's famous ^{14}C dating[1], discussed in § 12.6.

(b) *Reactor Irradiation of Be_3N_2 or AlN*

Any substance rich in nitrogen as beryllium or aluminium nitride, but free from impurities with high neutron capture cross section, when irradiated by the high flux slow neutrons of a reactor the same (*n,p*) reaction (as in above) occurs leading to the transmutation of ^{14}N into ^{14}C. The irradiated sample is dissolved in 65 per cent H_2SO_4 and H_2O_2 added when the radiocarbon is released as $^{14}CO_2$ together with smaller amounts of labelled CO, CH_4 and HCN. All these minor components are oxidised to $^{14}CO_2$ when the gas is passed over copper oxide heated to 750°C. Finally the $^{14}CO_2$ is absorbed in 60 per cent NaOH and precipitated as $Ba^{14}CO_3$, in which form it is usually marketed. The desired organic substance labelled with ^{14}C can be synthesized by releasing the $^{14}CO_2$ from the barium carbonate. A better absorber for CO_2 is methyl benzothonium hydroxide dissolved in methanol (1*M*).

Uses

^{14}C labelled compounds are widely used in research, in structure determination and for studying reaction mechanisms, and in the studies of photo and other biosyntheses, and particularly in rearrangement reactions. For the technical analysis of functional groups in organic substances, no more than 20 ng of ^{14}C labelled sample is needed while for the same analysis by most other spectroscopic methods some 10 mg of the sample would be necessary[11]. Typical applications of ^{14}C labelled compounds are considered in Ch.12.

(4) ^{13}N (decay : $^{13}N \xrightarrow[9.96\,min]{\beta^+} {}^{13}C$)

The isotope ^{13}N is formed by the (*d,n*) reaction on carbon. The target is pure graphite which is bombarded by high energy deuterons in a cyclotron. In the end the irradiated graphite is burnt up in oxygen and the CO_2 and $^{13}NO_2$ formed are liquefied and the condensed product is dissolved in CCl_4.

(5) (a) ^{22}Na (decay : $^{22}Na \xrightarrow[2.6\,y]{\beta^+} {}^{22}Ne$)

^{22}Na is a cyclotron product obtained by bombarding magnesium with 8.5 MeV deuterons when the reaction $^{24}Mg(d,\alpha)^{22}Na$ takes place. The product ^{22}Na may be recovered from the irradiated target in either of two ways.

(i) The target is dissolved in dilute HCl and the solution electrolysed at 12 V between a platinum anode and a revolving mercury cathode as in Fig. 11.3. The ^{22}Na amalgam formed is subsequently decomposed to yield ^{22}NaOH.

(ii) Alternatively, the HCl solution of the irradiated target material is neutralized and passed through a polythene chromatographic column packed with a cation exchange resin, using 0.1N HCl as the elutant. The magnesium is retained by the column.

(b) ^{24}Na (decay : $^{24}Na \xrightarrow[15.03\,h]{\beta^-} {}^{24}Mg$)

^{24}Na is a straight (n,γ) product obtainable by subjecting pure sodium carbonate to slow neutrons from a laboratory source or in a reactor.

Both ^{22}Na and ^{24}Na are useful radioisotopes of sodium, frequently used in research and in medicine. The shorter lived isotope (^{24}Na) is preferred for medical uses as it would be eliminated from the system in about three days' time.

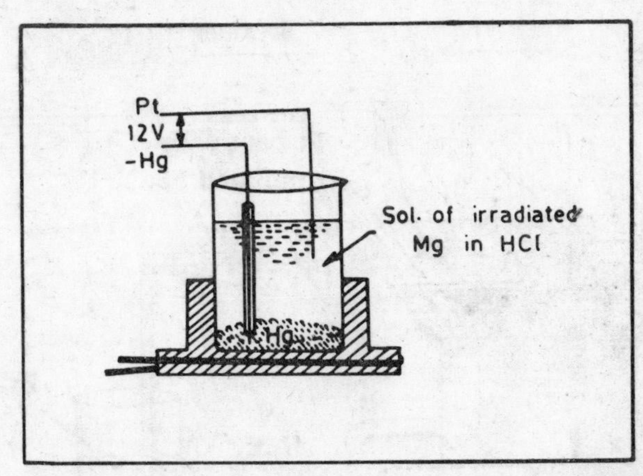

Fig. 11.3: **Separation of ^{22}Na formed in the (d, α) reaction on magnesium.**
(after M. Chemla et., J. Pauly, *Bull. Soc. Chim. France,* 1953, 432

^{24}Na is marketed as 0.9 per cent neutral NaCl solution of specific activity 1-20 mCi/ml.

(6) ^{32}P (decay : $^{32}P \xrightarrow[14.28\,d]{\beta^-} {}^{32}S$)

By (n,γ) reaction on red phosphorus (isotopic abundance 100 per cent), the resulting ^{32}P may be converted into any desired species as P_2O_5, PCl_3, etc. The process is less expensive, but the specific activity would be low as it would be mixed with excess of ^{31}P of the target.

(b) Alternatively, by (n, p) reaction on natural sulphur (^{32}S, abundance 95 per cent), the recovery of ^{32}P is effected by reacting the irradiated target in dilute NHO_3 in an autoclave. By boiling with *aqua regia* the initially formed $^{32}P_2S_5$ is oxidised to $H_3{}^{32}PO_4$. The phosphate, and impurity arsenate, ions are precipitated with ammonaical $La(NO_3)_3$. The lanthanum is removed by passing the HCl solution of the precipitate through a cation exchange resin. This method is preferred if high specific activity ^{32}P is desired.

(c) The collection of ^{32}P on charged plates of the (n, p) reaction on CS_2 was described in the last section.

^{32}P is marketed in a carrier-free state in the form of H_3PO_4 in 0.01 N HCl solution, or as Na_2HPO_4 in 0.9 per cent neutral NaCl solution with a specific activity of about 2 mCi/ml.

(7) ^{35}S (decay : $^{35}S \xrightarrow[87.5\,d]{\beta^-} {}^{35}Cl$)

The irradiation of pure NaCl by high flux neutrons in a reactor results in the reaction $^{35}Cl(n,p)^{35}S$. Side by side ^{24}Na, ^{38}Cl, and ^{32}P are formed by (n,γ) and (n,α) reactions on the Na and Cl of NaCl. The irradiated NaCl is allowed to age for a week when the short-lived ^{24}Na and ^{38}Cl decay. It is then heated in a platinum boat in a slow current of pure dry H_2 and the exit gas $H_2{}^{35}S$ is bubbled through bromine water which converts it into $H_2{}^{35}SO_4$

$$^{35}S \xrightarrow{H_2} H_2{}^{35}S \xrightarrow{\text{Br-water}} H_2{}^{35}SO_4$$

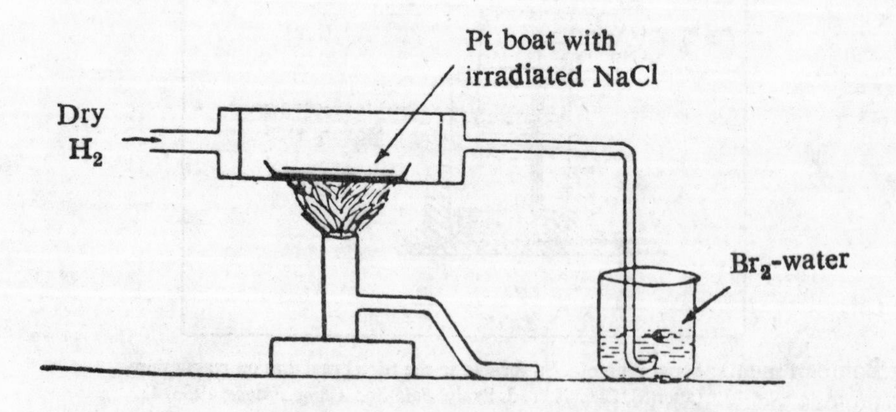

Fig. 11.4: **Separation of ^{35}S from neutron irradiated sodium chloride.**

Alternatively, the H_2S is reacted with cynamide which fixes it as thiourea.

$$H_2{}^{35}S + NH_2CN \longrightarrow C^{35}S(NH_2)_2$$

^{35}S is marketed either as carrier-free $H_2{}^{35}SO_4$ or as $Na_2{}^{35}SO_4$ in neutral 0.9 % NaCl solution.

(8) ^{42}K (decay : $^{42}K \xrightarrow[12.36\,h]{\beta^-} {}^{42}Ca$)

Metallic potassium on bombardment with deuterons yields ^{42}K by (*d, p*) reaction on ^{41}K (abundance 6.7 per cent). The irradiated target is dissolved in water, neutralized with HCl and potassium cobaltinitrite is precipitated. It is marketed as 1.2 per cent KCl solution of specific activity 0.5 - 2.0 mCi/ml.

(9) ^{51}Cr (decay : $^{51}Cr \xrightarrow[27.7\,d]{\beta^+} {}^{51}V$)

^{51}Cr results by (*p, n*) reaction when ^{51}V (abundance 99.75%) is bombarded by 0.5 -10 MeV protons.
The irradiated vanadium is dissolved in 6N HNO_3. To this are added $FeCl_3$ and boiling NaOH in excess when $^{51}Cr(OH)_3$ coprecipitates with $Fe(OH)_3$. If necessary the ^{51}Cr may be oxidized to the chromate state. It is marketed as $^{51}CrCl_3$ in HCl solution or as $Na_2{}^{51}CrO_4$ in 0.9 per cent neutral solution. The specific activity is high being around 10-50 mCi/mg Cr

(10) ^{54}Mn (decay : $^{54}Mn \xrightarrow[312.2\,d]{\beta^+} {}^{54}Cr$)

On deuteron irradiation of natural chromium the ^{53}Cr (abundance 9.5 per cent) yields ^{54}Mn by (*d, n*) reaction. The irradiated chromium is dissolved in 12 N HCl and evaporated nearly to dryness. The residue is dissolved in water and to this are added solid $FeCl_3$ and alkaline bromine solution. The chromium remains in solution as CrO_4^{2-} and the ^{54}Mn coprecipitates with $Fe(OH)_3$. The precipitate is dissolved in 6M HCl and on adding ether the ^{54}Mn stays in the aqueous phase while the iron goes into the ether phase. ^{54}Mn is marketed as carrier-free $MnCl_2$ solution in HCl.

(11) (a) ^{55}Fe (decay : $^{55}Fe \xrightarrow[2.7\,y]{EC} {}^{55}Mn$)

Bombardment of manganese ^{55}Mn (abundance 100 per cent) by accelerated protons yields ^{55}Fe by (*p, n*) reaction. The irradiated target is dissolved in 6M HCl. The iron is precipitated with cuperron. The precipitate is washed, dried and redissolved in HCl. These processes are repeated several times. The ^{55}Fe is marketed either as $FeCl_3$ solution in HCl or as ferric citrate in aqueous solution (pH = 5), the specific activity in either case being about 4 mCi/mg.

(b) ^{59}Fe (decay : $^{59}Fe \xrightarrow[44.6d]{\beta^-} {}^{59}Co$)

One way of obtaining ^{59}Fe would be by (*n, p*) reaction on natural cobalt ^{59}Co (abundance 100 per cent). This isotope is also marketed either as $FeCl_3$ in HCl solution or as ferric citrate solution (pH= 5), the specific activity is also the same as ^{55}Fe, *viz.* 4 mCi/mg.

(12) (a) ^{58}Co (decay : $^{58}Co \xrightarrow[70.79\ d]{\beta^+} {}^{58}Fe$)

One way of obtaining ^{58}Co would be by (*n, p*) reaction on ^{58}Ni (abundance 67.8 per cent). It is marketed as carrier-free $^{58}CoCl_2$ solution in HCl.

(b) ^{60}Co (decay : $^{60m}Co \xrightarrow[10.5\ min]{I.T.} {}^{60}Co \xrightarrow[5.27\ y]{\beta^-,\gamma} {}^{60}Ni$)

Pencils of cobalt ^{59}Co (abundance 100 per cent) on neutron irradiation in a reactor yield ^{60}Co, by (n,γ) reaction. This is an important source of γ radiation in the laboratory, in medicine and in industry. It is marketed for industrial applications in various forms and activities from a few curies to a megacurie, duly shielded in the form of gamma cells or chambers. Smaller activities are marketed as $^{60}CoCl_2$ solution in HCl of activity of 10 mCi/mg Co. For medicinal purposes, as an efficient substitute of very expensive radium, ^{60}Co is available in the form of small tubes or needles of length 5-40 mm, or seeds encapsulated in Pt-Rh alloy.

(13) (a) ^{128}I (decay : $^{128}I \xrightarrow[25\ min]{\beta^-/EC} {}^{128}Xe;/{}^{128}Te$)

^{128}I is a straight (n, γ) irradiation product of any iodine compound (natural abundance of ^{127}I being 100 per cent). Any neutron source can bring about this reaction.

(b) ^{131}I (decay : $^{131}I \xrightarrow[8.04\ d]{\beta^-} {}^{131}Xe$)

(i) *From Tellurium*
^{130}Te (natural abundance 34.5 per cent) on neutron irradiation leads to the formation of ^{131}I.

$$^{130}Te\ (n,\gamma)\ {}^{131}Te \xrightarrow[25\ min]{\beta^-} {}^{131}I)$$

The irradiated tellurium or tellurium oxide is dissolved in $HNO_3 + H_2SO_4$ mixture. A small amount of iodine is added as a carrier and the mixture distilled into CCl_4. The latter is shaken up with water containing sodium bisulphite when iodine is extracted into the water layer as NaI.

(ii) *From Fission Products*
However, the more important source of ^{131}I is the product of uranium fission

which contain several isotopes of iodine, as $^{127, 129, 131, 132}$ and ^{133}I. Of these ^{131}I forms about three per cent of the fission products. The recovery of the valuable isotopes from the irradiated uranium is carried out on a large scale by special remotely controlled equipment behind shielded walls. The spent uranium is dissolved in HNO_3 and on heating the mixture most of the iodine vaporizes and this is collected in a cold condenser at 3 °C. There are several alternative methods of recovering radioiodine. ^{131}I is marketed in various forms as carrier-free solution of NaI in $Na_2S_2O_3$ and in different radio-pharmaceutical forms as Rose Bengal, radioiodinated serum albumin, hippuran, diiodofluorescein and several others needed for special medical purposes.

(14) ^{137}Cs (decay : $^{137}\text{Cs} \xrightarrow[30.1y]{\beta^-} {}^{137}\text{Ba}$)

We shall conclude this chapter with a mention of another radioisotope ^{137}Cs of fission origin and of great industrial use as a long-lived source of γ radiation. The fission products are found to contain over ten isotopes of Cs besides the stable one ^{133}Cs. The recovery of ^{137}Cs depends on the procedure adopted for the separation of Pu from U (§ 10.3.2.). In the TBP process, the waste solution contains a large amount of $NaNO_3$ while in the hexone process the waste solution contains a large amount of $Al(NO_3)_3$. To these waste solutions hot K_2SO_4 is added to saturation. On cooling, double sulphates (alums) separate out which carry nearly the whole of Cs present. By repeated dissolution and recrystallization, pure Cs alum ($CsAl(SO_4)_2 12 H_2O$) separates out containing all the ^{137}Cs, while other alkali and ammonium ions remain in solution

11.5 PURITY AND STRENGTH OF RADIOISOTOPES

Radioisotopes may not always be *radiochemically pure* i.e. they may not be free from other radioisotopes formed from impurities in the target substance, and/or due to alternative nuclear reactions sometimes possible with the same substance. Similarly, a *carrier-free* radioisotope only means that it is free from the presence of *stable* isotope(s) of the same element, but may not be radiochemically pure. Lastly, it should be noted that all isotope marketed, except in the elementary form, are unavoidably associated with some other salts, acids or bases. For instance, carrier-free ^{131}I is in the form of NaI dissolved in Na_2SO_3 solution at a pH of 8, or in neutral phosphate buffer solution. Similarly, carrier-free ^{32}P is in the form of H_3PO_4 in 0.01*N* HCl, or as Na_2HPO_4 in neutral 0.9 per cent NaCl solution, with a specific activity of 2 mCi/ml.

References

1. H.J. Arnikar, *Essentials of Nuclear Chemistry,* Wiley Eastern Ltd, II Ed. 1987.
1a. H.J. Arnikar, and O.P. Mehta, *Current Science*, 1959, *28*, 400.
2. L. Szilard and T.A. Chalmers, *Nature*, 1934, *134*, 462.
3. An.N. Nesmeynov, *Radiochemistry*, Mir Publishers, Moscow. 1972

4. E. Amaldi, O.D. 'Agostino, E. Fermi, B. Pontecorvo, F. Rasetti and E. Segre, *Proc. Roy. Soc.*, (London), 1935, *A 149*, 522.
5. F. Paneth and J. Fay, *Nature*, 1935, *135*, 820; *J. Chem. Soc.*, 1936, 384.
6. P.C. Capron, G. Stokkink, and M. Van Meersche, *Nature*, 1946,*157*, 806.
7. S. Wexler, T.H. Davies, *J. Chem. Phys.*, 1952, *20*, 1688.
8. N.K. Saha and L.K. Rangan, *Ind. J. Phys.*, 1956, *30*, 80.
9. H.J. Arnikar, *J. Phys. Radium*, 1962, *23*, 578; *Artificial Radioactivity*. (Ed. K.N. Rao and H.J. Arnikar, Tata McGraw-Hill Publishing Co, 1985).
10. E.C. Anderson and W.F. Libby, *Phys. Rev.*, 1951, *81*, 64.
11. H.J.M. Bowen, *Chemical Applications of Radioisotopes*, Methuen and Co. Ltd., London, 1969.
12. S.G. Thompson, B.G. Harvey, G.R. Choppin and G.T. Seaborg, *J. Amer. Chem. Soc.*, 1954, *76*, 6229.

Chapter 12

USES OF RADIOISOTOPES

The fact that all isotopes of the same element stable or radioactive behave in a nearly identical manner*, makes it possible to use one of them, specially a radioactive one as marker to trace the movements of the atoms of this element, its distribution between the different phases and its displacement with time from one region to another of the system. In short, the tell-tale isotope reveals the history of the labelled element in its role in any reaction from the beginning to the end, from the reactant to the product state, through all the possible intermediate stages, as if the very atoms are visible to us! This is the basis of the use of isotopes as tracers in research and in diverse applied areas; typical examples of these applications are described in what follows.

In addition to the use of radioisotopes in tracer concentrations to label systems, certain isotopes in high concentration are used on account of their radiations, e.g. ^{60}Co, and ^{137}Cs and ^{192}Ir as sources of γ-radiation and ^{90}Sr and ^{63}Ni as sources of β radiation. These are described separately.

12.1 RADIOCHEMICAL PRINCIPLES IN THE USE OF TRACERS

Two molecules are isotopic with one another if they are identical in composition, structure and in every other respect, the only difference being that one of the atoms of one of the elements is present as an isotope of the same element in one of the molecules and in the same position e.g. $^{12}CH_3{}^{12}CH_2{}^{12}COOH$ and $^{12}CH_3{}^{12}CH_2{}^{14}COOH$. Let us look into a mixture of isotopic molecules, participating in a reaction, isotope effects being absent. Let the mixture consists of a very small trace amount of *a* moles of the radioisotope atom, mass number *A* and a large excess of *b* moles, of the stable isotope of mass number *B* (i.e. $a \ll b$). Whatever happens to molecules with the stable isotope (*B*), also happens to molecules with the radioisotope (*A*), precisely at every stage of the reaction, and to the same extent quantitatively, i.e. in strict proportion to their mole fraction *i.e.*

---

* This is on the assumption that *isotope effects* are absent, or are negligible. This agrees with the general observation that isotope effects are important only for hydrogen and a few light elements (Chapters 1 and 4).

$[a/(a+b)] \simeq a/b$. Let 1, 2,...*i*...*t* be the different steps in the transformation of the reactant molecule to the final product.

Table 12.1 Changes in the Amounts of Isotopic Molecules with the Progress of Reaction

Stage	(initial) 0	1	2	3	...	i	...	(final) ∞
Amount of stable isotope molecules	b_0	b_1	b_2	b_3	...	b_i	...	b_f
Amount of radioisotope molecules	a_0	a_1	a_2	a_3	...	a_i	...	a_f

The change in the stable molecule at the *i*th state is $\Delta b_i = (b_o - b_i)$, while the corresponding change in the radioisotope molecule is $\Delta a_i = (a_o - a_i)$. In the absence of isotope effects,

$$\frac{\Delta a_i}{\Delta b_i} = \frac{a_o}{b_0} = \text{constant} \tag{12.1}$$

Often, it is extremely difficult to determine the slight change in the amount of the stable isotopic molecules Δb_i at a given stage of the reaction but the corresponding amount of change in the labelled molecules Δa_i can be measured with precision, even upto changes of the order of 10^{-12} to 10^{-16} M. By measuring Δa_i, the quantity Δb_i is readily assessed by Eq. 12.1.

Following points should be taken care of in using radioisotopes as tracers.

(1) The tracer isotopic molecule and the stable one should be in the same chemical form, as sometimes isotopes in molecules of different chemical forms may not exchange rapidly.

(2) In case more that one radioisotope is available, the one to be chosen for labelling should have a convenient half-life, and the energy of the radiations emitted should be such that it permits easy measurement *e.g.* ^{24}Na (τ = 15h) is recommended for biological experiments, as it would be completely eliminated by decay in about three days, while ^{22}Na (τ:2.6y) would be preferable for investigations of physico-chemical problems, as diffusion.

(3) The total amount of the radioisotope should be just enough to provide measurable signals. Large excess should be avoided as otherwise the system under study may get chemically damaged by self-irradiation, besides increasing the health hazard to the worker.

12.2 TYPICAL APPLICATIONS OF RADIOISOTOPES AS TRACERS

We shall now present briefly some of the more important applications of isotopes as research tools. Besides radioisotopes, sometimes the less abundantly occurring stable isotopes also are used for labelling, as ^{2}H and ^{18}O. In such cases the monitoring is done by mass spectrometry. With the ready availability of radioisotopes innumerable applications are known and the number is on the increase; only some typical examples would be included here to give an idea of the vast scope of the technique. It may be emphasized here that the use of isotopes as labels is the *only* method of understanding some of the problems. The applications selected are grouped under the following heads.

12.3 Chemical investigations
12.4 Physico-chemical applications
12.5 Analytical applications
12.6 Age determinations
12.7 Medical applications
12.8 Agricultural applications
12.9 Prospecting of natural resources
12.10 Industrial applications.

12.3 CHEMICAL INVESTIGATIONS

Some typical applications of isotopes in elucidating (a) reaction mechanisms and (b) structure determinations, are described here.

12.3.1 Reaction Mechanisms

1. *Esterification - Hydrolysis*
Employing an alcohol containing oxygen enriched in the rare isotope ^{18}O (natural abundance 0.21 per cent), the mechanisms of esterification and hydrolysis, its reverse, have been studied. Results show that during esterification the water eliminated results from the OH of the acid and the H of the alcohol. In the reverse process of hydrolysis, the ester retains the carbonyl oxygen *i.e.* the C=O group and takes OH of the water to form the acid, the rest forming the alcohol.

$$R\,CO—OH + H—{}^{18}O\,R' \rightleftharpoons R\,CO—{}^{18}O\,R' + H—OH$$

The bonds suffering rupture are indicated by —

2. *Decomposition of hydrogen peroxide by MO_2*
In the decomposition of H_2O_2 by metal dioxides as PbO_2 or MnO_2, it was earlier believed that each reactant contributes one oxygen atom to form the oxygen liberated. The use of $H_2O^{18}O$, however, showed that the oxygen liberated consists purely of the oxygen of the H_2O_2, thus

$$O = Pb{=}O + H{\div}O{-}{}^{18}O{\div}H \longrightarrow PbO + H_2O + O{-}{}^{18}O$$

3. *Oxidation of fumaric acid by $KMnO_4$*
It is known that a molecule of fumaric acid on oxidation by acidified permanganate yields carbon dioxide, water and formic acid as per the reaction

$$\begin{array}{l} COOH \\ | \\ CH \\ \| \\ CH \\ | \\ COOH \end{array} + 5O \longrightarrow 3CO_2 + H_2O + HCOOH$$

It cannot, however, be known which of the four carbon atoms (two carboxylic and two methylinic) ends up as the formic acid. This was investigated by labelling the end carboxylic carbons with the radioisotope ^{11}C of 20 minutes' half-life. By this technique, Ruben and Allen[1] found that the activity is wholly associated with the CO_2 and none with the formic acid, showing thereby that the acid results from one of the methylinic carbons[10].

4. *Mechanism of Friedel-Crafts' reactions*

Anhydrous aluminium chloride is a useful reagent for bringing about a large number of reactions known as the Friedel-Crafts' reactions. Consider the reaction

$$C_6H_6 + ClCOCH_3 \xrightarrow{AlCl_3} C_6H_5COCH_3 + HCl$$

It was only by using $AlCl_3$ labelled with ^{36}Cl that the true mechanism was revealed as consisting of following steps.

First, the acetyl chloride and benzene dissociate into charged fragments.

(i) $CH_3COCl \longrightarrow CH_3CO^+ + Cl^-$

(ii) $C_6H_6 \longrightarrow C_6H_5^- + H^+$

followed by

(iii) $Al^*Cl_3 + Cl^- \longrightarrow Al^*Cl_4^-$

Finally, the appropriate fragments recombine

(iv) $C_6H_5^- + COCH_3^+ \longrightarrow C_6H_5COCH_3$

(v) $Al^*Cl_4^- + H^+ \longrightarrow Al^*Cl_3 + H^*Cl$

On this mechanism the HCl should carry 1/4 of the total activity and the catalyst the rest. Hence the ratio of activity of HCl to that of the residual catalyst should be 1/3. This is precisely what is observed experimentally; hence this mechanism is accepted as the valid one.

There are a large number of other instances where reaction mechanisms have been unequivocally established by the technique of using radioactive isotopes as tracers.

5. *Evidence for the formation of an intermediate precursor*

When a reaction is postulated to proceed in two or more stages as in

$$M \longrightarrow N \longrightarrow P$$

there may be some uncertainty regarding the formation of the intermediates.

Isotopic labelling of the reactant(s) and subsequent analysis of the products would, in general, be able to provide evidence in favour or otherwise of the intermediate. Two examples are cited here.

(i) *Cholesterol formation*

The hydrocarbon squalene ($C_{30}H_{50}$) is postulated as an intermediate in the transformation of pyruvic acid into cholesterol.

$$\begin{array}{ccccccc} \text{Pyruvic acid} & \longrightarrow & \text{Isoprene} & \longrightarrow & \text{Squalene} & \longrightarrow & \text{Cholesterol} \\ CH_3COCOOH & & (C_5H_8)_n & & C_{30}H_{50} & & C_{27}H_{45}OH \end{array}$$

To verify this, two groups of rats were experimented upon: one fed with ^{14}C labelled pyruvic acid and unlabelled squalene and the other fed with unlabelled pyruvic acid and labelled squalene. The observation of active squalene in the first group and active cholesterol in the second group demonstrated the formation of squalene as an intermediate precursor to cholesterol formation[2]. It is to be noted that the formation of the precursor hydrocarbon could not have been detected by known biochemical methods, without recourse to radioactive labelling.

(ii) *Electrodeposition of chromium*

A solution of chromate, Cr(VI), is used as the electrolyte bath in the electroplating of metals by chromium. The problem is to know if Cr(III) is an intermediate in the electroreduction process

$$\text{Cr(VI)} \longrightarrow \text{Cr(III)} \longrightarrow \text{Cr (0)}$$

Labelling by ^{51}Cr has provided the answer. A mixed electrolyte solution of CrO_4^{--} and Cr^{+++} ions was employed of which either alone was labelled with ^{51}Cr in a given experiment. This is possible as isotopic exchange between CrO_4^{--} and Cr^{+++} is known to be virtually absent at room temperature. The study revealed that the resulting metal deposit was active only when CrO_4^{--} was labelled and not when Cr^{+++} was labelled, establishing thereby that Cr(III) is *not* an intermediate in the electroreduction of the chromate.

6. *Periodate-iodide reaction*

The formation of iodine and iodate when periodate and iodide react in solution is well-known. That the iodine comes exclusively from the iodide ion and none from the periodate ion was shown by Magnier *et al*[3]. in a very elegant and simple way by employing the iodide labelled with the 25 minute ^{128}I in the reaction

$$IO_4^- + 2\,^{128}I^- + H_2O \longrightarrow {}^{128}I_2 + IO_3^- + 2OH^-$$

7. *Labile nature of the hydrogen in $C_6H_5NH_3^+$ ion*

Though the anilinium ion is stable at ordinary temperatures, one of the hydrogen atoms of the NH_3^+ becomes labile at 150° C as shown by Evans[4] employing tritiated anilinium hydrochloride $C_6H_5NH_2TCl$.

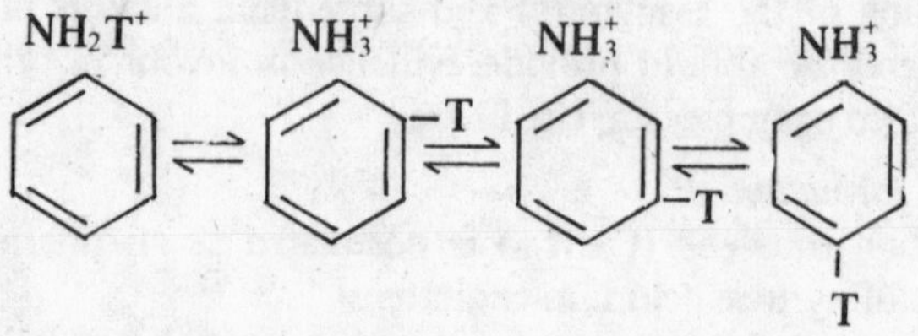

NH_2T^+ ⇌ NH_3^+ (—T) ⇌ NH_3^+ (—T) ⇌ NH_3^+ (T)

Other examples of reaction pathways studied by the use of radioisotopes are described by Bowen[5].

12.3.2 Structure Determination

When two or more atoms of the same element are present in a molecule, the question of their structural equivalence or otherwise arises which can be elegantly settled by the labelling technique. This is illustrated here with reference to four substances (1) phosphorus pentachloride, (2) phosphorous acid, (3) the thiosulphate ion and (4) ammonium disulphide.

1. *Phosphorus pentachloride*

To know whether all the five chlorine atoms in PCl_5 occupy structurally equivalent positions or not, the substance is synthesized using PCl_3, and Cl_2, the latter labelled with its radioisotope, ^{36}Cl.

$$PCl_5 + {}^*Cl_2 \longrightarrow P^*Cl_5$$

Subsequently the product is hydrolyzed when the following reaction takes place.

$$P^*Cl_5 + H_2O \longrightarrow POCl_3 + 2H^*Cl$$

Experimentally it was found that all the radioactivity remained with the HCl and none with $POCl_3$. It is obvious that two Cl atoms in PCl_5 occupy positions different from the rest of the three Cl atoms. This agrees with the trigonal bipyramidal structure accepted for PCl_5, with three Cl atoms in the equitorial plane and two along the vertices. Further, the equitorial P-Cl distance is known to be shorter than the apical P-Cl distance.

2. *Phosphorous acid*

That the three hydrogen atoms of phosphorous acid H_3PO_3 are not equivalent is shown by dissolving the acid in tritiated water and reisolating it. The final activity corresponds to just that of water, (2H atoms), and not 50 per cent more which would have been the value if the 3H atoms had been equivalent. This agrees with the structure[5]

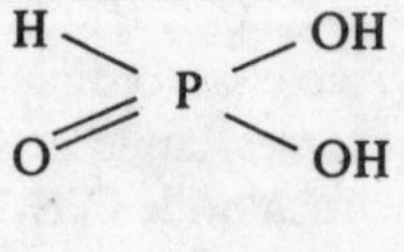

H, O= P (OH)(OH)

3. *Thiosulphate ion*

Two structures for the thiosulphate ion can be envisaged as

$$\left[S \begin{smallmatrix} \diagup O \diagdown \\ - O - \\ \diagdown O \diagup \end{smallmatrix} S \right]^{-2} \qquad \left[S = S \begin{smallmatrix} \diagup O \\ - O \\ \diagdown O \end{smallmatrix} \right]^{-2}$$

(a) Equivalent positions (b) Non-equivalent positions

The problem is to know which of the structures (*a*) or (*b*) is correct. This has been solved by synthesizing sodium thiosulphate by boiling a solution of sodium sulphite with sulphur labelled with ^{35}S

$$Na_2SO_3 + {}^*S \longrightarrow Na_2{}^*S_2O_3$$

The product is decomposed subsequently by adding to it a solution of silver nitrate in the presence of nitric acid. An examination of the final products, silver sulphide and sodium sulphate, shows the activity is exclusively on the former.

$$Na_2{}^*S_2O_3 + 2AgNO_3 + HNO_3 \longrightarrow Ag_2{}^*S + Na_2SO_4 + \text{products}$$

This points to non-equivalent structures for the two sulphur atoms in the thiosulphate ion.

4. *Ammonium disulphide*

Ammonium disulphide may be obtained by boiling a solution of $(NH_4)_2S$ with sulphur labelled with ^{35}S. The compound on treatment with HCl yields sulphur and H_2S both of which are nearly equally active.

$$(NH_4)_2S + {}^{35}S \longrightarrow (NH_4)_2^*S_2 \xrightarrow{HCl} {}^{35}S + H_2{}^{35}S$$

This experiment, due to Voge[6], shows the two sulphur atoms occupy equivalent positions in ammonium disulphide.

12.4 PHYSICO-CHEMICAL APPLICATIONS

These studies can well include, besides others, determinations of

1. rates of reactions, including isotope exchange reactions,
2. solubility and enthalpy of solutions,
3. surface area of powders, and
4. diffusion rates.

These subjects are dealt in detail in several books. We shall make only a brief mention of these to indicate where radioisotopes enter these studies.

12.4.1 Rates of Isotope Exchange Reactions

When two substances, *AX* and *BX*, molecules or ions, or a precipitate having an atom or an ion in common, come together in a uniform solution (homogeneous system), or along a surface (heterogeneous system), a *constant exchange* of the common species takes place between the two substances all the time, though the system be in equilibrium. Such exchanges cannot be detected, nor can their rates be measured, as the common species in the two substances are indistinguishable. However, if one of the species, AX or BX, be labelled with a radioisotope, the kinetics of the exchange reaction

$$AX + B^*X \rightleftharpoons A^*X + BX$$

can be followed by quenching the reaction at different intervals of time and measuring the proportion of $^*X$ in the two samples.

Adopting the following definitions for the concentration terms involved:

$a = [A^*X + AX]$: total AX species, active + inactive.
$b = [B^*X + BX]$: total BX species, active + inactive.
$x = [A^*X]$ and $y = [B^*X]$ for the active species.

McKay[7] arrived at the expression for the rate of gross exchange, R mol $l^{-1}s^{-1}$

$$\frac{dx}{dt} = \frac{R}{ab}(ay - bx) \tag{12.2}$$

which on integration yields

$$\frac{(a+b)}{ab}Rt = -\ln(1-F) \tag{12.3}$$

where F is the fraction exchanged in time t, i.e. $F = x/x_\infty$ where x_∞ is the final amount of active species A^*X after equilibrium is reached. The half-time of exchange τ is obtained by making $x = 1/2\, x_\infty$

$$\tau = \frac{0.693\,ab}{R\,(a+b)} \tag{12.4}$$

Half-times of exchange may vary very widely from micro-seconds to years at room temperature. Many examples of isotope exchange reactions are described in literature.[5,8-11]

12.4.2 Solubility and Enthalpy of Solutions

The enthalpy of a solution can be calculated from solubility data for different temperatures. The first step then is to determine the solubility. This is conveniently done by the tracer technique by comparing the activity of an aliquot of a saturated solution of the labelled substance with that of a known amount of the labelled substance.[10] If S_1 and S_2 be the solubilities at temperatures T_1 and T_2, the enthalpy ΔH is given by

$$\ln\frac{S_2}{S_1} = \frac{\Delta H}{R}\left[\frac{1}{T_1} - \frac{1}{T_2}\right] \tag{12.5}$$

where R is the gas constant.

12.4.3 Surface Area of Powders

Information regarding the surface area of a powder, or a precipitate, is of great importance in surface chemistry in the study of the processes of adsorption and catalysis. Isotopic labelling provides a convenient method for computing surface areas of powders. This is based on the principle that when a precipitate is in contact with its saturated solution, there is a rapid exchange of ions between the surface of the solid phase and the solution. The extent of exchange depends on the surface area of the powder and this is readily measured if the solution is initially labelled with a radioactive isotope. On attainment of equilibrium, we have

$$\frac{\text{Total active atoms on surface}}{\text{Total active atoms in solution}} = \frac{\text{Total amount of substance on surface}}{\text{Total amount of substance in solution}} \tag{12.6}$$

The technique of this method will be described with reference to the determination of the surface area of a given preparation of barium sulphate.

A clear saturated solution of sodium sulphate free from solid particles is prepared and to it is added between 0.1 to 0.5 ml of tracer $Na_2{}^{35}SO_4$ of high specific activity. Let the solution thus prepared possess a specific activity of s_1 counts min^{-1} ml^{-1}. 1 g of the given $BaSO_4$ precipitate sample is taken and to it is added say V ml of the labelled Na_2SO_4 solution. A rapid isotopic exchange occurs between the SO_4^{--} ions on the surface of the precipitate with the $^{35}SO_4^{--}$ ions in solution till an equilibrium is reached.

$$SO_4^{--}\,(\text{solid}) + {}^*SO_4^{--}\,(\text{solution}) \rightleftharpoons {}^*SO_4^{--}\,(\text{solid}) + SO_4^{--}\,(\text{Solution})$$

We now have

$$\frac{\text{Total } SO_4^{--} \text{ on surface}}{\text{Total } SO_4^{--} \text{ in solution}} = \frac{\text{Active } {}^*SO_4^{--} \text{ on surface}}{\text{Active } {}^*SO_4^{--} \text{ in solution}}$$

Here the total SO_4^{--} stands for active + inactive SO_4^{--}. The precipitate is centrifuged and the reduced specific activity of the solution is determined ($= s_2$ counts min^{-1} ml^{-1}).

The activity transferred from the solution to the surface of the precipitate is $V(s_1 - s_2)$ and that remaining in the solution is Vs_2. Since each SO_4^{--} ion corresponds to one $BaSO_4$ entity in precipitate and one Na_2SO_4 in solution,

$$\frac{\text{Total } BaSO_4 \text{ entities on surface}}{\text{Total } Na_2SO_4 \text{ entities in solution}} = \frac{s_1 - s_2}{s_2}$$

Here all the quantities except the numerator of the left hand side are known. This last is readily calculated. Let it be = *a*.

Dividing the formula weight of $BaSO_4$ (= 233) by its density (d = 4.5 g cm^{-3}) and by the Avogadro constant (L), one obtains the *surface area b* of one $BaSO_4$ entity

$$b = \left(\frac{M}{dL}\right)^{2/3} = 1.95 \times 10^{-15}\ \text{cm}^2$$

The surface area of the $BaSO_4$ precipitate sample is simply ab cm^2 g^{-1} (= $1.95 \times 10^{-15} . a$ cm^2 g^{-1})

12.4.4 Diffusion Rates

Diffusion is an irreversible process by which a difference in concentration, if present between different regions of a medium or across two phases, is reduced by the spontaneous flow of matter from the region of higher to that of lower concentration. This is a direct consequence of Brownian motion present in all matter at all temperatures above 0 K. The rate of net flow of matter is approximately proportional to the difference in concentrations between the two volume elements in contact.

A second order partial differential equation known as the *Fick's second law*, relating concentration with distance and time, is stated in the form:

$$\frac{\partial c}{\partial t} = D \frac{\partial^2 c}{\partial x^2} \tag{12.7}$$

One of the integral forms of this equation is :

$$C_{(x,t)} = \frac{0.5\ C_{(0,0)}}{\sqrt{\pi D t}} \exp\left(-x^2/4Dt\right) \tag{12.8}$$

where $C_{(x,t)}$ is the amount of the diffusant at distance x from the origin, at time t of diffusion, and $C_{(o,\ 0)}$ is the initial concentration or the amount at x=0 and t=0. A plot of log C *versus* x^2 is linear whose slope gives the diffusion coefficient. Where the diffusant is labelled with a radioisotope, the concentration $C_{(x,\ t)}$ is replaced by the activity $a_{(x,\ t)}$ which is more precisely and conveniently measured.

(1) *Diffusion in solids*

Experiments to study diffusion in solids have been variously designed with different boundary conditions. In the early days, a thin plating of a radioactive metal was electrodeposited on the plane end of a metal, or solid, and allowed to diffuse into the solid at a constant temperature for a given period of time. In the end, the bar was cut into short sections of thin layers and the distribution of activity along the length of the bar measured. The slope of the plot of log C *vs* x^2, as per Eq. 12.8, gave the diffusion coefficient.

A simple method was developed by Chemla[12] into what has come to be known as the *zone diffusion* technique. A thin zone of the labelled ion is sublimed on to a (100) face of a single crystal of NaCl or KCl and the same is placed face to face with another similar crystal such that the labelled zone is sandwiched between the two crystals. The crystal pair is next placed in an electric muffle maintained at the desired temperature and diffusion allowed to proceed for about 24 hours. The diffusion of the zone of negligible thickness in the midst of semi-infinite column of length $2l$ of the crystal, corresponds to the boundary conditions:

$$\begin{aligned} a &= a_0 \text{ for } x = t && = 0 \\ a &= 0 \text{ for } x \lesseqgtr 0 \text{ at } t && = 0 \\ a &= 0 \text{ for } x = \pm l \text{ for } t > 0 && \end{aligned} \qquad (12.9)$$

After the diffusion is over, the crystals are separated and sectioned into 0.1 mm thick layers with a precision microtome with a glass knife and the activity in the powder collected corresponding to each section is measured. As before D is determined from a plot of log activity *vs* the square of the distance.

The results of Chemla and Arnikar[13] for the tracer diffusion of $^{22}Na^+$ in a single crystal of KCl at 710° C are shown in Fig. 12.1, while the variation of log D with $1/T$ for the tracer diffusion of $^{22}Na^+$ and of $^{137}Cs^+$ in KCl crystal and the self-diffusion of $^{42}K^+$ in the same crystal are shown in Fig. 12.2 over the temperature range 570-750° C. The data of the latter provide values for the energies of activation for the corresponding diffusion processes.

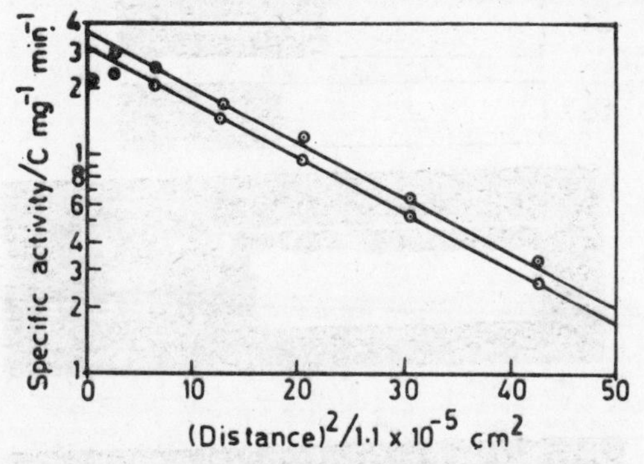

Fig. 12.1: **Tracer diffusion of $^{22}Na^+$ in single crystal of KCl** for 4 h 20 min at 710° C. (Arnikar and Chemla[13]).

(2) *Surface migration*

Migration of atoms and ions on the surface of metals, crystals and in adsorbed layers has been known and studied by different techniques. Surface migration in adsorbed layers of iodide ions on copper and silver was studied by Arnikar *et al.*[14, 14a] by adsorbing the ions on a strip of the matal to a definite layer thickness, one half area of which had been labelled with ^{131}I, with a sharp boundary. The surface self-diffusion coefficient was determined by a modified form of Fick's law applicable to the present case:

$$D^* = \frac{\pi a^2}{b^2 a_0^2 t} \qquad (12.10)$$

where a is the activity crossing the boundary in time t, a_o is the total activity initially deposited per unit area and b the width of the boundary. Fig 12.3 is

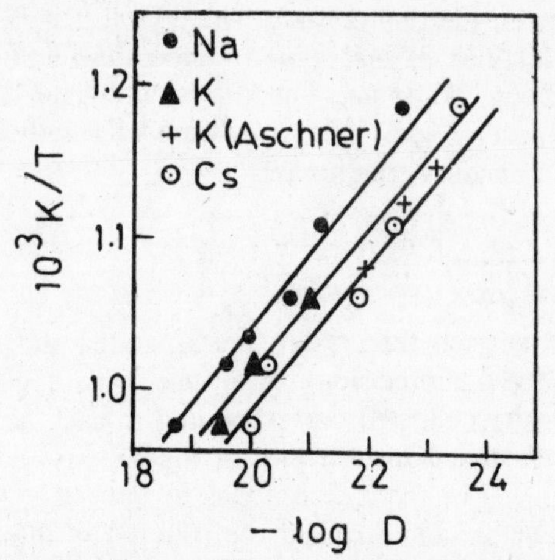

Fig. 12.2: **Tracer diffusion of $^{22}Na^+$ and $^{137}Cs^+$ and self-diffusion of $^{42}K^+$ in single crystals of KCl as a function of temperature.** (Arnikar and Chemla[13]).

an autoradiograph of $I^-(+^{131}I)$ deposited on silver before and after surface self-diffusion for 4 hours at 440° C.[14]

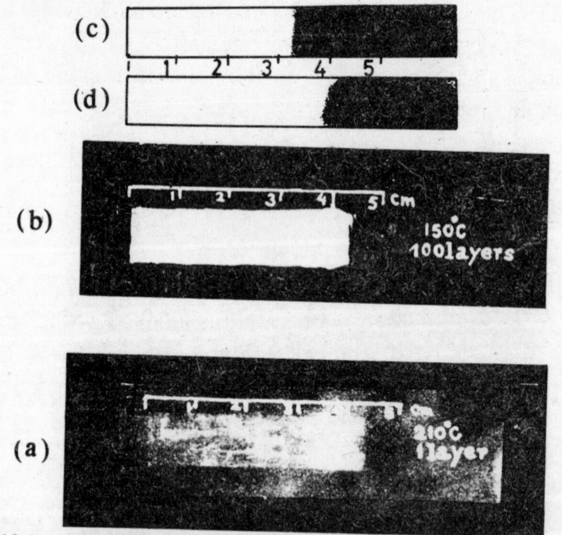

Fig. 12.3: **Autoradiographs representing the surface self-diffusion** of I^- in its (*a*) mono- and (*b*) multilayers adsorbed on copper[14]. The initial boundary line between the labelled and unlabelled halves was at 4 cm. (*c*) and (*d*) in 10 layer thick I^- ions on silver at 400°C[14a]; initial line at 3.2 cm.

Studies of diffusion in solids, liquids, aqueous solutions, including gels and surface migration on solids, have been facilitated by the use of radioisotopes for labeling the diffusant phase[12–15].

12.5 ANALYTICAL APPLICATIONS

Radioisotopes have been used in every branch of analytical chemistry.

Almost every process and technique beginning from simple titrations to sophisticated techniques as polarography and gas chromatography have been extended and adopted to permit the use of radioisotopes and radiation detectors. We shall limit to a brief description of the principles of following techniques which have come to be recognized as the core of radioanalytical chemistry.

1. Radiometric titrations,
2. Radiochromatography,
3. Isotope dilution analysis and
4. Neutron activation analysis.

12.5.1 Radiometric Titrations

Radiometric titrations employ radioisotopes to indicate the end point by a sudden release to, or absorption from the solution of activity, examples of which are given below:

1. *Titration of Ag^+ by Cl^- ions*

In the titration of silver by chloride ions labelled with ^{36}Cl, the solution remains inactive till the end point is reached, as all the activity is removed as AgCl precipitate. The end point is indicated by a sudden release of activity into the solution which keeps increasing thereafter. On the contrary, if the silver ion had been labelled with ^{110}Ag, the solution remains active, though decreasingly so, and the end point corresponds to a constant minimum of activity. In either case, the end point corresponds to a sharp inflection in the activity *vs* titre curve. It is necessary to ensure in every case that the silver chloride formed precipitates out without remaining colloidally suspended.

2. *Titration of metals by $^{32}PO_4^{3-}$ ions*

Metals as, Be, Mg, Al, Zr, Pb, Th and U, whose phosphates are insoluble may be titrated by sodium phosphate solution labelled with ^{32}P. Only after the end point, the clear solution shows activity.

3. *Titration of metals by $^{59}Fe\,(CN)_6^{4-}$ ions*

The ferrocyanides of metals like Co, Cu and Zn being insoluble may be titrated by potassium ferrocyanide labelled with ^{59}Fe. All the end points are sharp and microgram amounts can be detected.

In another type of radiometric titration, the indicator remains an insoluble solid till the end point is reached when it starts dissolving and releasing activity into the solution. As an example may be cited the complexometric titration of calcium by EDTA with solid $^{110}AgIO_3$ as the indicator. The added EDTA is all removed by Ca^{++} and it is only after the end point that the EDTA reacts with the solid $AgIO_3$ to form a soluble Ag-EDTAc complex, thereby releasing activity into the solution.

4. *Release of radioactive gases*

Several radioreagents are known which on reaction release a radioactive gas whose amount can be measured and correlated to the amount of the reactant*. As examples, we have:

a. *Anhydrous aluminium chloride*: $Al^{36}Cl_3$
This reacts quantitatively with water in an organic solvent the amount of which can be determined by measuring the activity of HCl liberated.[16]

$$Al^{36}Cl_3 + 3H_2O \longrightarrow Al(OH)_3 + 3H^{36}Cl$$

b. *Kryptonated calcium carbide*: $CaC_2\,^{85}Kr$
Calcium carbide imbibes a large amount of krypton (^{85}Kr). On reacting with water present in an organic solvent the ^{85}Kr is released which is carried by the acetylene or by a stream of nitrogen[5,17]

$$CaC_2.^{85}Kr + 2H_2O \longrightarrow Ca(OH)_2 + H_2C_2 + {}^{85}Kr$$

c. *Kryptonated copper*: $Cu\,^{85}Kr$
Metallic copper imbibes krypton and when the foil is heated it reacts with oxygen or ozone present and releases ^{85}Kr

$$Cu\,^{85}Kr \xrightarrow[500°C]{+O_2} CuO + {}^{85}Kr$$

$$Cu^{85}Kr \xrightarrow[200\ °C]{+O_3} CuO + {}^{85}Kr$$

These reactions permit the detection and estimation of less than one part per million of oxygen.

d. *Kryptonated platinum oxide*: $PtO_2\,^{85}Kr$
Similarly, kryptonated PtO_2 has been used in detecting trace amounts of hydrogen and fluorine.

12.5.2 Radiochromatography

There are innumerable applications of radioisotopes in all forms of chromatography, paper, column, or gas, as well as in electrophoresis. Almost any chromatographic separation can be carried out using labelled samples. No other developer is needed to reveal the movement of the initial spot and its resolution in the final chromatogram; as the activity distribution is easily scanned.

(1) *Discovery of trans-plutonium elements*
The most spectacular use of radiochromatography has been in the detection and discovery of some trans-plutonium elements by Seaborg and collaborators (1950-54)[18]. They were able to separate from a mixture, six 5*f* elements, americium, curium, berkelium, californium, einsteinium and fermium, all radioactive. The mixture was placed on a column of a cation exchanger Dowex-50 of 2 mm diameter and 5-6 cm length and eluted with a solution

* For details see H.J.M. Bowen—Ref 5.

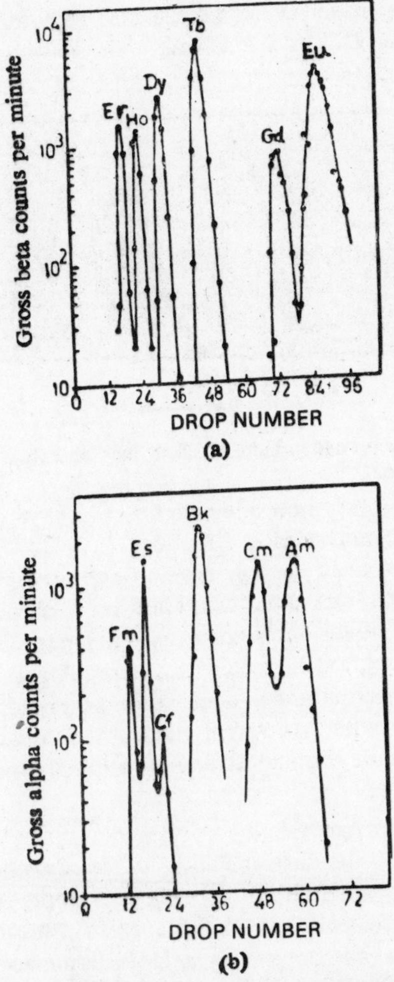

Fig. 12.4 Radiochromatography of (a) lanthanides: Eu to Er, (b) actinides: Am to Fm (from S.G. Thompson, B.G. Harvey, G.R. Chopin and G.T. Seaborg[18] reproduced with authors' permission).

of ammonium citrate or ammonium acetate, at 87° C. An automatic drop fraction collector received the eluant and the α and/or fission activity was recorded for each fraction.

(2) *Separation of alkali and alkaline earth elements*

There are many other instances of the use of radiochromatography in analytical work. A complete separation of the alkali metal ions Li^+, Na^+, K^+, Rb^+ and Cs^+ was shown to be possible by Arnikar and Chemla[19,20] using an acid washed asbestos paper and dilute hydrochloric acid as solvent, by labelling the ions with their radioisotopes, ^{22}Na, ^{42}K, ^{86}Rb and ^{137}Cs (Fig. 12.5).

The same asbestos paper could be used to separate $^{90}Y^{+++}$, $^{90}Sr^{++}$ and $^{131}Ba^{++}$ also with dilute HCl as the solvent[21], the R_f values being Y^{+++}(0.18), Sr^{++} (0.76) and Ba^{++}(0.90).

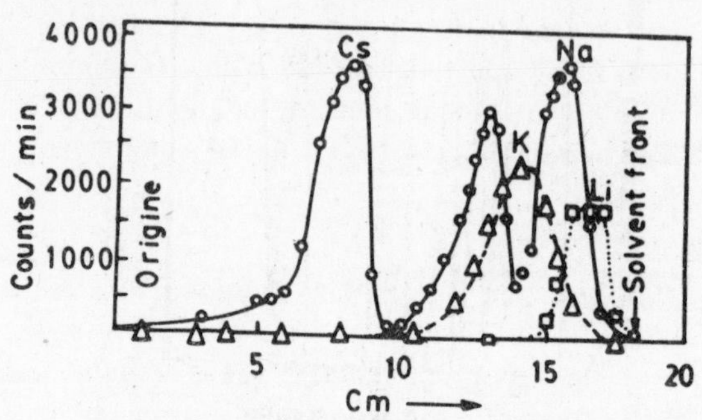

Fig. 12.5 : **Radiochromatography of alkali ions on asbestos paper** (Arnikar[19,20])

(3) *As chromatographic reagents*
Radioisotopes have also been used to develop chromatograms at the microgram level. Heavy metal ions after separation on a chromatographic paper are exposed to $H_2{}^{35}S$ which fixes them as sulphides. On driving off the excess H_2S, the chromatogram is revealed by autoradiography, or by scanning the activity. Similarly, unsaturated acids after separation on exposure to iodine vapour labelled with ^{131}I reveal the chromatogram. Specific radioreagents are available for developing different functional groups[5].

12.5.3 Isotope Dilution Analysis

This technique concerns the determination of an unknown amount of a given species of matter mixed up in a large sample which cannot be otherwise analysed conveniently, as for instance in assessing the amount of an active component as aureomycin in a large fermentation broth, or the volume of blood in a living being.

The technique consists in adding to the sample containing say x g (unknown) of the species, y g of a radioisotopic form of the species of initial specific activity s_i counts min^{-1} mg^{-1}. After ensuring that thorough mixing has taken place, a small amount of the mixture is taken out, the species isolated and its final specific activity s_f is determined. Obviously s_f would be less than s_i. Just with these two measurements, the unknown amount x is calculated on the principle of the conservation of total activity. The initial total activity should equal the final total activity, *i.e.*

$$(x + y)\, s_f = y s_i$$

Hence

$$x = y\,\frac{(s_i - s_f)}{s_f} \qquad (12.11)$$

Some Applications

(1) *Aureomycin in a fermentation broth*
Suppose to a kg of the broth containing x mg (unknown) of aureomycin, one adds a milligram of aureomycin labelled with ^{14}C whose initial specific activity is s_i. After thorough mixing, one mg of the aureomycin is isolated. Let its specific activity be s_f. The amount x mg of aureomycin per kg of the broth is found from Eq. 12.11.

(2) *Blood volume*
The simplest way of assessing the volume of blood in a person is by the technique of isotope dilution. One cm^3 of the patient's blood is withdrawn and it is labelled with a solution of ^{24}Na as NaCl. The initial specific activity of the labelled blood is measured in 0.1 cm^3 of it. Let this be s_i cm^{-3}. The rest of the labelled sample containing 0.9 cm^3 of blood ($=y$) is reinjected intraveinously. After about 15 min the time needed for the circulation and homogenization of the blood, once again 1 cm^3 of the blood is withdrawn and its specific activity s_f determined. The unknown volume of the blood in the patient's body x cm^3 is calculated from Eq. 12.11.

12.5.4 Substoichiometric Isotope Dilution

The principle of the so-called *substoichiometric* isotope dilution consists in separating the same mass of the substance A from the original labelled tracer and from the isotopically diluted sample after mixing. To both these are added a second substance B with which A reacts and forms AB a precipitate, or in some other separable form. The amount of B to be added is substoichiometric *i.e.* not enough to react with all of A present. By the same principle that the resulting relative activity is inversely proportional to the dilution factor enables the calculation of the amount of the substance present in the mixture. This method has been well developed by Ruzicka and Stary[22]

12.5.5 Neutron Activation Analysis

In this technique the element whose amount in a sample is to be determined is transformed by (n,γ) reaction into its radioactive isotope and from the activity of the product, the amount of the target element is computed. Since the amounts of radioactive isotopes of half-lives between hours and years can be determined with a precision of the order of 10^{-16} g, the technique is well-suited and extremely efficient for a large number of elements, specialy those with high neutron capture cross sections. Elements which permit neutron activation analysis include Na, P, Ca, Sc, Cr, Mn, Co, Cu, Zn, Ge, As, Se, Ag, In, W, Ir, Pt, Au and some of the rare earth elements as Sm, Eu, Gd, Dy, Ho and Yb, having high neutron capture cross sections. Analysis of trace elements in minerals and ores, in soils and in special preparations has been possible by neutron activation. The greatest advantage of this technique is that it is nondestructive, *i.e.* the sample remains unchanged at the end of the analysis*, and hence is well adapted for the analysis of rare and precious samples as jewels, precious stones, ancient coins, paintings and other archaeological specimens as pottery sherds, etc.

Suppose the problem is to determine the amount of a given element in particular sample. Let w g of the sample be placed in a uniform flux of thermal neutrons for a sufficiently long period of time t to produce measurable activity. The activity A_t produced is given by

$$A_t = N \phi \sigma (1 - e^{-\lambda t}) \qquad (12.12)$$

where N is the number of atoms of the element in the target (to be determined), ϕ is the neutron flux (*i.e.* number of neutrons cm^{-2} s^{-1}), σ the neutron capture cross section in cm^2 for the given target nuclide, obtainable from tables, and λ is the decay constant in s^{-1} of the radioactive product. In the above (Eq. 12.12), A_t is the activity of the product if measured *immediately at the end of irradiation* (time t). Often this is not possible and some t' lapses between the end of irradiation and the start of counting, during which time some of the activity would have decayed and what one observes is the activity A'_t, given by

$$A'_t = A_t e^{-\lambda t'} = N\phi\sigma (1 - e^{-\lambda t}) e^{-\lambda t'} \qquad (12.12a)$$

In actual practice, two samples are irradiated simultaneously under identical conditions, one containing a known amount of the element W^o and the other the unknown W. If the resulting activities are A_t^o and A_t respectively, we have

$$\frac{W}{W^0} = \frac{A_t}{A_t^0} \qquad (12.13)$$

Better still, a series of known samples containing W_1, W_2... grams of the element and the unknown are irradiated together and from the linear plot of W *versus* A_t, the content of the unknown is determined.

Generally, the target sample contains several elements some or all of which may be neutron activated at the same time, but their yields would depend on their amounts in the sample, their cross sections and half-lives. Thus, more than one activity results often, and the activity of the particular isotope in question has to be distinguished. Usually a large difference in the energy of the principal γ and/or β component and/or the difference in half-lives help in differentiating the relevant product isotope from the others. With the availability of high resolution γ ray spectrometers, using solid state detectors (as Si-Li or Ge-Li or intrinsic Ge), the analysis by neutron activation has today become a highly perfected technique for detecting and estimating a very large number of elements present in trace amounts in a

* Except for the nuclear transformation of atoms in tracer concentration.

variety of samples. In fact with modern techniques, the limit of detection by neutron activation has reached a new dimension, close on one part per trillion!

Some Applications

(1) *Chromium content of a ruby*
The ^{50}Cr in natural chromium becomes by (n,γ) reaction the 27.7 day ^{51}Cr. By irradiating with slow neutrons under identical conditions, a chromium bearing ruby with a series of $Al_2O_3 + Cr_2O_3$ mixtures of varying, but known Cr-content, it is found that the rubies have 0.1 to 0.4 per cent Cr. The precision and the fact that the precious ruby remains undamaged at the end of the analysis reveal the value of this method of analysis.

(2) *Manganese content of tea leaves*
The monoisotope ^{55}Mn becomes by (n,γ) reaction the 2.6 h ^{56}Mn. Neutron irradiation of a known weight of calcined tea leaves, along with a series of samples of known Mn content, reveal that tea leaves contain around 0.13 per cent of Mn, about half of which passes into the brew and the rest goes to waste.

(3) *Archaelogical specimens*
Neutron activation analysis has helped in determining the precise composition of some ancient coins non-destructively and the results have thrown light on their historical and geographic origin. Similarly, the analysis of coloured pottery sherds of archaeological discoveries go to establish the possible age and geographic proximity to available minerals, also throwing light on the development of ceramics in that region and in that epoch. All this is achieved without destroying the valuable specimens.

(4) *Ancient paintings*
An interesting and valuable application of neutron activation analysis is that it permits distinguishing ancient paintings of pre-1850 era from forged ones of a later date. This has become possible because the paints available in the pre-1850 period differed in the microcomposition at the ppm level in respect of several trace elements from the paints of later origin. Presence of appreciable Ag, Cu, Mn and Cr and absence of Zn, Sb and Ti characterize ancient paints. Table 12.2 shows these differences.

Table 12.2 Microcomposition in *ppm* of Paints of pre - and post -1850 Periods[26‡]

	Ag	Cu	Mn	Cr	Hg	Zn	Sb	Ti
pre-1850	18-27	15-220	70-110	225-500	3-7	0-60	0.01	0
post-1850	0-5	0-60	0-12	0-35	0-1	600-6000	30-110	*

* There was no Ti in paints before 1921.

‡ From *Principles of Radiochemistry*, by G.W.A. Newton and V.J.Robinson, MacMillan Education Series, 1971. Reproduced with authors' permission.

All that is necessary is a trace amount of the paint as may be picked up on a needle point for neutron activation analysis. The paintings remain undamaged.

(5) *Arsenic in hair : slow arsenic poisoning*
The monoisotope ^{75}As becomes by (n,γ) reaction the 26.4 h ^{76}As.

If arsenic is administered in slow doses over a period of time, a definite fraction of the element tends to accumulate at the root of hairs and nails. As the hair grows on the average about 0.35 mm per day, the arsenic also moves forward along the length of the hair at the same rate. If now some of the hair (in case they have not been destroyed) of a person suspected of death by slow arsenic poisoning, be neutron irradiated, along with some hair of a normal person, and the distribution of ^{76}As, along the length of the hair compared, the pattern and the schedule of arsenic poisoning would be revealed. In the case of hair from a normal person, the ^{76}As content is much lower and remains nearly constant around 0.8 ppm all along the length except towards the tip, while the pattern in the case of arsenic poisoning reveals distinct peaks corresponding to days of poisoning, each day corresponding to about 0.35 mm length of hair. The technique was first developed by Dr. Hamilton Smith[23] of the Glasgow University in 1959 and described in detail by Savel[24].

The technique has obviously a great importance in criminology and medical jurisprudence. An examination in 1960 of Napoleons's hair by this technique by Dr. Smith himself revealed an abnormal amount of arsenic in excess of 10 μg per g of hair the distribution pattern suggesting slow arsenic poisoning*[10,25]

12.6 AGE DETERMINATIONS

Each radioactive isotope decays with a characteristic rate which is invariable under all conditions of temperature, pressure and chemical environment. However, radioactivity being a statistical phenomenon, the invariability of the decay constant is assured only as long as a large number of the atoms are present; with this provision, radioisotopes can be looked upon as nature's atomic clocks recording the passage of time since the birth of the Universe in terms of the number of atoms decayed. Measurement of residual radioactivity has been one of the most reliable methods of dating samples.

The basis of age determination by a radioactive method depends on the sample containing a radioisotope in equilibrium with stable product and of an age commensurate with the half-life period of the isotope, *i.e.* within about 3-4 spans of the half-life. Thus, the ratio of ^{3}H/H in aqueous samples could be depended upon for samples of 0-40 years' age, the half-life of ^{3}H being 12.3 years. For historic and archaeological ages upto about 25,000 years one

* At this distance of time, it was not possible to identify the guilty person(s): (for details *see The Murder of Napoleon* by B. Weider and D. Hapgood. Congdon and Laties Inc., New York, 1982).

resorts to ^{14}C with a half-life of 5730 years in the form of $^{14}C/^{12}C$ ratio of the sample, by a method first developed by Libby. On the basis that the ^{14}C content of all living beings (plant or animal) corresponds to 16 disintegrations per minute per gram of total carbon, the age of a dead sample can be computed if its ^{14}C content can be measured. If this latter value be s disintegrations min^{-1} g^{-1} (total C), we have

$$s = 16\,(1/2)^n \tag{12.14}$$

where n is the age of the sample in units of half-life of ^{14}C, viz., 5730 years. In other words, the sample has remained dead for 5730.n years. For computing the age of minerals and rocks or the earth, the very long lived radioelements as U and Th in equilibrium with radiogenic Pb, the end product, are depended upon, in a variety of combinations[27].

12.7 MEDICAL APPLICATIONS

Some of the earliest uses of radioisotopes were in medicine as they promised to be a new powerful tool in an area of vital importance as the understanding of metabolic reactions, the mechanisms of the action of drugs, the location of cancerous growths, obstructions in the flow of blood, interphase seepages between the different body fluids, etc. Besides their use in diagnosis, radioisotopes in larger concentrations have also been used for therapeutic purposes. A few typical applications are described here without details. Specialized publications, listed in the bibliography[35–40], may be consulted for details of dosage calculations, instruments used in the measurement of activities *in vivo* and *in vitro*, radiation hazards involved and protective devices to be used to counter them. Modern techniques include whole body scanning by gamma cameras and autofluoroscopes with provision for computer analysis. Isotope scintigraphy enables visualization of the size and shape of internal organs with structural damages clearly demarcated.

12.7.1 Thyroidisis (Goitre)

Most of the iodine we obtain from food is known to accumulate in the thyroid gland which plays a vital role in our well being as it controls the growth and proper metabolism. As against a normal thyroid, some 20 g in weight, the gland in some people becomes over-active (*hyperthyroidisis*) while in some others it becomes sluggish or underactive (*hypothyroidisis*). Both conditions are unhealthy and call for treatment.

The condition of the thyroid is clearly revealed by administering a light dose of radioiodine to the patient. Usually, a glass of orange juice containing about 10 μCi of ^{131}I in the form of NaI is given to the patient and it has no unpleasant taste and the patient is hardly aware of it. The counting of the γ activity emitted by the patient's thyroid is started immediately with a scintillation counter fixed at a distance of 20 cm from the thyroid. The counts are compared with those from a phantom (*dummy plastic thyroid*) injected with the same amount of ^{131}I and counted at the same distance. The counts are continued every hour for the first six hours and at larger intervals thereafter.

The ratio of the counts, dummy/patient (D/P), is plotted as a function of time and compared with the curve for a normal. Fig. 12.6 shows the characteristic patterns of D/P for a hyperthyroidisis, a hypothyroidisis and a normal person. Throughout the period of diagnosis, the patient feels no unpleasant effect.

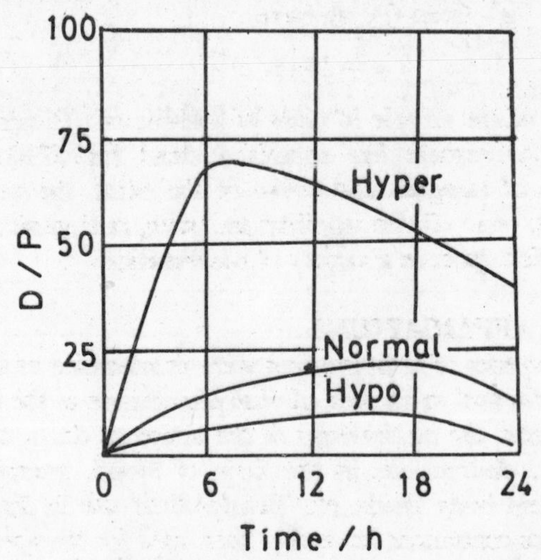

Fig. 12.6: **Diagnosis of abnormal thyroidisis**

Once a case is diagnosed to be hyperthyroidisis, the treatment is possible, by radioiodine ^{131}I or better still ^{125}I. The dose has to be much higher, usually around 200 μCi. The progress is followed by the autoradiography of the gland and by point collimated scintillation counter.

12.7.2 **Brain Tumour Location**

Brain tumours are difficult to locate. It is known that certain dyes as fluorescein, or Rose Bengal are preferentially absorbed by cancerous cells. The technique is to lable the dye with ^{131}I as diiodofluorescein or Rose Bengal with ^{131}I and scan the entire space around the skull by special counters. Sometimes, a solution of albuminate of iodine, labelled with ^{131}I is administered and the region of the brain where the radioiodine accumulates is located.

For radiation therapy, the very expensive radium is replaced by ^{60}Co which emits gamma radiation of 1.17 and 1.33 MeV, with a half-life of 5.27 years.

12.7.3 **Defective Blood Circulation**

Any imperfection in the circulation of blood is detected by injecting in the left forearm a saline solution labelled with ^{24}Na and noting the times needed for the activity to reach different parts of the body and comparing the same with the pattern for a normal. Any obstruction by way of a clot or thrombosis is indicated by the delay in the activity arriving there.

12.7.4 Studies of Blood Volume

The radiotracer method of measuring the total blood volume using ^{24}Na was described under isotope dilution analysis (§ 12.5.3). Similar procedures are used for measuring the red blood cell volume and their survival period by ^{51}Cr as sodium chromate, or by ^{54}Fe as ferric chloride. Pernecious anemia is studied by ^{57}Co labelled cyanocobalamine (vitamin B-12).

12.7.5 Other Medical Uses

Besides the above, a very large number of radioisotopes are finding increasing uses in medicine. Some of these are listed in Table 12.3 only to indicate the wide scope. For details, special books[22,28–30,32] have to be consulted.

Table 12.3 Some Uses of Radioisotopes in Medicine

Purpose	:	*Isotope used*
Bone imaging	:	^{18}F(as *NaF*), ^{85}Sr (as chloride), ^{99m}Tc (as methylene diphosphate)
Cerebrospinal fluid	:	^{111}In (as chloride)
Kidney disorder	:	^{99m}Tc (as dimercaptosuccinic acid) ^{131}I (as hippuran)
Liver/Spleen	:	^{89m}Tc (as sulphur colloid) ^{131}I (as Rose Bengal) ^{198}Au (as colloidal gold)
Lung ventilation	:	^{67}Ga (as citrate), ^{131}I (as macroaggregated albumin), ^{99m}Tc (as macroaggregate)
Metabolism (general)	:	^{24}Na, ^{42}K (as chlorides)

It may be noticed that the radioisotope ^{99m}Tc finds multiple applications. This versatile isotope is marketed in the form of a ready to use kit, along with its parent the "cow" ^{99}Mo, from which it can be "milked" once every 24 hours, for a week or so. The details of the cow-milk systems were described earlier (11.3.1 (2))

12.7.6 Radioimmunoassay

The technique of *radioimmunoassay* (RIA) was first developed by Yalow and Berson[31] in 1959. This has since then proved to be a versatile techinique for assessing the concentration levels of vitally important biological ingredients in the body fluids, such as hormones, vitamines, steroids, drugs and antigens of extraneous origin. As an example consider the determination of the concentration of the antigen hormone-insulin in a sample of blood serum. An excess of the labelled antigen is added to the sample and it is incubated with a known amount of the corresponding antibody (in this case anti-insulin serum). Becasue of the highly specific reaction of an antigen with its antibody the two get bound as an antigen-antibody complex which precipitates out. The excess of the antigen not bound to the antibody is removed by centrifuging in the presence of an immuno adsorbent as dextran coated charcoal. The radioactivities of the supernatant (free antigen) (*F*) and of the precipitate (bound) (*B*) are determined. The concentration of the antigen in the original sample is obtained by interpolating the observed *B/F* value against a standard curve of *B/F vs* antigen concentration.

The RIA technique is being used increasingly in the estimation of the human placentol lactogen (HPL) in the early stages of pregnancy. The information is of vital importance as it enables a clear differentiation of normal pregnancies from abnormal ones involving risks.

The use of radioisotopes in the study of absorption, metabolism and secretion of steroid hormones and vitamins, as well as in ascertaining the safety limits of toxic drugs, has become a routine practice in modern clinical physiology[33].

The Isotope Group of the Bhabha Atomic Research Centre provides ready to use RIA kits for the assay of insulin, HPL and thyroid hormones.

12.8 AGRICULTURAL APPLICATIONS

Ever so many applications of radioisotopes in agriculture have become routine procedures[34]. The principles involved in some of the applications are outlined here.

12.8.1 The Optimum Use of Fertilisers

Using ^{45}Ca as a tracer, it has been found that the uptake by plants of calcium from the soil is nearly the same both for CaO and $CaCO_3$ in acidic soils. However, the uptake is distinctly less if the calcium is present as $CaSO_4$, unless the soil is markedly acidic. In respect of phosphate fertilizer, the farmer would like to know how much of it he needs to add in addition to what is present in the soil, and at what stage of the growth of the plant would it be the best to add the same. By adding ammonium phosphate labelled with ^{32}P of known specific activity, the uptake of phosphorus is followed by measuring the radioactivity as the activity reaches first the lower parts of the plant, then the upper parts, branches, leaves, etc. The *total* phosphorus uptake by the whole plant is determined by chemical analysis and that of the added fertilizer by the activity measurement. The difference is the natural phosphorus present in the soil. By a series of experiments it has been established that in the case of crops it is best to add the phosphorus fertilizer very early at the time of sowing itself, when over 60 per cent is taken up. On the contrary, if the fertilizer is added at a later stage, the uptake is less than 35 per cent. Similar studies in respect of the optimum uptake of nitrogen are now possible employing nitrogen fertilizers labelled with ^{15}N (§ 7.3.6).

12.8.2 Irradiation of Seeds

Beneficial effects of exposing seeds to X or γ radiation on their growth has been well known. This is more easily effected with a 1-2 kCi ^{60}Co source arranged to irradiate panoramically all plants in a large area to a dose of 1-50 grays. However doses beyond 100 grays would be lethal.

12.8.3 Control of Predatory Insects

It is not always possible to effectively combat predatory insects with pesticides only. It is possible to obtain to some extent data in this regard by labelling the insects themselves with ^{32}P or ^{60}Co. This is done by dipping a collection of the insects in a solution of cobalt chloride labelled with ^{60}Co. Usually each insect absorbs around 0.4 μCi of activity which provides it a dose of about 3 Gy over a period of 6-8 months, which is very much less than the lethal dose. With an elaborate system of a number of detectors spread over a large area, it is possible to follow the migration of the labelled insects. Sometimes the insects have been observed to move about a kilometre per day and scatter around 10 km in a season[30, 34]. They have also been observed to prefer old decayed wood to settle down for hybernation. These spots once located, are sprayed with a suitable insecticide and thus the predators are destroyed.

12.9 PROSPECTING OF NATURAL RESOURCES

Because a large number of elements can be activated by neutrons and in the process they emit radiation whose energy is characteristic of the element, the technique of identifying some of the elements and their compounds by a neutron probe is well developed. Some examples are presented here.

12.9.1 Water and Petroleum

A drilling is made in the terrain to be examined and a neutron probe, usually a (Po + Be) source of flux around $10^7 n/s$ is inserted, as shown in Fig. 12.7. As the probe is lowered to different depths, the neutrons induce radioactivity in the elements present in the various layers of the earth and in the process of activation each element emits its characteristic γ-rays. By proper collimation, these γ-photons reach a scintillation detector placed in the probe and well shielded by lead to cut off stray radiation. The signals after amplification are analyzed by a γ-ray spectrometer and recorded. From the energy of the photons the nature of the elements present at the given depth is known. This technique is widely used in locating the presence of large amounts of water and petroleum underground by the fingerprint gamma radiations as : H : 2.2 MeV : O:6.7 MeV and C : 4.4 MeV.*

12.9.2 Uranium and Thorium

Being naturally radioactive, minerals bearing uranium and/or thorium are readily detectable with a sensitive *GM* counter. Granite rocks contain usually 4 g of U and about 13 g of Th per tonne of the rock. This should be borne

* Some others: Mg: 1.37; K: 1.53; Cl: 2.15; S: 2.25; Ca: 3.73 MeV, etc.

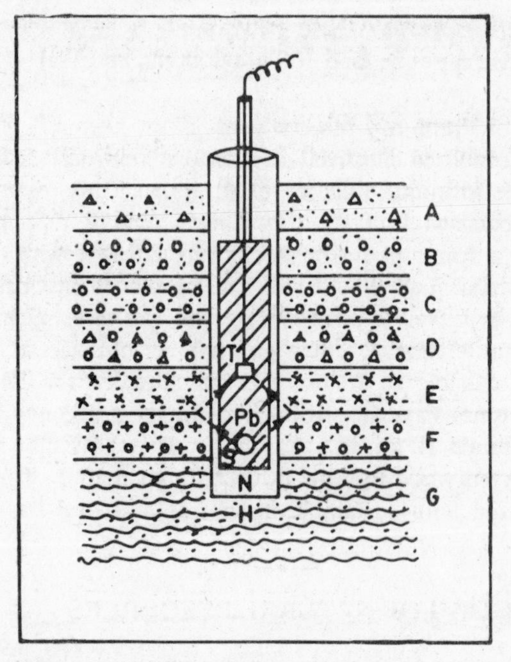

Fig. 12.1: **Neutron probe for prospecting natural resources**
A, B,....: different rock formations
N: neutron probe
S: neutron source; T: detector
Pb: lead shielding; H: outerpipe
(from *Radioisotopes and their Industrial Applications*,
By H. Piraux, N.V. Philips Gloeilampenfabrieken, Eindhoven, 1964,
reproduced with permission)

in mind as constituting the background. It is only when the detector registers an activity well in excess of the background that the material may be considered as a source of these elements.

12.10 INDUSTRIAL APPLICATIONS

Numerous applications of radioisotopes and their radiations, mainly β and γ in industry are known. Some of these are described below.

12.10.1 Friction and Wearout

When two surfaces of same, or different metals or alloys, rub against each other, as during the motion of one relative to the other, friction develops and trace amounts of material are transferred and the surfaces wear out. The presence of a lubricant reduces the friction and wearout. It is difficult by normal chemical analysis to know the trace amounts of matter lost by friction. It is however desirable to know this, so that the period after which the machine parts became unserviceable can be anticipated, besides being able to evaluate the efficiency of a given lubricant in minimizing friction. Radioisotopes find a valuable application in this context. To know this, one of the moving parts is neutron activated so that one of the elements in it becomes radioactive. The parts are repositioned and after a certain period of operation, without a lubricant, the other surface is autoradiographed and this indicates the amount of matter transferred from the first surface during that period of time. When worked with a lubricant, the used oil after a given period is counted for radioactivity. Transfer of matter of less than a microgram can be measured by this technique. For example, a wearout of one per cent of a steel component can be determined after working it for just a minute. The technique finds routine application in automobile industries for assessing the rate of wear out of piston rings. In modern practice, the entire piston or the ring is no longer activated, but only a selected spot on it, say of area of 1 cm^2 or less. This is realized, not in a reactor, but by focussing on to the spot a beam of neutrons from a neutron-generator. The irradiated component is repositioned and worked normally. The wearout is measured continuously by monitoring the activity carried down by the lubricant. The method is also used in knowing in advance when the holes of a draw-plate used for drawing out wires of a fixed diameter, enlarge by friction.

12.10.2 Wearout of a Furnace Lining

The rate of wearout of the refractory lining of a steel (or any other) furnace is monitorable from the outside if a few ceramic pellets labelled with ^{60}Co are incorporated in the lining. The extent of wearout is indicated by the fall in the activity as measured from the outside. This permits the continuous operation of the furnace without a shut-down for examination of the lining. The technique is widely used in the steel industry.

12.10.3 Thickness Measurement and Control

A non-destructive method of measuring the thickness of coatings, or layers, levels of liquids in containers (static), or of moving sheets or layers of textiles, paper, rubber sheets, moving on a conveyor belt and controlling their thickness has been in vogue in many industries, using usually a ^{90}Sr beta source below the sheet or coating and a GM detector above it (Fig. 12.8). The 0.54 MeV betas are well suited for measuring thicknesses over the range of 60 to 500 mg/cm^2 of matter (0.2 to 1.85 mm thick).

Sometimes only one side may be accessible, as for instance, the paint on a solid body, or lining of the walls of a furnance, or corrosion growth in a reactor vessel, etc. In such cases, the *back-scattering* technique is used, the source and the detector are placed on the same side. The signal strength due to radiation scattered by the layer can be calibrated against its thickness. In this case the problem is only to know the thickness of the paint, as it peels off or as the corrosion layer grows.

The radiations get reflected from the coating and the detector is placed at the right position and direction to receive the back-scattered radiation. Thicker the coat, stronger the back scatter, upto a limit.

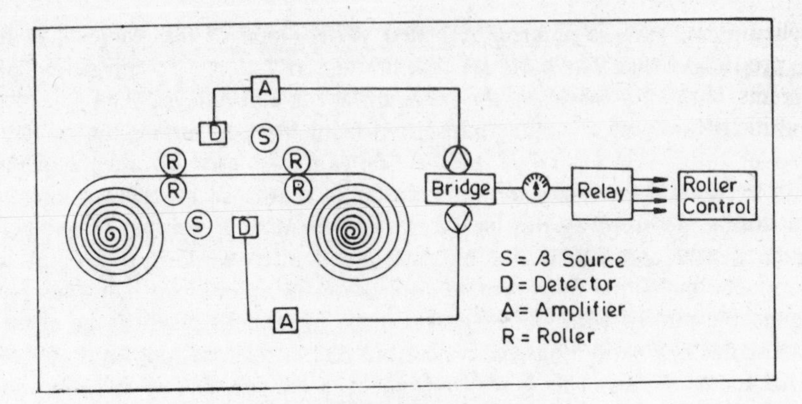

Fig. 12.8: **Thickness control of sheets**
S: β source (usually ^{90}Sr); D: GM detector
A: amplifier; R: roller
B: bridge for anticoincidence circuit.
(from *Radioisotopes and their Industrial Applications,* by
H. Piraux, N.V. Philips Gloeilampenfabrieken, Eindhoven, 1964.
Reproduced with permission)

12.10.4 Gammagraphy

Suppose a large metallic casting is to be examined for defects. A photographic film sensitive to γ radiation is placed behind it in close contact with it and a gamma source is placed in front of it at different positions. Any denser blackening at certain parts of the film on development corresponds to the presence of a crack or defect in the casting. Some convenient sources for gammagraphy are listed in Table 12.4.

Table 12.4: Sources for Gamma Radiography

Source	Half-life	γ Energy/Mev	Half-thickness*/mm steel
^{55}Fe	2.7 y	0.0059	0.01
^{60}Co	5.27 y	1.17; 1.33	34
^{137}Cs	30.17 y	0.662	16.5
^{170}Tm	128.6 d	0.084	1
^{192}Ir	74.2 d	0.3 to 2.0	10-50

The Diesel Locomotive Works, Varanasi, have been using a high strength ^{192}Ir for testing by gammagraphy steel plates of thickness upto 76 mm used in the manufacture of Diesel locomotives for the Indian Railways, since the mid sixties.

In metallurgy, valuable information has been obtained about inclusions and distributions of small amounts of other elements in alloys by neutron

* The half-thickness reduces the intensity of the radiation to half the initial value. This is given by $0.693/\mu$ where μ is the linear absorption coefficient.

activation followed by autoradiography, or by adding the impurity element labelled with its radioisotope during metallurgy and then autoradiographing the product. The general finding in all cases has been that the impurity elements tend to concentrate along the surface, grain boundaries or inter-dendritic sites.

12.10.5 Radiation Sterilization

It is found more convenient and effective to sterilize surgical instruments, sutures, gloves, and ampules, by exposure to a high gamma dose of $\sim 10^4$ Gy, than by the conventional way of steaming in an autoclave, or by chemicals. Similarly, the radiation sterilization of drugs and pharmaceuticals has also been realized with success.

All foodstuffs, vegetables, milk, eggs, and meat have a limited life after which they decay and become inedible. Sometimes large amounts of food-stuffs are lost this way. Usually this happens when the number of micro-organisms exceed 10^8 per gram of matter. An exposure to an appropriate dose of γ radiation at room temperature is known to destory the harmful micro-organisms and the shelf-life of the food is thereby considerably increased, without affecting the taste, colour or appearance, so that it remains as acceptable for consumption as a fresh sample.

Water itself can be adequately sterilized at room temperature by a dose of the order 2 kGy. A dose of 50 to 200 Gy given to potatoes prevents sprouting for a few months longer than unirradiated samples. Similarly, γ-irradiated onions and meat keep fresh for long periods without any adverse effect. However, there is always an optimum dose for each food material, which if exceeded, may lead to the destruction of the vitamins in the food and to its ultimate decay, even faster.

12.10.6 Radiation Induced Synthesis

A large number of chemical syntheses employing γ radiation has been shown to be technically feasible and economically viable. It would appear to be an excellent use of the energy of large (few kGy/h) γ-sources, as the spent fuel rods which emit 20-30 kGy/h while they are "cooling"; in any case all this energy would go waste, otherwise. There are other advantages besides, such as the use of metallic reactors on account of the high penetrability of γ radiation and the more uniform irradiation of the bulk of the reaction space. Since radiation-induced reactions take place at room temperatures and without a catalyst, the products are likely to be purer. Some of the reactions found to be feasible are mentioned below.[35]

(i) *Ethyl Bromide*

Passing a mixture of hydrogen bromide and ethylene at room temperature around a high intensity ^{60}Co source leads to the formation of ethyl bromide

$$C_2H_4 + HBr \rightarrow C_2H_5Br$$

The purity is around 99.5 per cent and the yield corresponds to a low energy absorption of 96 J/mol.

(ii) *Other syntheses*
Some other industrially useful products synthesized by nuclear radiation are gammexane and cyclohexane sulphonyl chloride.[10,35]

12.10.7 Radioisotopes as a Source of Electricity

Because of the large amount of energy released in the decay of each radioactive atom, it has been a tempting and a challenging proposition from the early times to convert this energy into electricity. However, only low power generators have been possible so far. One of these, the SNAP is described below.

The SNAP (strontium-ninety auxiliary power), developed in US, represents an efficient source of power, rather auxilliary power, as it is used mainly for charging batteries used in submarines, space craft, unmanned Arctic stations, and in nuclear devices coupled to high frequency transmitters for continuous signalling of weather conditions, and of sudden large bursts of radiation release, as in nuclear explosions. The SNAP is reported to use the 0.54 MeV β emiter ^{90}Sr in the form of strontium titanate of strength around 20 kCi in a highly compact packing which gets heated to temperatures of the order of 500 °C, which provides the thermoelectric power through a large number of efficient thermojunctions as of Ag-Bi, or Te-Bi, the cold junctions outside may be at a temperature around 50 °C (Fig. 12.9). Sometimes the 5.4 MeV β emitter ^{238}Pu (86.4 years' half life period) is stated to be used in place of ^{90}Sr.

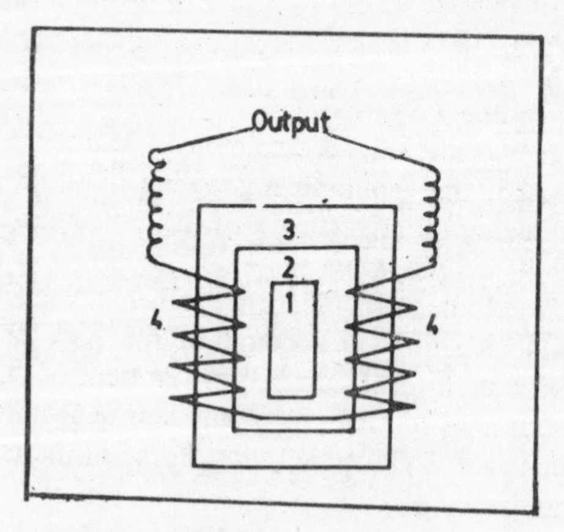

Fig. 12.9: **Radisiotope ($^{90}SrTiO_3$) as a source of thermoelectricity**
1: Column of $^{90}SrTiO_3 \simeq 20$ kCi; 2: Insulator; 3: Shielding
4: Thermocouples
(from *Radioisotopes and their Industrial Applications*, by H. Piraux, N.V. Philips Gloeilampenfabrieken, Eindhoven, 1964, reproduced with permission)

The generation of electricity on a commercial scale, by burning nuclear fuels in a reactor, does not come under the purview of this subject.

References

1. M.B. Allen and S. Ruben, *J. Amer. Chem. Soc.*, 1942, *64*,948
2. K. Bloch and F. Lynen—*Nobel Lectures in Medicine and Physiology*, 1964, Nobel Foundation, Stockholm.
3. P. Magnier, A. Khramoff, M. Martin, P. Daudel, *Bull. Soc. Chim. France*, 1947, *14*, 626.
4. E.A. Evans, *Tritium and its compounds*, Butterworths, 1966.
5. H.J.M. Bowen, *Chemical Applications of Radioisotopes*, Metheun, London, 1969.
6. H.Voge, *J. Amer. Chem. Soc.*, 1939, *61*, 1032.
7. H.A.C. McKay, *Nature*, 1938, 142, 997; *J. Amer. Chem. Soc.*, 1943, *65*, 702: *Principles of Radiochemistry*, Butterworths, 1971.
8. A.C. Wahl and N.A. Bonner, (Ed) *Radioactivity Applied to Chemistry*, Wiley, New York, 1951.
9. R.T. Overman and H.M. Clark, *Radioisotope Techniques*, McGraw-Hill Book Co., New York, 1960.
10. H.J. Arnikar, *Essentials of Nuclear Chemistry*, Wiley Eastern Ltd., New Delhi, II Ed. 1987.
11. J.F. Duncan and G.B. Cook, *Isotopes in Chemistry*, Oxford Univ. Press, 1968.
12. M. Chemla, *Thesis*, Paris, 1954; *Ann. de. phys.*, 1956, *13*, 959;
13. H.J. Arnikar, *et* M. Chemla, *Comptes Rendus* (Paris) 1956, *242*, 2132.
14. H.J. Arnikar and O.P. Mehta, *J. Phys. and Chem.*, of Solids, 1963, *24*, 1633.

14.a H.J. Arnikar, E.A. Daniels and S.V. Kulkarni, *Ind. J. Chem.*, 1979, *18A*, 13.

15. M. Chemla, *Proc. Internat. Symp. on Isotopes*, 1957-58, 288,

15.a H. London, *Separation of Isotopes*, G. Newnes Ltd., London, 1961.

16. C.H. Wallace and J.E. Willard, *J. Amer. Chem. Soc.*, 1950, *72*, 5275.
17. D. Chleck *et al. Chemist Analyst*, 1965, *54*. 81.
18. S.G. Thompson, B.G. Harvey, G.R. Chopin and G.T. Seaborg, *J. Amer. Chem. Soc*, 1954, *76*, 6229.
19. H.J. Arnikar and M. Chemla, *Compt. Rendus* (Paris) 1957, *244*, 68.
20. H.J. Arnikar, *Nature*, 1958, *182*, 1230.
21. H.J. Arnikar and O.P. Mehta, *Current Science*, 1959, *28*, 400.
22. J. Ruzicka and J. Stary, *Talanta*, 1964, *11*, 697.
23. H. Smith, *Anal. Chem.* 1959.
24. P. Savel, *Ann. Pharm. France*, 1963, *21*, 303.
25. B. Weider and D. Hapgood, *The Murder of Napoleon*, Congdon and Laties, Inc., New York, 1982.
26. G.W.A. Newton and V.J. Robinson, *Principles of Radiochemistry* (MacMillan, 1971).
27. K. Rankama, *Isotope Geology*, McGraw-Hill, Book Co., New York, 1954.
28. S. Silver, *Radioisotopes in Medicine and Biology*, Henry Kimpton, London, 1962.
29. G.Wolf, *Isotopes in Biology*, Academic Press, New York, 1964.
30. M.D. Kamen, *Isotopic Tracers in Biology*, Academic Press, New York, 1967.
31. R.S. Yalow and S.A. Berson, *Nature*, 1959, *184*, 1648.
32. E.H. Quimby, S. Feiterberg, and M. Gross, *Radioactive Nuclides in Medicine and Biology*, Lea Febiger, Philadelphia, 1970.
33. E.R. Habermann, *Radioimmunoassey and Related Procedures in Medicine*, IAEA, Vo1.2, Vienna, 1974, 1977.
34. H. Piraux, *Radioisotopes and their Industrial Applications*, N.V. Philips Gloeilampenfabrieken, Eindhoven, 1964.
35. J.W.T.Spinks and R.J. Woods, *An Introduction to Radiation Chemistry*, John Wiley & Sons, New York, 1964.

APPENDIX

Characteristics of some of the more commonly used radioisotopes*

Radio-isotope	*Facility*	*Mode of formation*	*Half-life*	*Principal radiation energy/ MeV*		*Usual chemical form*
				β	γ	
1	2	3	4	5	6	7
^{3}H	*N*	^{14}N$(n,t)^{12}$C	12.33 y	0.018		various
	R	^{6}Li$(n,\alpha)^{3}$H				
^{14}C	*N*	^{14}N$(n,p)^{14}$C	5730 y	0.155		various
	R					
^{22}Na	*A*	^{24}Mg$(d,\alpha)^{22}$Na	2.6 y	β^{+}0.54−1.83	0.51	
^{24}Na	*R*	(n,y)	15.02 h	1.39	1.38-2.76	NaCl aq.
^{31}Si	*R*	(n,y)	2.6^{2} h	1.47		
^{32}P	*R*	(n,y)	14.28 d	1.71		H_3PO_4 in
		^{32}S($(n,p)^{32}$P				HCl soln.
^{35}S	*R*	(n,γ)	87.4 d	0.167		H_2SO_4aq.
		^{35}Cl$(n,p)^{35}$S				
^{36}Cl	*R*	(n,y)	3.0 x 10^{5} y	0.714		HCl aq.
^{38}Cl	*R*	(n,y)	37.3 m	4.9	2.17	
^{40}K	*N*	0.012%	1.28 x 10^{9} y	1.3		
^{42}K	*R*	(n,y)	12.36 h	3.5	1.53	KCl aq.
^{45}Ca	*R*	(n,y)	165 d	0.25		$CaCl_2$ in HCl
^{46}Sc	*R*	(n,y)	83.8 d	0.36	0.89 1.12 X	HCl_8 in HCl
^{49}V	*A*	^{48}Ti$(d,n)^{49}$V	330 d			
^{51}Cr	*R*	(n,y)	27.7 d		0.005	$CrCl_3$ in HCl
^{56}Mn	*R*	(n,y)	2.58 h	2.9	0.85 X 0.0059	
^{55}Fe	*A*	^{55}Mn	$(p,n)^{55}$Fe	2.7 y		
	R	(n,y)				$FeCl_3$ in HCl
^{59}Fe	*R*	(n,y)	44.6 d	0.46	1.1	$FeCl_3$ in HCl
^{58}Co	*R*	^{58}Ni$(n,p)^{58}$Co	70.8 d	β^{+}0.49	0.51	$CoCl_2$ in HCl

* Symbols used are explained at the end of the Appendix.

Contd.

1	2	3	4	5	6	7
^{60}Co	*R*	(*n*,*y*)	5.27 y	0.366	1.17, 1.33	metal or $CoCl_2$ in HCl
^{63}Ni	*R*	(*n*,*y*)	100 y	0.067		$NiCl_2$ in HCl
^{64}Cu	*R*	(*n*,*y*)	12.7 h	β^+0.66	0.51, 1.34 0.0075 X	$CuSO_4$ aq.
^{65}Zn	*R*	(*n*,*y*)	244.1 d	β^+0.325	0.51, 1.11 X 0.008	$ZnCl_2$ aq.
	A	$^{65}Cu(p,n)^{65}Zn$				
^{76}As	*R*	(*n*,*y*)	26.3 h	2.97	0.55	$HAsO_2$ in HCl
^{75}Se	*R*	(*n*,*y*)	118.5 d		0.024 0.402 X 1.0105	Na_2SeO_3 in Na_2SO_4 aq.
^{80m}Br	*R*	(*n*,*y*)	4.42 h		0.4	NaBr aq.
^{80}Br		from ^{80m}Br	17.6 m			
^{82}Br	*R*	(*n*,*y*)	35.34 h	0.04	0.78	
^{85}Kr	*F*		10.7 y	0.67	0.52	
^{86}Rb	*R*	(*n*,*y*)	18.8 d	1.77	1.08	RbCl in HCl
^{89}Sr	*R*	(*n*,*y*)	50.5 d	1.46	0.39	$Sr(NO_3)_2$ in HNO_3
^{90}Sr	*F*		28.8 y	0.54		$Sr(NO_3)_2$ in HNO_3
^{90}Y	*F*	from ^{90}Sr	64.1h	2.26		YCl_3 aq.
^{99}Mo	*R*,*F*	(*n*,*y*)	66 h	1.23	0.14 impurity	$(NH_4)_2MoO_4$ in NH_4OH
^{108}Ag	*R*	(*n*,*y*)	2.4 m	1.7		
^{111}Ag	*R*	$^{110}Pd(n,y)^{111}Ag$	7.45 h	1.05	0.34	$AgNO_3$ in HNO_3
^{115m}Cd	*R*	(*n*,*y*)	44.8 d	1.61	0.94	$Cd(NO_3)_2$ in HNO_3
^{116}In	*R*	(*n*,*y*)	54.1 m	1.0	1.27	
^{113}Sn	*R*	(*n*,*y*)	115.1 d		0.26, 0.39	$SnCl_2$ in HCl
^{124}Sb	*R*	(*n*,*y*)	60.3 d	0.61	0.60	$SbCl_3$+ SbOCl in HCl
^{125}I	*A*	^{124}Te	$(d,n)^{125}I$	60.14 d	0.035	
^{128}I	*R*	(*n*,*y*)	24.99 m	2.12	0.45	
^{131}I	*F*		8.04 d	0.61	0.36	NaI in Na_2SO_3 sol.
^{133m}Xe	*F*		2.19 d		0.23 X	
^{131}Cs	*F*		9.688 d		0.36	

Contd.

1	2	3	4	5	6	7
^{137}Cs	*F*		30.17 y	0.52	0.662 from ^{137}Ba	$CsCl$ in HCl
^{131}Ba	*R*	(*n*,*y*)	12.0 d		0.5 X 0.031	$BaCl_2$in HCl
^{140}Ba	*F*		12.79 d	1.02	0.03-0.54	$BaCl_2$ in HCl
^{144}Ce	*F*		284 d	0.31	0.134	$CeCl_3$ in HCl
^{147}Pm	*F*		2.62 y	0.223		
^{170}Tm	*R*	(*n*,*y*)	128.6 d	1.0	0.084	
^{192}Ir	*R*	(*n*,*y*)	74.2 d	0.7	0.316	
^{198}Au	*R*	(*n*,*y*)	2.696 d	1.0	0.41	$HAuCl_4$ aq.
^{203}Hg	*R*	(*n*,*y*)	46.8 d	0.21	0.279 X 0.011-0.073	$Hg(NO_3)_2$ in HNO_3
^{204}Tl	*R*	(*n*,*y*)	3.77 y	0.77		Tl_2SO_4 aq.
^{210}Pb (Ra-D)	*N*		22.3 y	0.017	0.047 X 0.01	

Symbols in the above Table

Facility (Column 2)
A: Accelerator product; F: Fission product;
N: Naturally occurring and/or being formed in nature by neutrons of cosmic origin;
(n, γ): a product of (n, γ) reaction;
R: Nuclear reactor product.
Half-life (column 4)
y: year; d: day: h: hour; m: minute; s: second
Radiation (columns 5 and 6)
β always β , except where marked β^+
X: X radiation

BIBLIOGRAPHY

Arnikar, H.J., *Essentials of Nuclear Chemistry*, Wiley Eastern, New Delhi, II Ed., 1987.

Aston, F.W., *Mass Spectra and Isotopes*, Longmans Green & Co., 1933

Bowen, H.J.M., *Chemical Applications of Radioisotopes*, Methuen & Co. London, 1969

Chemla, M. and Périé, J., *La Séparation des Isotopes*, Presses Universitaire de France, 1974.

Duncan, J.F. and Cook, G.B., *Isotopes in Chemistry*, Oxford Univ. Press, 1968

Ehrlich, H.L., *Geomicrobiology*, Marcell Dekker, New York, 1981

Evans, E.A., *Tritium and its Compounds*, Butterworths, 1966

Habermann, E.R., *Radioimmunoassey and Related Procedures in Medicine*, IAEA, Vol 2, Vienna

Haissinsky, M., *Nuclear Chemistry and its Applications*, Addison-Wesley, Reading, Mass., 1964

Kirshenbaum, I., *The Physical Properties of Heavy Water*, McGraw-Hill Book Co., New York 1954

Lederer, C.M. and Shirley, V.M. *Table of Isotopes*, Wiley Inter-Science, New York, 1978

London, H., *Separation of Isotopes*, George Newnes, London, 1961

Nash, L.K., *Elements of Statistical Thermodynamics*, Addison-Wesley, Reading Mass., 1968

Piraux, H. *Radioisotopes and their Industrial Applications*, N.V. Philips Gloeilampenfabrieken, Eindhoven, 1964

Proceedings of the International Symposium on Isotope Separations, North Holland Press, 1957, 1958.

Quimby, E.H., Feitelberg, S. and Gross, W., *Radionuclides in Medicine and Biology*, Lea and Febiger, Philadelphia, 1970

Silver, S., *Radioisotopes in Medicine and Biology*, Henry Kimpton, London, 1962

Spindel, W., *Isotope Effects in Chemical Processes*, American Chemical Society, Washington, D.C., 1969

Villani, S., *Isotope Separation*, American Nuclear Society, 1976

Wahl, A.C. and Bonner, N.A., *Radioactivity Applied to Chemistry*, Wiley, New York, 1951

Wolf, G., *Isotopes in Biology*, Academic Press, New York, 1964.

AUTHOR INDEX

SUBJECT INDEX

INDEX OF ISOTOPE SEPARATIONS